# 2nd International Seminar for Fire Safety of Facades 2016



MATEC Web of Conferences Volume 46 (2016)

Lund, Sweden
11 - 13 May 2016

**Editors:**

**Stephanie Vallerent**

ISBN: 978-1-5108-2406-5

**Printed from e-media with permission by:**

Curran Associates, Inc.
57 Morehouse Lane
Red Hook, NY 12571

CURRAN ASSOCIATES INC. proceedings.com

**Some format issues inherent in the e-media version may also appear in this print version.**



Printed by Curran Associates, Inc. (2016)

For additional information, please contact EDP Sciences – Web of Conferences at the address below.

EDP Sciences – Web of Conferences
17, Avenue du Hoggar
Parc d'Activité de Courtabœuf
BP 112
F-91944 Les Ulis Cedex A
France

Phone: +33 (0) 1 69 18 75 75
Fax: +33 (0) 1 69 28 84 91

contact-edps@webofconferences.org

**Additional copies of this publication are available from:**

Curran Associates, Inc.
57 Morehouse Lane
Red Hook, NY 12571 USA
Phone: 845-758-0400
Fax: 845-758-2633
Email: curran@proceedings.com
Web: www.proceedings.com

# TABLE OF CONTENTS

MATEC Web of Conferences
Open access conference proceedings in Materials science, Engineering and Chemistry
A flexible, fast and efficient publication service,
for an open access recording of your event
and a wide dissemination of your proceedings
www.matec-conferences.org
edp open Open Access Publishing from EDP Sciences

# Aims & scope

*MATEC Web of Conferences* is an open-access journal devoted to the publishing of conference proceedings dealing with Materials science, Engineering and Chemistry.

The journal aims to make conference proceedings visible, citable and widely usable by the scientific community. It offers outstanding services for the publication and dissemination of contents in order to ensure long-term archiving and maximum exposure of scientific events.

Coverage includes all engineering disciplines: civil, naval, mechanical, chemical, and electrical engineering as well as nanotechnology and metrology. The journal also deals with the physical-chemical characterization of materials, their implementation, their resistance in their environment, etc. Others subdisciplines of chemistry, such as analytical chemistry, petrochemistry, organic chemistry, etc., and even pharmacology, are also welcome.

# Why publish with MATEC Web of Conferences?

- **Open access:** conference papers available worldwide.
- **Fast publication:** less than 3 months.
- **Total flexibility:** no page restrictions, free unlimited use of colour in the online version, any kind of scientific document welcomed (traditional articles, posters, abstracts and slideshows).
- **High visibility:** all documents indexed in main databases and easily found online.
- **Fair author rights:** CC-BY Creative Commons license.
- **Range of additional publishing options:** printed proceedings, CD-ROMs, copy-editing, etc.

**Contact: Isaline Augusto**
**matec@webofconferences.org**

ISSN (electronic edition): 2261-236X

**www.matec-conferences.org**

edp open Open Access Publishing from EDP Sciences

MATEC Web of Conferences **46**, 00001 (2016)
DOI: 10.1051/matecconf/20164600001


# Preface

The international seminar for fire safety of facades (FSF) was created in 2013 to gather international experts in the field of fire safety for façade, to provide them the opportunity to present their work and research activities, to receive feedback from their colleagues and to establish collaborations and discussion with their peers.

The first international seminar for fire safety of façades was successfully held on November 14–15 2013 in Paris (France). The second edition of the seminar was held on May 11–13, 2016 in Lund (Sweden). For this second edition, we received again excellent abstracts among them the scientific committee selected 24 for oral presentations and 20 for posters. All authors were invited to submit full papers that are grouped into the following topics: standardization and regulations, test and assessment methods, fire safety engineering, facade systems, case studies, products and material.

For this second edition, we wanted to foster information exchange and technical discussions. Therefore we organized two preliminary workshops on the following topics:

- Workshop 1: Fire safety of facades on multi-storey buildings
- Workshop 2: Test methods – can we rely on the test methods available?

We hope that every participant will find during the seminar knowledge and contact that will be useful to their work and that will increase fire safety of facade.

Finally we would like to thank the members of the scientific committee who did a wonderful work in reviewing the submitted abstracts, the members of the organizing committee and the authors.

Stéphanie Vallerent
*Chairman of the organizing committee*

Lars Boström
*Chairman of the scientific committee*

edp sciences

MATEC Web of Conferences

# Statement of Peer review

In submitting conference proceedings to MATEC Web of Conferences, I certify to the Publisher that I adhere to the **Policy on Publishing Integrity** of the journal in order to safeguard good scientific practice in publishing.

1. All articles have been subjected to peer review administered by the proceedings editors.
2. Reviews have been conducted by expert referees, who have been requested to provide unbiased and constructive comments aimed, whenever possible, at improving the work.
3. Proceedings editors have taken all reasonable steps to ensure the quality of the materials they publish and their decision to accept or reject a paper for publication has been based only on the merits of the work and the relevance to the journal.

Title, date and place of the conference

Fire Safety of Facades

11-13 may 2016

Lund, Sweden

Proceedings editor(s):

Stéphanie Vallerent

Date and editor's signature

08/08/2016

MATEC Web of Conferences **46**, 01001 (2016)
DOI: 10.1051/matecconf/20164601001


# Regulative and standard requirements on facades in Slovakia and their mutual comparison with selected European countries

Róbert Leško and Martin Lopušniak

Technical University of Košice, Faculty of Civil Engineering, Department of Building Physics, Vysokoškolská 4, 042 00 Košice, Slovak republic

**Abstract.** Increasingly stricter requirements on the energy effective and environment-friendly construction of buildings are projected even to the other parts of project design including fire safety. Resulting requirements for fire safety can mutually differ in spite of stricter requirements in individual European countries. The subject of this paper is represented by description of the solution of façade fire safety in Slovakia with subsequent comparison of requirements on class of reaction to fire and fire resistance of façades between selected European countries on two structures differing by their height. From the viewpoint of class of reaction to fire, the Slovak system appears to be strictest one, on the contrary, from the viewpoint of fire resistance as the least strict one. The necessity of modification of fire safety solution in Slovakia follows from analysis carried out taking into account wider application of products with class of reaction to fire class other than A1, or A2-s1, d0.

## 1. INTRODUCTION

The design of façades has been considerably changed during last several years. Nowadays, the façade forms not only means of expression of the structure, but requirements of most areas of design solution are put on it. They are represented mainly by the architectonic-structural design, thermal and acoustic protection, daylight, technological and technical solutions, and last but not least by fire safety. In general, the façade is a sandwich type building structure consisting of the mixture of heterogeneous building products. Just heterogeneity of used building product multiplies demandingness of its final design even due to possible propagation of fire on its outer surface. According to O´Connor [1], the height of flames can achieve the level of more than 5 m depending on the size of fire load and surfaces opened to fire. If the façade also contains combustible components, in addition to quicker development of fire, it also could exercise an influence on the reduction of resulting value of building structure fire resistance [2]. Therefore it is necessary to find optimum (safe or acceptable) level of the use of building products and building structures within the framework of the façade structure.

Requirement on the uniform solution within the framework of requirements on the façade structure for all European countries appears to be unreal. This follows mainly from different national approaches to the solution of fire safety. This fact has also been confirmed by study [3] of requirements for fire safety of façades within Nordic countries. In spite of the fact that fire risk and traditional nature of the construction are very similar in Nordic countries, resulting requirements are different. Differences in the solution across European countries are manifested not only within the framework of requirements on reaction to fire, but also in requirements on fire resistance [4].

Continual development in the field of material engineering is projected into innovative proposals of design solution. Fire safety, as a profession part of the design solution should be operatively responded to the aforementioned fact. Therefore it is very questionable, whether the existing method of the solution of fire safety in Slovakia (developed almost 40 years ago) is able to reflect really demands of the existing progress of the building industry. Therefore a summary of the solution of fire safety of façades in Slovakia has been created within the framework of this paper with the comparison of requirements in selected European countries in order of creating the basis for possible optimizing of the existing solution of fire safety in Slovakia.

The design of façades of multi-storey buildings in Slovakia is not so variable from the point of view of materials used as it could be possible with the existing progress of building industry. In comparison with the other European countries, building products with different class of reaction to fire than A1 or A2-s1,d0 can be used within the framework of the construction of façades only in very limited degree. Therefore substantiation of this state and its mutual comparison with the solution used in European countries on model cases is the subject of this paper.

## 2. SOLUTION OF FIRE SAFETY IN SLOVAKIA

The solution of fire safety of buildings in Slovakia (former Czechoslovakia) in the form as it is known today was described by REICHEL [5–8] at the turn of the 70 s and 80 s. Structure of the solution consists of fields that are assessed in the most of countries of Europe and the world. The determination of requirements for building structures is one of these fields. Requirements on fire safety within the framework of assessment of building structures are focused mainly on risk of the origin and propagation of fire, physiological effects of smoke, heat and the like [9]. Taking into account the fact that the building industry has passed through gradual changes since the 70 s, it was necessary to transfer this development even into requirements of fire safety. During almost 40-year solution of fire safety in Slovakia, the method of classification of building structures can be divided from the point of view of fire safety into two development stages.

### 2.1 Classification of building structures – elements of structure

In early stages of the solution of fire safety in Slovakia, no reliable method existed for unambiguous verification of individual building structures [5]. Therefore building structures were evaluated separately according to fire resistance and fire-technical properties (combustibility) of the building materials used. Classification of the structure depending on the type of building materials used was preceded with relative toilsome and lengthy calculation of their overall weight. In spite of criteria set in the aforementioned way, according to REICHEL [5] this method did not represent the most suitable method for assessment of fire safety, therefore it has been optimized.

It has been shown, after years of the designing practice that assessment of individual elements of structure and subsequently assessment of the structure as a whole is more suitable method than to assess the structure from the point of view of its useful properties [6]. On the basis of this method of assessment of building structures that is used up to the present time, impact of combustible building products on fire intensity and bearing capacity of building structure was taken into account. Mutual position of building products in the building structure also plays an important role in this method of assessment. Depending on reaction to fire and position of building products, structures are identified as an element of structure. Elements of structure are divided into three basic types, D1, D2 and D3 (Table 1, Fig. 1) according

**Table 1.** Criteria for sorting of elements of structure [5].

| Criteria for sorting of elements of structure | Type of element of structure | | |
|---|---|---|---|
| | D1 | D2 | D3 |
| Influence of combustible materials on fire intensity | NO | NO | YES |
| Influence of combustible materials on stability and bearing capacity | NO | NO | YES |

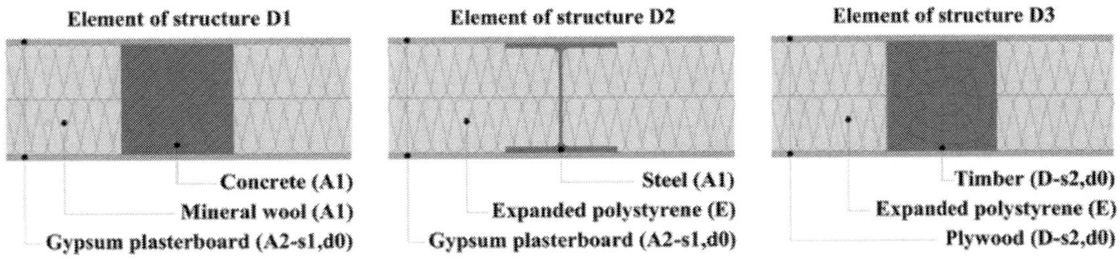


**Figure 1.** Elements of structure.

to impact of combustible products on intensity of fire, stability and bearing capacity of the building structure.

Element of structure of the D1 type is the building structure that does not increase intensity of fire, because it consists exclusively of building products with class of reaction to fire maximum A2-s1, d0. Element of structure of the D2 type is the building structure that does not increase intensity of fire, because its building products with class of reaction to fire other than A1 or A2-s1, d0 are completely closed between building products with class of resistance to fire of A1 or A2-s1, d0. Bearing function of this element of structure must be provided exclusively by building products with class of reaction to fire of A1 or A2-s1, d0. Element of structure of the D3 type is the building structure that increases intensity of fire and that does not meet aforementioned requirements, while it can be made of building products without any limitation of class of reaction to fire [STNEN] (Fig. 1).

## 2.2 Classification of structures – type of construction

As a set of building structures becomes the construction system from the construction point of view, a set of building structures creates, from the point of view of fire safety, the construction system (type of construction). The nature of type of construction is determined on the basis of type of building structures (elements of structure) used. Like in the case of the classification of building structures, the classification of types of construction is one of main factors for layout of elements of structure (Fig. 2).

From the point of view of fire safety, buildings are divided on the basis of their utilization of purpose (for example buildings intended for housing and accommodation, buildings intended for health services and the like), but requirements on material-construction design of building structures (class of reaction to fire and fire resistance) are determined on the basis of a type of construction of the building (Table 2).

## 2.3 Façade as an element of fire safety in Slovakia

Designing of buildings has been developing in various stages since half of the 70 s, when effort for reduction of heat consumption for heating was asserted. Gradual reduction of thermal losses through thermal exchange structures (including façades) become the primary goal. From this point of view, it was necessary to make requirements on individual coefficients gradually stricter expressing the allowed quantity of heat that is transmitted to exterior through individual building structures. The development of requirements from the point of view of heat transmission coefficient through the façade is stated in Fig. 3 on the left [11]. Figure 3 on the right states the quantity of heat that can be released from the outer

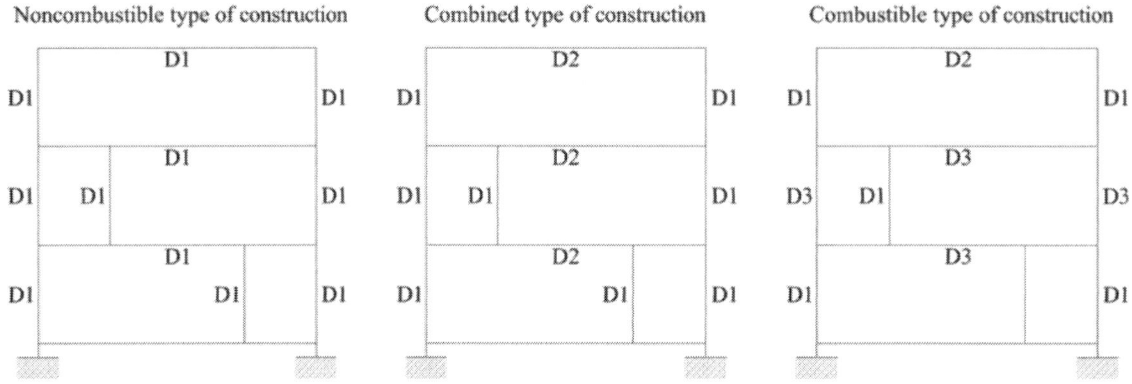


**Figure 2.** Type of construction.

**Table 2.** Fire height limit and requirements on fire resistance of building structures [10].

| Type of building (except family houses) | Type of construction | | |
|---|---|---|---|
| | **Noncombustible** | **Combined** | **Combustible** |
| Buildings for living and accommodation (dwelling houses, hotels, guest-houses etc.) | FH = NO LIMIT R(EI) 30 – 120 | NoS = 1 – 3 storeys R(EI) 30 – 45 | NoS = 1 – 2 storeys R(EI) 30 – 45 |
| Buildings for medical services (hospitals, health care, nursery, etc.) | FH = NO LIMIT R(EI) 30 – 180 | NoS = 1 – 2 storeys R(EI) 30 – 120 | NoS = 1 – 2 storeys R(EI) 30 – 180 |
| Non-industrial buildings (schools, office buildings, libraries, services, shops, etc.) | FH = NO LIMIT R(EI) 30 – 180 | FH = 0 – 22,5 m R(EI) 30 – 120 | FH = 0 – 9 m R(EI) 30 – 120 |
| FH – Fire height, NoS – Available number of fire storeys R(EI) – Fire resistance of loadbearing and fire separation building structures (fire wall, fire ceilings, etc.). | | | |

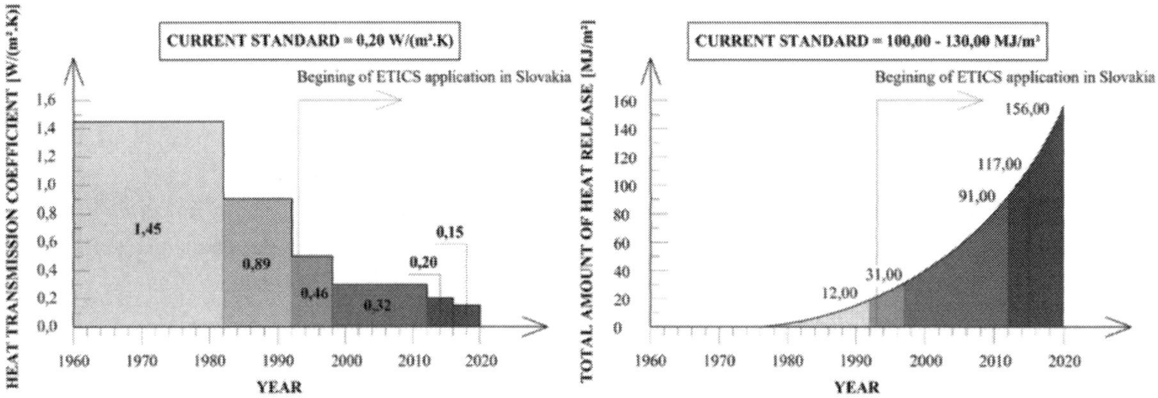


**Figure 3.** Increase in demands on the heat transmission coefficient value and total amount of heat release from façade [11].

surface of façade during fire corresponding to values of heat transmission coefficient of heat insulation system from thermal insulation material on the basis of foam polystyrene (this is the most frequently used form of the thermal insulation system in Slovakia).

By 1980, the presence of building products with class of reaction to fire other than A1 or A2, s1, d0 within the framework of façade structure was minimal (they were represented, for example, by primary forms of disruption of the heat bridge and the like) causing that the classification of construction elements appeared to be suitable at that time. This state had been gradually changed with the development of building products, mainly on the basis of plastics and composite materials. The fact that this progress has not been caught within the framework of determination of structural elements,

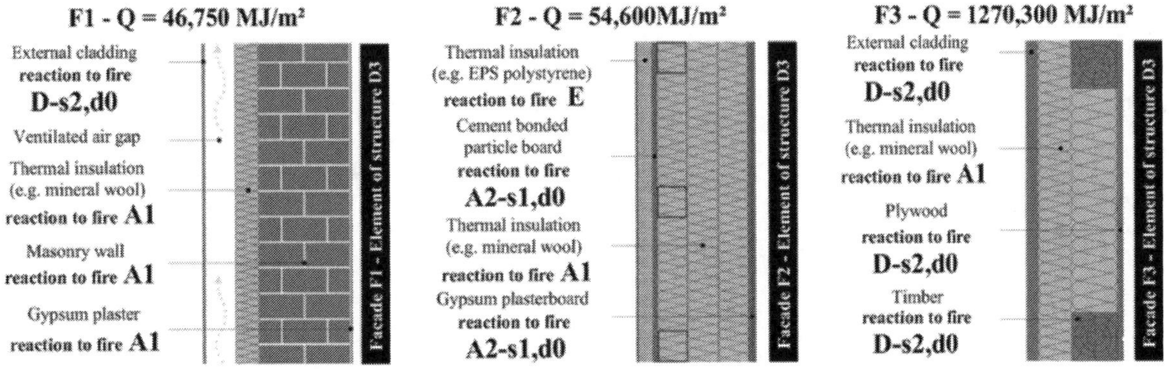


**Figure 4.** Total amount of heat release from façade in various material composition.

the design of façades of multi-storey buildings can become problematic with increasing demands on energy efficiency from the point of view of fire safety by 2020. The situation, when three apparently different variants of façade are classified as the identical structural element in spite of different material composition, is an example of this state (Fig. 4).

Differences between compared alternatives are visible not only in the material composition, but also in the quantity of heat that can be released from their surface. Alternative F3 is able, in comparison with alternative F2 or F1, to release from its surface almost 30-multiple larger quantity of heat during fire. The use of the existing method of classification of building structures causes that any sandwich building structure (for example façade) consisting of building products with class of reaction to fire other than A1 or A2-s1, d0 is defined as the D2 or D3 element of structure (Fig. 1). Actual quantity of heat released from its material during fire (Fig. 4), or degree of risk that can be represented by it in the case of fire for its vertical development, are not taken into account. This state, taking into account increasing demands on energy effectiveness and permanent sustainability of buildings appears to be unsustainable.

## 3. REQUIREMENTS ON FIRE SAFETY IN SELECTED EUROPEAN COUNTRIES

Each country of Europe and the world solves fire safety of buildings individually. But the basic scheme of the solution of fire safety is not different in most of countries. Assessment of designing building structures is one of parts of this scheme. Taking into account the fact that designing façades from the point of view of fire safety is discussed at the present time not only in Slovakia, but across Europe, the list of requirements of selected European countries was prepared within the framework of this paper on two model cases.

### 3.1 Areas of assessment

Requirements on outer surface of façade and on bearing structure providing its stability (in cases of façade with bearing function) are important from the point of view of the solution of fire safety of façades. Requirements on their designing are changed depending on the height of buildings, or the number of its storeys. Two height variants were taking into account for the purposes of this analysis. The first variant takes into account height level of the floor of the highest storey at 10.0 m, and the second one at 22.0 m. These are heights corresponding to 4 or 8 overhead storeys. The building is intended for residential purposes (housing or housing with polyfunction). Within the framework of aforementioned requirements (Table 3) on the outer surface of façade and bearing structures providing its stability the element preventing the propagation of fire on façade (fire cut-offs, fire barriers and the like) had not been

**Table 3.** Fire safety requirements on facades and buildings structures in European countries [3, 4, 10, 12–19].

| European country | Fire safety requirements (h ≤10,0 m) | | Fire safety requirements (h ≤22,0 m) | |
|---|---|---|---|---|
| | Facades | Building structures | Facades | Building structures |
| **Slovak republic** | **A2-s1,d0** | **A2-s1,d0 – R 45** | **A2-s1,d0** | **A2-s1,d0 – R 45** |
| Czech republic | B-s1,d0 | A2-s1,d0 – R 45 | B-s1,d0* | A2-s1,d0 – R 60 |
| Austria | D-d0 | R 60 | B-d1 | R 90 |
| Germany | DCM[e] | R 60 | DCM[e] | A2-s1,d0 – R 90 |
| Great Britain | No provisions [a] | R 60 | B-s3,d2, C-s3,d2[d] | R 90 |
| Finland | D-s2,d2[c] | R 60 | B-s1,d0 | A2-s1,d0 – R 120 |
| Norway | D-s3,d0[b] | A2-s1,d0 – R 60 | B-s3,d0 | A2-s1,d0 – R 90 |
| Estonia | D-s2,d2 | R 30 – R 60 | D-s2,d2 | A2-s1,d0 R60 – R120 |
| Belgium | D-s3,d1 | R 60 | B-s3,d1 | R 90 |

Notes:

*Till the 12 m building fire height acceptable B-s1, d0 with barriers (max. A2-s1, d0), over 12 m to the top of the building max. A2-s1, d0 (see Fig. 5).

DCM[e] Difficult combustible material – reaction to fire B – C.

[a] No provisions in respect of the boundary indicated.

[b] Maximum 4 storeys – depending on risk class and hazard class and if the fire risk in the facade is limited and the risk of fire spread to other buildings is low.

[c] Accepted in 2 – 4 storey building depending on its building class.

[d] Till the 18 m building height acceptable C-s3, d2, over 18 m to the top of the building C-s3, d2 (see Fig. 5).

taken into account. Reference buildings are designed as separately standing ones with sufficient distance separations for the prevention of transfer of fire to adjacent buildings. Fire load of these buildings was taken into account at the level of 60 kg/m$^2$, or 600 MJ/m$^2$.

Visual comparison of differences in requirements on the outer surface of façade from the point of view of class of reaction to fire and fire resistance of bearing elements providing its stability is stated in Fig. 5.

## 4. RESULTS

The following facts can be stated on the basis of analysis of requirements on outer surfaces of façades and fire resistance of bearing elements providing its stability:

- Strictest requirements are put in Slovakia from the point of view of class of reaction to fire on the outer surface of façades,
- Slovakia requires, as the only country among compared countries, full-area non-combustible design of facades in both assessed height alternatives of buildings, while this fact will not be influenced by possible equipment of building with active elements of fire protection (for example sprinklers, smoke detection and the like),
- The application of building products with class of reaction to fire other than A1 or A2-s1, d0 is higher in the other assessed countries than in Slovakia, but it depends on particular requirements on fire safety,
- From the point of view of requirements on fire resistance of bearing elements providing stability of the façade, Slovakia is the country with least strict requirements (value of R 45 represents national criterion of fire resistance used in the Czech Republic and in the Slovak Republic).

## 5. DISCUSSION

An approach is applied in Slovakia in the existing solution of fire safety, within which the “optimal state” between requirements of fire safety (both active and passive one) and resulting level of protection against possible fire is found only with difficulties (Fig. 6).

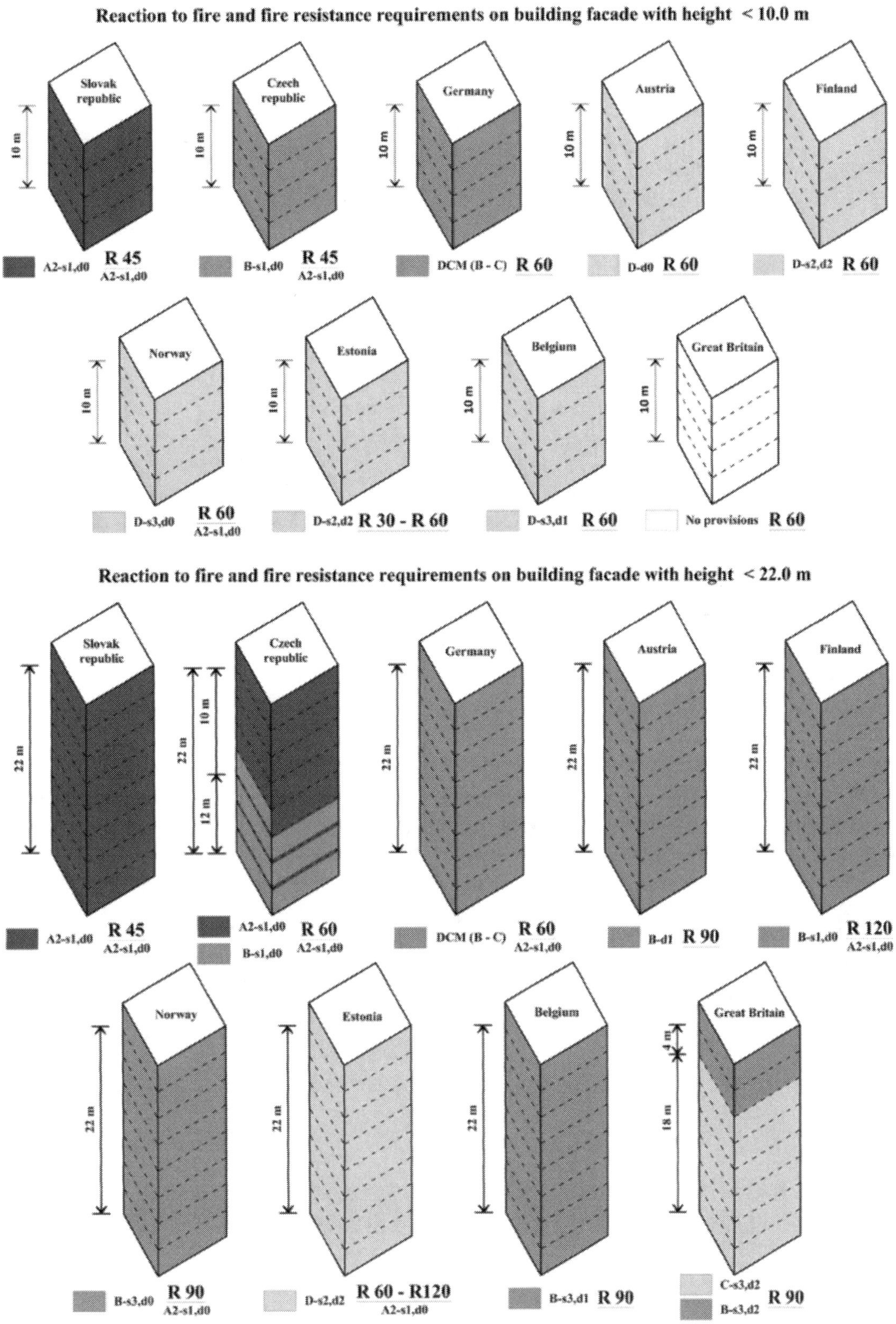


**Figure 5.** Mutual comparison of façade requirements in selected European countries on buildings with height till 10 and till 22 m.

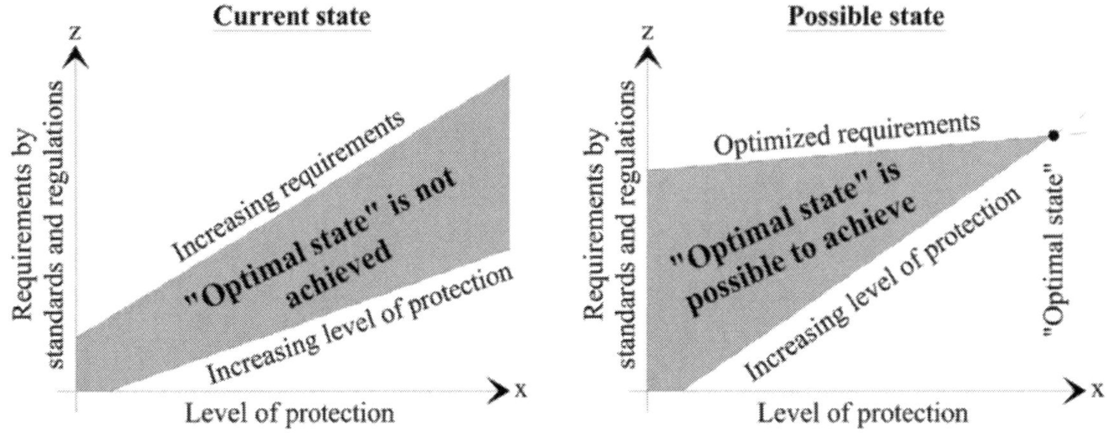


**Figure 6.** Current and possible state of fire safety in Slovakia.

Carried out analysis pointed out the fact that Slovakia appears as one of European countries, in which one of strictest requirements on fire safety are applied from the point of view of material-construction requirements. The more intensive application of active fire protection systems appears to be the solution of this situation. They are currently applied (mainly due to costs on their implementation) only in buildings, where they are required by the legislation. The reduction of requirements on fire resistance of building structures in Great Britain [16], or the reduction of risk of propagation of fire on the outer surface of façade made of combustible building products in Sweden using sprinklers can be stated as examples.

Not only energy efficiency, but also environment-friendly construction with outlook to 2020 requires modification of the existing concept of solution of fire safety in Slovakia. According to [20], the increase of allowed height to five and more overhead storeys is expected by 2020 in European countries, which limits at the present time the use of buildings with bearing system on the basis of wood up to two overhead storeys (including Slovakia). In [20], it is also stated that increased awareness of designing wooden structures from the point of fire safety, in combination with systems of active fire safety (for example sprinklers, smoke detection and the like) was the reason why some European countries allowed the use of higher wooden structures by modification of their regulations. Similar process, built on the basis verified by system and case studies, should also be implemented in Slovakia for the development of the building industry and dire safety.

## 6. CONCLUSION

The paper deals with different approach to requirements on fire safety of façades. Results have shown that requirements on fire safety of façades are considerably different between Slovakia and the other European countries assessed. Omitting the classification of elements of structure that is not used in any of assessed countries (except the Czech Republic) appears to be one of variants of optimizing the existing solution. On the basis of the aforementioned results, it is necessary to change the existing classification of fire safety of façades in Slovakia in order it would reflect the development and innovations in the building industry.

This paper was created by support of grant project 1/0835/14 "Experimental research of physical fragment properties and construction details of building envelopes in non-stationary hygro-thermal conditions".

## References

[1] O Connor, D. J. „Building Facade or Fire Safety Facade?", *Council on Tall Buildings and Urban Habitat Journal*. **9**, 30–39, 2008.

[2] White, N., Delichatsios, M., "Fire Hazards of Exterior Wall Assemblies Containing Combustible Components", The Fire Protection Research Foundation, Quincy, MA, 2014.

[3] Strömgren, M., Albrektsson, J., Johansson, A., Almgren, E., "Comparative Analysis of Façade Regulations in the Nordic Countries" Proceedings of the International seminar for fire safety of facades, Paris, France, 14-15 November, 2013.

[4] Loebus, S., Friquin, K., Tulamo, T. S., SmartTES – Innovation in timber construction for the modernization of the building envelope, Book 5 – Fire safety, WoodWisdom-Net, 2014.

[5] Reichel, V., *Navrhovaní požární bezpečnosti staveb I.*, Nakladatelství technické literatury ve Středisku interních publikací, Praha, 1978.

[6] Reichel, V., *Navrhovaní požární bezpečnosti staveb II.*, Nakladatelství technické literatury ve Středisku interních publikací, Praha, 1979.

[7] Reichel, V., *Navrhovaní požární bezpečnosti staveb III.*, Nakladatelství technické literatury ve Středisku interních publikací, Praha, 1980.

[8] Reichel, V., *Navrhovaní požární bezpečnosti staveb*, Nakladatelství technické literatury ve Středisku interních publikací, Praha, 1984.

[9] Ďurica, T., "Horavosť stavebných materiálov – faktor požiarnej bezpečnosti stavieb", *Stavebné hmoty*, 12–14, 2010.

[10] STN 92 0201-2, Structural fire protection, Common regulations, Part 2: Building constructions, 2007.

[11] Sternová, Z. a kol., *Energetická hospodárnosť a energetická certifikácia budov*, JAGA Group, Bratislava, 2010.

[12] ČSN 73 0802, Fire protection of buildings – Non-industrial buildings, 2009.

[13] ČSN 73 0810, Fire protection of buildings – General requirements, 2009.

[14] The National Building Code of Finland – Fire safety of buildings, Regulations and guidelines, 2002.

[15] Yves, M., "Fire regulations, research and challenges for fire safe use of bio-based building products – State of the art of Belgium", Proceedings of the Bio-based Building Products and Fire Safe Design of Buildings – Recent Developments – Book of Abstracts, 3-4, Barcelona, Spain, 20–21 April, 2015.

[16] The Building Regulation, Fire Safety, Approved Document B, Volume 2 – Buildings other than Dwellinghouses, 2010.

[17] Decree no. 315, Fire safety of building and a building part. 2005.

[18] Werther, N., Hofmann, V., Merk, M., Winter, S., "Fire safety regulations in Germany and limitations for the use bio-based building products – State of the art report", Proceedings of the Bio-based Building Products and Fire Safe Design of Buildings – Recent Developments – Book of Abstracts, 15-18, Barcelona, Spain, 20-21 April, 2015.

[19] Österreichisches Institut für Bautechnik OiB-Richtlinie 2, Brandschutz, 2011.

[20] Gerrard, R., Barber, D., Wolski, A., "Fire safety Challenges of Tall Wood Buildings", The Fire Protection Research Foundation, Quincy, USA, 2013.

MATEC Web of Conferences **46**, 01002 (2016)
DOI: 10.1051/matecconf/20164601002


# Brief summary of the Hungarian test method (MSZ 14800-6:2009) of fire propagation on building façades

István Móder, Ádám Varga, Péter Geier, Bálint Vágó and Edit Rajna

Non-Profit Limited Liability Company for Quality Control and Innovation in Building, Szentendre, Hungary

**Abstract.** When a fully developed fire from one compartment of a building spreads from a window to a façade, it is considered to be one of the most dangerous causes of fire propagation on façades. The Hungarian test method described in standard MSZ 14800-6:2009 focuses on determining the effect and risk evaluation of combustible façade materials by modelling a heat release of a compartment/office fire scenario in a model building with realistic construction and detail types. The method provides actual fire spread limit ($T_h$) in minutes and an evaluation method as well.

## 1. INTRODUCTION

The main causes of fire spread on façade can be led back to one of the following three main reasons. A fully developed fire from at least one compartment of a building spreads from a window to a façade. Materials and goods placed in front of a façade get on fire causing fire propagation on the façade. Fire spreads from one building to another one. The first one is considered to be the most dangerous and common one. The Hungarian test method focuses on determining the effect and risk evaluation of combustible façade materials in case of fire.

Origin of the current test method dates back to 1974. During the last four decades the test standard got improved on modelling actual room/office fire. The test method described in standard MSZ 14800-6:2009 [1] models the heat release of a compartment/office fire scenario in a model building with realistic construction types and detail solutions. The method provides actual fire spread limit ($T_h$) in minutes and an evaluation method as well.

So far only one laboratory (Non-Profit Limited Liability Company for Quality Control and Innovation in Building) has got accredited for this test standard since 2009, up to this date over one hundred tests were carried out forming a solid experience and knowledge about the standard within the Laboratory.

During the current standardization work in Europe the outcome of work and effort which were put into developing this standard needs to be shared.

## 2. TEST METHOD

The scope of this test method is determining the vertical and horizontal fire propagation on building façades which contain openings (window, French balcony etc.) where the wall is finished with combustible materials (such as ETICS), or ventilated or not ventilated cladding systems. Other aim is to determine the effect of façade design solutions that do not satisfy the fire barrier criteria of 1.30 (m) stated in the Hungarian Fire Code [2].

The outdoor test rig is a 3-storey high building, the ground floor is the fire chamber, the second and third floors are the observing levels. The façade is made according to the end use application of the product to be tested; ETICS are usually tested on aerated concrete walls. The façade has standardized window openings with dimensions of 1.20 (m)×1.20 (m). On the level of fire chamber an openable timber frame window is installed before every test.

One of the main concepts of this test method is to model a fully developed fire in one compartment of a building and control the fire propagation on the façade to one level above. Several heating curves are used in fire testing. Using standardized internal exposure for modelling one compartment fire is an unquestionable fact. This is the reason why a standard heating curve is used in this test standard for testing fire propagation on façades.

The fire chamber has a volume of 48.76 ($m^3$) in which fire source of a 650 (kg) timber crib is placed 0.5 (m) away from the test chamber's front wall. Timber cribs shall be made out of first class pinewood roof battens with the dimensions of 25×50×1500–2000 (mm) while the moisture content shall be controlled within 12 ± 2 (%). Elements have to be placed one element away from each other during forming the crib in the test chamber. This timber crib provides ~3.25 MW energy during the test which is meant to be modelling a fully furnished room (taking into count the calorific values of several combustible materials). This provides the following heating curve in the test chamber for a maximum test period of 45 (min):

$$T - T_0 = 345 \times \lg(8t + 1)[\mathrm{K}] \quad (1)$$

where

$T$ is the average temperature in the test chamber, in [°C]
$T_0$ is the initial temperature in the test chamber, in [°C]
$t$ is the time, in minutes.

The standard gives tolerance criteria for the heating curve that the Laboratory shall fulfill. For the first 5 (min) of the test no tolerances are given, while for the test period $5 < t \leq 10$ (min)± 15 (%) and for the test period $t > 10$ (min) ± 10 (%) deviation from the standard heating curve shall not be exceeded.

To achieve a proper combustion of a timber crib and heating curve the timber window is opened after 300±10 (sec) from the start of a test and the air ventilation within the fire chamber is controlled by the Laboratory by several air grills.

Temperature data shall be recorded at the following locations: five places in the test chamber ($T_{tt}$), between the test chamber and the first observation floor 10 (cm) and 50 (cm) away from the parapet at 9 – 9 locations ($T_{lz}$) and in the observation floor 10 (cm) away from the window opening at 16 locations ($T_{any}$) by using a measurement panel of 1.20 (m)×1.20 (m). Additional thermocouples can be used on locations where higher temperature might be expected. Temperature rise is measured due to failure criteria.

## 3. CRITERIA AND EVALUATION METHOD

The fire spread limit on façade ($T_h$) is determined in minutes by reaching at least one of the following criteria. Flame spread reaches the highest horizontal line of the parapet wall. Flame spread reaches

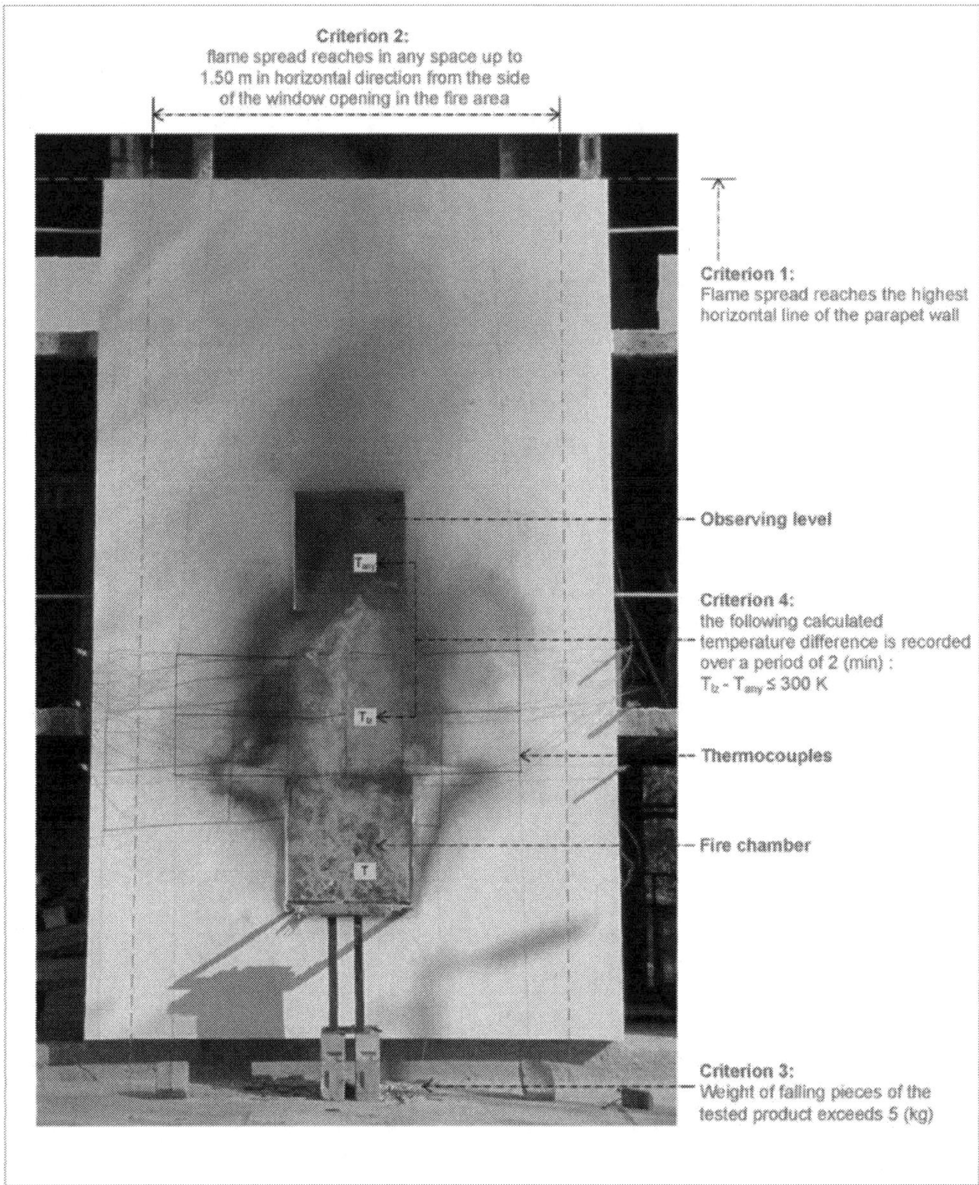


**Figure 1.** Criteria of MSZ 14800-6:2009.

1.50 (m) width from both sides of the fire chamber's window opening. Weight of falling pieces of the tested product exceeds 5 (kg). The following calculated temperature difference is recorded over a period of 2 (min):

$$T_{lz} - T_{any} \leq 300[\mathrm{K}] \tag{2}$$

The aim of Eq. (2) is to verify that the upstand and downstand (with the kit and cladding system applied on) between the two window opening successfully decreases the effect of flame zone in the observing floor.

The products' fire spread limit on facades can be classified as 15 (min), 30 (min) or 45 (min). According to the current National Fire Code [2], the following minimum requirements are held against fire propagation (reaction to fire requirements are not dealt in this paper). Buildings with ground floor and further maximum 2 upper floors $T_h \geq 15$ (min) requirement shall be met, buildings with ground floor and at least 3 maximum 4 floors $T_h \geq 30$ (min) while buildings with ground floor and more than 4 floors $T_h = 45$ (min) requirements shall be fulfilled.

Testing and evaluating facades with French balcony without significant fire resistance is out of the scope of this paper. In this case a special hybrid of EN 1634-1 [3] and MSZ 14800-6 shall be used, so far only few tests and evaluations were carried out.

## 4. CORRALATION WITH OTHER TEST METHODS

Currently over 10 national test methods are available in Europe. With each test method numerous amount of tests were carried out therefore a heavy base of experience is gained by each testing laboratory. Each test method has its own concept, ideology and evaluation method behind it, furthermore requirements hold against fire propagation on facades vary from country to country.

Several comparisons of test methods were carried out on very solid – input – aspects [4]. However finding correlation with other test methods' results is almost impossible since the classification – output – is not uniform for all test method. MSZ 14800-6:2009 can only be compared to the "large scale" test methods like Lepir II [5] or SP Fire 105 [6]. Small scale test methods (like the ones described in EOTA draftTR073 [7]) will not be compared to the Hungarian test method in this paper due to significant differences in the aspect of test rig and fire exposure among many.

Considering the inputs of Lepir II have major similarities to the Hungarian test method by applying actual window openings on the ground floor of the test rig and using almost the same amount of fire source (10 (%) less than at MSZ 14800-6:2009). But the fire exposure time is not the same in the two test methods, and the test rig of Lepir II test method has two window openings on the ground floor, while MSZ 14800-6:2009 has only one. Criteria of SP Fire 105 are almost the same as the ones in the Hungarian test method, such as falling parts and flame spread, however, the Swedish method uses dead windows on the test rig and the fire source is different as well.

ZAG Ljubjana carried out a large scale experiment [8] where the test rig shows big similarities to the Hungarian test rig. The Slovenian test rig consisted of two floors with openable windows installed on each floor. The fire source was higher by 10 (%) than in MSZ 14800-6:2009 but the test duration was shorter. Both methods applied the standard heating curve (1) during the fire exposure. The temperature measurements were carried out just like in MSZ 14800-6:2009, and even more thermocouples were used and during the experiment, temperature rise were recorded in the insulation material as well.

## 5. CONNECTION WITH ACTUAL FIRE CASES

The aim of all fire testing is to determine products' and kits' response to fire in controlled conditions. Real fire exposures can really vary from the standardized heating curves. During real fires neither fire source, nor combustible material or oxygen amount are controlled – clearly – no measurements are made.

One of the most tragic and widely publicized fire propagation on facades in Hungary was the one in Miskolc at 15 August, 2009. The 10-storey high prefabricated panel building had been rehabilitated in 2007 when external insulation was installed, new insulated windows were installed, and the heating system was modernized as well. The fire broke out in one apartment's kitchen on the $6^{th}$ floor causing 3 deaths and 12 smoke poisoning. Expert judgements and fire investigations were published on this topic stating that several causes led to the tragedy.

Concerning the ETICS system the following statements were made [9]. The adhesive was not used properly, the leading and overlapping of fiberglass was not adequate, the number of anchors were less than required and the positioning were randomly made, the rendering thickness was half size of

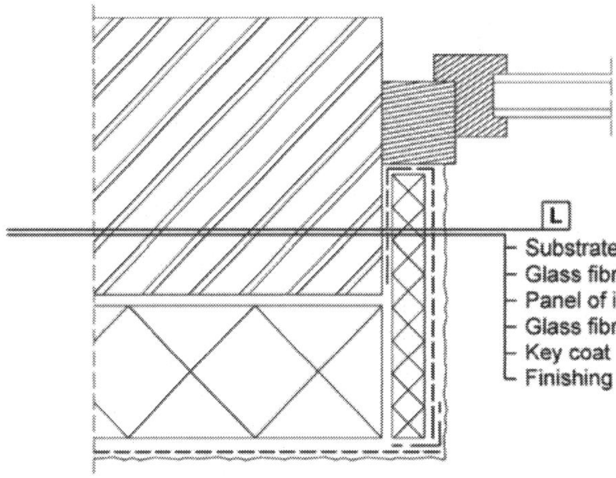


**Figure 2.** Embrasure detail.

the required amount. Furthermore, the expert judgements highlighted that problems were not with the components themselves but with the system as one and the details of it.

## 6. IDEOLOGY

The test standard's ideology is to test combustible insulation materials or cladding system as one 'structure'/ one system and control the details of it. The fire is modeled just like in real fire cases, flames come out from (at least) one window opening to the façade [10]. The second floor is not considered to be necessarily lost, the damage can be controlled. The test method screens whether the ETICS or cladding system contributes significantly to the fire spread or not.

Test results and real fire investigations show that the details around window openings play a significant role in the systems' fire performance. Closing the insulation core around the window opening and at the edges of the structure is the key which can be achieved by reinforcing the right amount of render by fiberglass mesh and filling the gap between the windows' frame and the panel of insulations by adhesive.

## 7. CONCLUSION AS HIGHLIGHTS OF MSZ 14800-6:2009

The test method provides

- 1:1 construction model with realistic details
- evaluation of the effect of fire spread in vertical and horizontal plane
- standardized temperature-time curve (connection to EN 1363-1 [11])
- sophisticated criteria and evaluation method
- results in minutes
- classification range 15 (min), 30 (min) or 45 (min).

### References

[1] MSZ 14800-6:2009 Fire resistance tests. Part 6: Fire propagation test for building facades.
[2] National Fire Code; Issued by Ministry of Internal Affairs 54/2014. (XII. 5.).
[3] MSZ EN 1634-1:2014 Fire resistance and smoke control tests for door and shutter assemblies, openable windows and elements of building hardware. Part 1: Fire resistance test for door and shutter assemblies and openable windows.

[4] Smolka, M. et al., "Semi-natural test methods to evaluate fire safety of wall claddings", Proceedings of the 1st International Seminar on Fire Safety of Facades, Paris, France, 14–15 November, 2013.
[5] LEPIR Test. Large scale Fire Performance testing of External wall cladding systems. LNE, 2015.
[6] SP FIRE 105. Issue 5. Large scale testing of facade systems. SP Boras Sweden, 1994.
[7] Technical Report TR073 (draft). Large Scale Fire Performance Testing of External Wall Cladding Systems. EOTA PT4 Task Group, 2013.
[8] Dr. Nataśa Knez (speaker), "Experimental large scale facade test", EGOLF Workshop on Facades Szentendre, Hungary, 29 October, 2015.
[9] Dr. Tamás Bánky, Sándor Mezei "Tényvázlat a tűzkárt szenvedett miskolci panelház homlokzati hőszigetelő rendszerének viselkedéséről", (translation: Representations of behaviour of the external insulation system of the panel building in Miskolc) *Védelem* (ISSN: 1218–2958), XVI. 6, 2009, p. 20–21.
[10] Dr. Tamás Bánky (speaker), "A full-scale test method for determination of fire propagation parameters on facades with openings in Hungary", Fire Safety Seminar, University of Tokyo, Tokyo, Japan, 26 August, 2015.
[11] MSZ EN 1363-1:2013 Fire resistance tests. Part 1: General requirements.

MATEC Web of Conferences **46**, 01003 (2016)
DOI: 10.1051/matecconf/20164601003


# Semi-natural test methods to evaluate fire safety of wall claddings: Update

Miroslav Smolka[1], Eleonora Anselmi[2], Tony Crimi[3], Bruce Le Madec[1], István Ferenc Móder[4], Kye Won Park[5], Roman Rupp[1], Yong-Ho Yoo[6] and Hideki Yoshioka[7]

[1] Rockwool International A/S, Denmark
[2] L.S.Fire Testing Institute srl, Como, Italy
[3] A.C. Consulting Solutions Inc., Canada
[4] ÉMI Budapest, Hungary
[5] Fire Insurers Laboratories of Korea
[6] Korea Institute of Civil Eng. and Building Technology
[7] National Institute for Land and Infrastructure, Japan

## 1. INTRODUCTION

There are a number of test methods worldwide to evaluate fire safety of facades. An overview of available test methods implemented in fire safety codes was presented at the 1$^{st}$ Conference of Fire Safety of Facades in 2013. [1] Since then, a number of changes and developments occurred. The purpose of this paper is to present the updated global overview of facade fire spread test methods made for building regulations.

On top of that, work is going on developing new test methods or amending existing ones. The paper presents major developments and principles of revisions to existing standards.

## 2. UPDATES ON NEW FAÇADE TEST METHODS AND REVISIONS

### 2.1 Germany: New test for façade base (bottom part/edge)

An internal procedure of the German building institute (DIBT) [2] has been in use since 2015. Its purpose is to demonstrate the suitability of fire protection measures against a fire exposure from outside a building in the immediate vicinity of the facade ("base fire exposure") for External Thermal Composite Insulating Systems (ETICS) with polystyrene insulation.

The test wall is a reentrant corner 4 wide, 2 m deep, and 9.8 m high (minimum dimensions). Fire load is made up by a wood crib 200 kg in total. The test requires quite substantial instrumentation: more than 200 thermocouples placed inside the tested system, and 20 mm from its outer surface.

The test procedure does not contain a set of pass/fail criteria; test evaluation is reserved by the German building institute. The measured values and observations include temperatures, visible flame spread, falling parts, burning droplets, and any damage observed on the façade system after the test. The

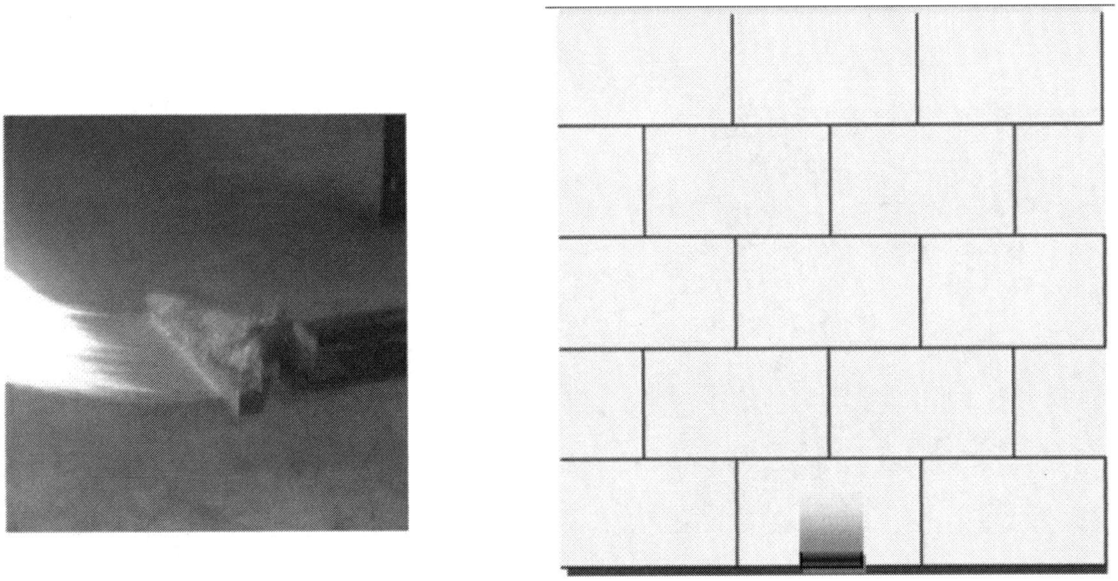

**Figure 1.** The 1st (insulation material) test. Left: the burner. Right: test configuration. Author: Eleonora Anselmi.

only criteria mentioned as reasons to terminate the test prematurely (prior to 60 minutes into the test) is the flame reaching the top edge of the specimen or a safety threat to personnel.

Although the test is reportedly intended as an internal procedure for establishing safety levels and building regulations by authorities with a temporary applicability, the current occupation of laboratories performing this test suggest that this test method continues to be used on commercial basis.

The other German test method DIN E 4102-20 [3] is expected to be published as a full DIN standard in 2016.

## 2.2 Japan: New façade fire test

New test method was published as the Japanese industrial standard in 2015 as JIS A 1310: 2015 [4]. The test method has been developed on the basis of experiments presented in 2013 [5]. The method originated from ISO 13785-1:2002 [6] (the screening test accompanying the large-scale test ISO 13785-2 [7]. Similar to ISO13785-1, the new Japanese standard is intended as "a screening method for determining the fire propagation of products and constructions of a building façade when exposed to flames ejected from the opening" [4].

The method differentiates from ISO 13785-1 quite substantially, however. It lacks the lateral wing, and the fire load is 6x higher. Instead of a linear burner placed directly under the lintel or the bottom edge (base) of the façade system, it has a square burner placed inside the elevated combustion chamber, in order to avoid the situation where molten droplets from the burning façade specimen covers the top surface of linear burner.

The standard follows the pattern of ISO façade fire testing standards in focussing on the incident heat flux in the first place. Surface temperatures and temperatures 50 mm from the surface are measured as well. There is no measurement of internal temperature measurment inside insulation products, but observation of melting droplets and falling parts is recommended.

The method is not aimed at any particular product or construction technology; its scope are all types of facades even including those with ventilated air gaps.

Japan is currently looking at developing a classification system based on this standard.

**Figure 2.** The 2nd (ETICS system) test. Left: the burner. Right: test configuration. Author: Eleonora Anselmi.

## 2.3 Italy: New façade fire tests using the Room Corner measuring equipment

The Mistry of Interior has commissioned developing a relatively-low-cost method for evaluating fire safety of rendered façade systems [8]. The proposal has two parts:

- **Construction work fire scenario:** mounting face fire scenario in which only the uncoated insulation element of the system is exposed to an ignition source.
- **Finished work fire scenario:** fire scenario in which the ETICS system in its end use condition is exposed to an ignition source.

Both tests are carried out using the room corner ISO 9705 [9] equipment in order to measure the heat release rate and to compare it considering different insulation components.

For both tests, the size of the specimen is 2950 mm×2950 mm. The damaged height (vertical flame spread) must not exceed 2.5 m, and lateral flame spread must not reach the specimen edges; if the 1st test passes it is possible to proceed with the 2nd test.

The 1st test is performed with the specimen consisting in the insulation product of its real thickness installed on the wall without the top covering layers (render).

In Italy, majority of fire accidents occurred in the construction site scenario, where the uncoated insulation element is exposed directly to the fire; the main ignition sources are uncontrolled use of the flame during the application of the bituminous sheathing. As a result, the test is carried out with a ribbon-type propane gas-burner that reaches 30 kW [8].

Total heat release, rate of heat release, and smoke optical density are also measured and reported. The specimen of the 2nd test consists of the finished ETICS system but looks at the same measured values and criteria; the main differences are:

- The power of the burner that reaches 300 kW.
- The type of the burner (linear spreading burner) and its position (under the bottom edge of the test specimen, 50% deep).
- The fact that the ignition of the specimen due to burning droplets is not evaluated.

In its current version, the test scope and test procedure description claim that it is only the system base that is subjected to examination. In fact, in Italy, the historical fire events occurred in such a scenario – i.e. with a ETICS system protected on its surface by a glass fiber mesh and cement-based mortar – because of ignition sources like burning garbage bins adjacent to the walls of the building or burning cars parked along its external perimeter. The test carried out on this scenario thus required a much more powerful fire model than the fire model used for the construction work fire scenario [8].

The burner power and position are based on the assumption that they represent the window-plume attack fire scenario as well.

## 2.4 EOTA: Draft test procedures TR073

A task group within the European Organisation for Technical Assessments (EOTA) has been drafting TR073 [9], a document containing 2 different test methods which are very similar to DIN E 4102-20 [3] and BS 8414-1 [10]. The two test methods described therein follow the usual test setup of re-entrant corner and a combustion chamber. The test methodologies do not address incident heat flux but otherwise follow the usual pattern of temperature measurements and observations of the tested facade performance.

## 2.5 CEN: New draft test method for cavity barriers

One of the European routes to dealing with fire safety of wall claddings is testing fire performance of cavity barriers [11]. The new test method designated prEN 1364-6 [12] actually deals with all the applications where hidden fire spread is a safety risk: "Cavity barriers are provided within buildings to restrict the spread of smoke and fire in concealed spaces. Cavity barriers may be "closed" e.g. in roof voids, under raised floors, between the inner and outer leaves of walls such as in timber framed construction; or they may be "open" e.g. as used in ventilated facade systems, roof eaves, rain screen cladding systems etc." [12]

The test proposed is an indicative or benchmark test carried out in fire resistance furnace utilising the standard temperature/time curve. The cavity barrier under examination is installed within an artificial cavity made of aerated concrete, built on the fire resistance furnace ceiling. The test is looking at performance of the barrier in terms of integrity (i.e. flame penetration through the barrier) or insulation (maximum temperature above the barrier). The test is intended to be evaluated to European fire resistance criteria.

As it is obvious from the test description, the test method cannot replace a façade fire spread test; it can be used to complement such a test so that the large-scale façade test does not have to be repeated every time when a cavity barrier is replaced for an alternative product. The purpose of the test fits well in the European concept of Extended Application of Test Results (EXAP) where it can be used in conjunction with a future European façade spread test.

## 2.6 ISO: Proposal for ISO 13785-2 revision

ISO 13785-2 [7] was developed in ISO TC92 as a combination of large-scale facade tests available in 1990´s and experience available at that time. Since then, significant experience has been gathered using the method as well as similar test methods worldwide. The need for revising ISO 13785-2 is obvious, and in particular concerns the following elements:

- Fire load and its calibration: the 3 types of fire load allowed need to be equalized;
- Heat flux measurement: plate thermometer can successfully replace heat-flux meters;
- Testing construction details of the tested facade system.

To serve the current and future needs, attention could also be given to the following areas:

- Measurements and observations, including new items like falling parts and continuous smouldering;
- Presentation of results so that they are compatible with most used regulatory requirements;
- Use of the method for Fire Safety Engineering;
- Relation of the method with the intermediate-scale screening method ISO 13785-1 [6] which can be used as an approval test for specific areas of use.

ISO 13785-2 test facilities exist in Korea, Japan, Czech Republic, Poland. ISO 13785-1 and/or -2 are referred to in regulations of Czech republic and Korea, and have been used on voluntary basis in Poland and Slovakia; Australia and China are considering using the standard. The test method itself has recently been subjected to research [13, 14].

## 2.7 USA: NFPA 285 revision

The test method, which today is titled NFPA 285 [15], has been applied nationally in the US via adoption in the model building codes, and has resulted in an existing building stock with exterior walls that are more resistant to self-propagating fires. The NFPA 285 Standard is currently in its revision cycle. A new edition will be published in 2016. Several proposed changes have been considered, the most important of which is a revision to require at least one horizontal joint or seam to be located between 1 ft (305 mm) and 2 ft (610 mm) above the top of the window opening, and at least one vertical joint or seam upward from the center of the window opening. Combustible cladding materials and systems that have been tested together to show code compliance, such as various insulation or barrier materials, without joint/seams located as defined would also require each existing NFPA 285 test to show compliance with the new protocol.

## 2.8 France: LEPIR II revision

The test Lepir II [16] has been used in France since 1970´s and improved ever since. The current set of measured values and observations is the most complete of all current façade fire spread test methods including visible flame spread, temperatures at different levels, droplets, falling parts, and observation of glowing and smouldering combustion after the test.

Since 2013, a revision of the test protocol is leading to a series of tests in order to justify the possible compliance of façade solutions with the French regulation.

## 2.9 Hungary: MSZ 14800-6 revision

A minor revision to MSZ 14800-6 [17] is under preparation. It includes the extension of the test duration from 45 minutes to 60 minutes, specification of falling parts in the normative text of the standard, and more precise specification of phenomena to be observed during the test including burning droplets, ignition due to falling droplets, smoke, high radiation-level etc.

The test remains unique among the façade fire spread test methods in that it requires a controlled combustion following the standard temperature/time curve in the combustion chamber, and opening of the standard glazed window 5 min into the test simulating a sudden impingement of flames onto the façade.

**Figure 3.** Large-scale façade tests. Left: Lepir II (France). Author: Bruce Le Madec. Right: CAN/ULC-S135 (Canada). Author: Tony Crimi.

## 3. FAÇADE FIRE SPREAD TEST METHODS – EUROPEAN AND GLOBAL OVERVIEW

Following are the updated table originally published in [1]. Apart from adding new test methods, minor additions and/or corrections were made.

## 4. DISCUSSION AND CONCLUSION

Due to the growing awareness about the importance of façade fire spread phenomenon supported by real fire cases, a number of countries has developed or revised their fire testing protocols. Despite attempts for harmonisation, the amount of diverse test methods is on the increase (even in European countries). The European Commission has started its own initiative to harmonise the methodology and classification of façade fire spread; the topic has drawn attention of both member states and industry. The obvious end result of such an initiative should be a harmonised system of façade fire performance testing and classification, based on a test that reflects real safety risks. There is no reason why such initiatives should be limited to different parts of the world when the problem of façade fire performance and its evaluation is truly global.

**Table 1.** Main parameters of major international and national test methods for fire spread on external wall cladding. Published standards and technical documents in Europe.

| **Standard** | **BS 8414-1** | **LEPIR II** | **MSZ 14800-6** | **SP FIRE 105** | **Önorm B 3800-5** | **PN-90/ B-02867** |
|---|---|---|---|---|---|---|
| **Country** | UK | F | H | S, DK | A | PL |
| **Fire exposure** | Wood crib, peak heat 3.5 MW, 4500 MJ | 600 kg wood crib | 650 kg wood crib, 3 MW approx. (controlled) | 60 l heptane | 25 kg wood / 320 kW propane | 20 kg wood crib + wind towards the wall (2 m/s) |
| **Max. heat flux on surface** | 70 kW/m² at 1 m height | Not specified | Not specified | 15 or 80 kW/m² | Not specified | Not specified |
| **Max. tem-perature on surface** | 600°C / 20 min | Average 500°C, peak 800°C (read at 150 mm from the surface) | 600°C 0.5 m high / 50 min | 450°C / 12 min | Not specified | 800°C peak |
| **Test duration** | 30 min | 60 min 30 min fire exposure | 60 min | Min. 12 min | 30 min | 30 min |
| **Test specimen** | Corner 2.5 m × 8.0 m + 1.5 m × 8.0 m | Flat wall 5.0 m × 6.6 m | Flat wall 6.0 m × 7.0 m | Flat wall 4.0 m × 6.7 m | Corner 3.0 m × 6.0 m + 1.5 m × 6.0 m | Flat wall 2.3 m high |
| **Substrate** | Masonry or light frame | Any | Masonry | Aerated concrete | Aerated concrete | Masonry |
| **Criteria**[a] | Temp. limits | Flame spread, system integrity, temp. rise, falling parts[b] | Temp. rise, fire spread, falling parts[b] | Flames 2 floors above; Falling parts | Temp. rise, fire spread, falling parts[b] | Temp. limits; Burning particles |

a) Additional criteria may be contained in other standards or regulations.
b) Temperature rise, vertical and/or horizontal flame spread, burning droplets and particles, falling parts of certain size or mass.

**Table 2.** Main parameters of major international and national test methods for fire spread on external wall cladding. Draft standards and technical documents in Europe.

| **Standard** | **DIN E 4102-20** | **DIBT procedure** | **Procedure No. 281 – construction phase** | **Procedure No. 281 – finished phase** | **EOTA draft TR073 Exp. 1** | **EOTA draft TR073 Exp. 2** | **prEN 1364-6** |
|---|---|---|---|---|---|---|---|
| **Country** | D | D | I | I | EU | EU | EU |
| **Fire exposure** | 25 kg wood / 320 kW propane | 200 kg wood crib | 30 kW propane burner | 300 kW propane burner | 30 kg wood crib | 382.5 kg wood crib | Standard temp./time curve |
| **Max. heat flux on surface** | 70–95 kW/m² at 1 m height | Not specified | 55–60 kW/m² | Not specified | Not specified | Not specified | N/A |
| **Max. temperature on surface** | Not specified | Not specified | Not specified | Not specified | Not specified | Not specified | Defined by standard temp./time curve |
| **Test duration** | 21 min gas, 30 min wood | 60 min 25 min fire exposure | 15 min 10 min fire exposure | 15 min 10 min fire exposure | 60 min 30 min fire exposure | 60 min 30 min fire exposure | Customer declaration |
| **Test specimen** | Corner 3.0 m × 5.2 m + 2.0 m × 5.2 m | Corner 4 m × 9.8 m + 2 m × 9.8 m | 3 m × 3 m insulation product uncovered | 3 m × 3 m finished ETICS system | Corner 2.8 m × 5.5 m + 1.5 m × 5.5 m | Corner 2.8 m × 9.0 m + 1.5 m × 9.0 m | 10 × nominal width of cavity barrier, max 3 m |
| **Substrate** | Aerated concrete | Aerated concrete | According to "end-use condition" | According to "end-use condition" | Any substrate | Any substrate | Aerated concrete |
| **Criteria**[a] | Temp. rise, fire spread, falling parts[b] | Not included | height of the damaged area < 250 cm and not extending to the edges; droplets not igniting the specimen | height of the damaged area < 250 cm and not extending to the edges | Temp. rise, fire spread, falling parts[b] | Temp. rise, fire spread, falling parts[b] | Integrity and insulation (180 K max temp. rise) |

a) Additional criteria may be contained in other standards or regulations.

b) Temperature rise, vertical and/or horizontal flame spread, burning droplets and particles, falling parts of certain size or mass.

**Table 3.** Main parameters of major international and national test methods for fire spread on external wall cladding. International and Asian tests.

| Standard | ISO 13785-2 | ISO 13785-1 | GOST 31251 | JIS A 1310 |
|---|---|---|---|---|
| **Country** | CZ, SK | CZ, SK | AZ, AM, KG, KZ, MD, RU, TJ, UZ, UA | JP |
| **Fire exposure** | Propane set by calibration | Propane 100 kW | Soft wood 700 MJ/m$^2$ | Propane 600 kW or more |
| **Max. heat flux on surface** | 55 kW/m$^2$ at 0.6 m height | Not specified | 12.5 kW/m$^2$ at 2 m height | Not specified |
| **Max. temp. on surface** | Min. 800°C | Max. 150°C at 0.5 m height | 750°C at lintel | Not specified |
| **Test duration** | 23–27 min | 30 min | 35 min | 25 min<br>20 min fire exposure |
| **Test specimen** | Corner 3.0 m × 5.7 m + 1.2 m × 5.7 m | Corner 1.2 m × 2.4 m + 0.6 m × 2.4 m | Flat wall 3.0 m × 5.1 m | Flat wall 1.8 m × 4.1 m |
| **Substrate** | Any | 12 mm Ca-Si board | Masonry, concrete | Two layers of 12 mm Ca-Si board |
| **Criteria**[a] | Not included | Not included (CZ version: internal temp.) | Not included | Not included |

a) Additional criteria may be contained in other standards or regulations.

b) Temperature rise, vertical and horizontal flame spread.

**Table 4.** Main parameters of major international and national test methods for fire spread on external wall cladding. *North American tests.*

| Standard | NFPA 285 | NFPA 268 | CAN/ULC-S134 | CAN/ULC-S101 + CAN/ULC-S114 |
|---|---|---|---|---|
| **Country** | USA | USA | CAN | CAN |
| **Fire exposure** | Gas burner | Radiant panel 12.5 kW/m$^2$ | Propane set by calibration | Defined time/temp curve |
| **Max. heat flux on surface** | 38–40 kW/m$^2$ at 0.6–1.2 m height | 12.5 kW/m$^2$ | 45 kW/m$^2$ at 0.5 m height | Not specified |
| **Max. temperature on surface** | 538°C at 3 m height or in cavities; 260°C on the 2nd storey | Not specified | Approx. 415°C | Not specified |
| **Test duration** | 30+10 min | 20 min | 25 min | Min. 15 min |
| **Test specimen** | Flat wall 4.1 m × 5.3 m | Flat wall 1.2 m × 2.4 m | Flat wall 10 m high | Flat wall min. 2.8 m × 2.8 m |
| **Substrate** | 150 mm th. concrete wall | Steel tube | Concrete | None |
| **Criteria**[a] | Flame spread, cavity fire spread, temp. rise[b] | Ignition | Vertical flame spread, heat flux | Integrity of top layer[c] |

a) Additional criteria may be contained in other standards or regulations.

b) Temperature rise, vertical and horizontal flame spread on or within the wall.

c) Protection of foam plastics on external facade –protective layer must remain in place, no openings developed, does not disintegrate. S114 is additional non-combustibility requirement.

## References


[1] Smolka, M. et al., "Semi-natural test methods to evaluate fire safety of wall claddings", Proceedings of the 1st International Seminar on Fire Safety of Facades, Paris, France, 14–15 November, 2013.

[2] Principles of the assessment procedure, the experimental procedure and the evaluation of tests to prove the suitability of fire protection measures against a fire exposure from outside a building close to the facade ('base fire exposure') for ETICS with polystyrene insulation – Revision 2 by DIBt internal issue 01/30/2015.

[3] DIN E 4102-20 (draft) Brandverhalten von Baustoffen und Bauteilen, Teil 20: Besonderer Nachweis für das Brandverhalten von Außenwandbekleidungen. DIN 2009.

[4] JIS A 1310 Test method for fire propagation over building façades. Japanese Standards Association, 2015.

[5] Nishio, M. et al., "Experimental study on fire propagation over combustible exterior facades in Japan", Proceedings of the 1st International Seminar on Fire Safety of Facades, Paris, France, 14–15 November, 2013.

[6] ISO 13785-1 Reaction-to-fire tests for façades –Part 2: Intermediate-scale test. ISO, 2002.

[7] ISO 13785-2 Reaction-to-fire tests for façades –Part 2: Large-scale test. ISO, 2002.

[8] Internal Form 281. Ministry of Interior, Department of National Fire Corps, Public and Civil Defense, General Directorate for Prevention and Technical Safety, Italy, 2015.

[9] Technical Report TR073 (draft). Large Scale Fire Performance Testing of External Wall Cladding Systems. EOTA PT4 Task Group, 2013.

[10] BS 8414-1 Fire performance of external cladding systems. Test methods for non-loadbearing external cladding systems applied to the face of a building. BSI, 2002ISO 9705 Fire Tests-Full-scale Room Test for Surface Products. ISO, 1993.

[11] Jensen, G., "Fire spread modes and performance of fire stops in vented façade constructions – overview and standardization of test methods", Proceedings of the 1st International Seminar on Fire Safety of Facades, Paris, France, 14-15 November, 2013.

[12] prEN 1364-6 (draft). Fire resistance tests for non-loadbearing elements – Part 6: Cavity Barriers. CEN, 2015.

[13] Yoshioka H. et al., "Large-scale Facade Fire Tests Conducted Based on ISO 13785-2 with Noncombustible Facade Specimens", Fire Science and Technology, **31**, No.1, 2012.

[14] Yong-Ho Yoo et al., "Study on Prevention of Spread of Vertical Fire along Finishing Materials for External Wall of High-Rise Buildings", Proceedings of the 1st International Seminar on Fire Safety of Facades, Paris, France, 14–15 November, 2013.

[15] NFPA 285. Standard Fire Test Method for Evaluation of Fire Propagation Characteristics of Exterior Non-Load-Bearing Wall Assemblies Containing Combustible Components. NFPA, 2012.

[16] LEPIR II Test. Large scale Fire Performance testing of construction systems for façade – CECMI (French committee for the evaluation and the classification of products and elements of construction as regards to fire hazard – under the French Ministry of Internal Affairs directives), 2013.

[17] MSZ 14800-6 Fire resistance tests. Part 6: Fire propagation test for building facades. Hungarian standardisation institute 2009.

[18] SP FIRE 105. Issue 5. Large scale testing of facade systems. SP Boras Sweden, 1994.

[19] Önorm B 3800-5 (draft) Fire behaviour of building materials and components – Part 5: Fire behaviour of facades – Requirements, tests and evaluations. Austrian Standards Institute, 2003.

[20] GOST 31251-2008. Facades of buildings. Fire hazard test method. Interstate council for standardisation, metrology and certification 2010.

[21] NFPA 268. Standard Test Method for Determining Ignitibility of Exterior Wall Assemblies Using a Radiant Heat Energy Source. NFPA, 2012.

[22] CAN/ULC-S134. Standard Method of Fire Test of Exterior Wall Assemblies. Underwriters Laboratories of Canada, 2013.

[23] CAN/ULC-S101 Standard Methods of Fire Endurance Tests of Building Construction and Materials. Underwriters Laboratories of Canada, 2007.

[24] CAN/ULC-S114. Standard Method of Test for Determination of Non-Combustibility in Building Materials. Underwriters Laboratories of Canada, 2005.

MATEC Web of Conferences **46**, 01004 (2016)
DOI: 10.1051/matecconf/20164601004


# Recent progress on test evidence, standardization and design of protection for exterior openings

Ruth Astrid Strøm

Securo AS, Norway

**Abstract.** Any opening in the exterior of structures are critical when exposed to fire. Windows and doors may be open, ventilation openings may be unprotected, wall cladding and roof tiles may allow flames or embers to enter cavities. Gaps may have developed over time and allow embers to settle in crevices or cavities in the structure. In fact, rarely does initial fire penetration occur right through cladding, sheets, insulation or solid walls. Penetration is most often related to openings, weak joints or cracks.

There is no doubt these weak points to fire spread in construction or indoor are crucial to the outcome of exterior fire exposure whether caused by wildfire or fire in outdoor trash or neighbouring buildings. Therefore, this has been addressed by authorities, standardization bodies and industry world-wide and much progress have been made during the last decade. This paper provide overview on new standards, on codes and on various pass/fail criteria set by code or classification standards. The paper describes how new designs of fire resistant ventilating facades and roofs apply dedicated cavity fire barriers and how new standard test methods helps to improve performances.

A number of different fire scenarios that elements in vented facades/constructions can be exposed to have been studied. New test standards have been developed for the last few years and are explained. Finally, this paper evaluate how different fire stop products perform in different fire scenarios and which test standards are applicable.

## 1. INTRODUCTION

Over the last decade, increasing attention has been drawn to the fire risk that openings in vented construction represent and how these should be protected.

It is crucial that fire spread in building cavities is prevented. Several different fire barriers, solid and ventilated, are available but apply for different types of fire exposures. Standard furnace test methods for exterior vents might provide limited proof of fire resistance. These tests apply the ISO 834 standard heating curve, a situation very different from the realistic worst case fire scenarios. Another concern is that temperature readings for the first 10 minutes of the fire are omitted from the tests while, according to studies done, most combustible debris ignites within 1–5 s when hit by flame [1]. For ventilating fire barriers which are open at start and then sealed upon heat exposure, this means that flames can penetrate the vent in the initial open state before it is completely sealed and spread through the building cavities.

### 1.1 Pass/fail criteria and test evidence vented construction

Exterior vents are inherently open at the time of fire starting and subject to harsh environmental exposures. An exterior fire challenges performance of vents by multiple exposures:

Embers
Flaming
Radiation
Sustained fire

of which scenarios vary from nearby building fires, wildfire, arson to auto ignition incidents. Secondary effects of exposure may be metal expansion against supporting construction, delamination and deflection of sandwich panels that create gaps for fire to pass cavity barriers etc.

## 2. OVERVIEW OF EXTERIOR FIRE SCENARIOS

### 2.1 Flaming fire through window

When window glass breaks in a fire, an immense smoke plume at 500–900°C is released. The plume's circumference will be flaming as it contacts oxygen in air. Part of the non-flaming hot smoke will spread into cavities where they rise as in a chimney. At first, there may not be enough air in the cavity to sustain flaming but as the buoyancy force creates substantial pressure differential more hot smoke and fresh air is drawn from all sides of the facade cavity. This may cause flaming inside the cavity or an immediate ignition of accumulated hot, unburnt smoke to a flashover. Another flaming phenomenon occurs when unburnt, hot smoke leaves the cavity from opening at the top. The unburnt gases will ignite as they get in contact with oxygen in the fresh air. This is the same principle as oil refinery torches or chimney fires as gas streams out of pipes or ducts to ignite, flare-up and sustain as a steady flame. It is assumed that this flame further increases the buoyancy force to draw even more unburnt smoke into the cavity. An assumed third effect of this condition we find no research on: The make-up air to the flaming on top of the cavity should cause an upward draft and likely extend the flame height on the outer surface as well. All this adds up to a possible explanation as to why these scenarios often escalate very fast and may explain why eaves are known as the most vulnerable part of the envelope in exterior fires.

Full scale research on the complete scenarios is lacking. Although the single effects are reported by research and observations in product and ad hoc testing the full understanding and quantification remains a hypothesis that should be investigated.

Fire through windows are most frequently applied worst credible scenario in FSE design of facades.

### 2.2 Wildfire or fire in nearby structures

Wildfire spread almost exclusively via openings, by penetration of embers or flames and to a less degree by radiated heat. The protective means are usually designed and tested to 10 min since a typical wildfire attach to structures last for about 3 min only and few more than 6 min. Extending the burn-through times are easily achieved by increasing thickness of wood, other material, sealants or intumescents. Therefore, the groundbreaking results of wildfire protection make an invaluable base of knowledge base when designing protection against building-to-building fire spread as well.

Wildfire or fire in neighboring structures challenges facades the most. Therefore, research has led to standards for testing, fire resistive designs and test methods to verify the fire resistance or burn-through times of most parts of facades.

### 2.3 Incidental fire

Exterior fires may develop incidentally by processes, maintenance work, play, vehicles, auto ignition and quite often in trash cans. Most of these represent slow development but are insidious as they often start at night or when unseen.

### 2.4 Liquid fire

Spilled flammable liquid fire at ground level of facade cladding may develop faster from ignition than enclosure fires in rooms. Direct flame impact from flashover by enclosure fires typically occurs after 5 min so these conditions cannot be compared.

### 2.5 Flashover

Occurs when enclosed unburnt gas/smoke ignite or get exposed to oxygen in air. It typically occurs in a room when opening a door. It may happen with cavities of facades as well when gaps develop by deflection from heat exposure or when fire fighters make access by breaking off cladding panel board.

### 2.6 Direct flame impingement by hot work, pyrotechnics, lightning strikes, gas or dust explosions

Direct flame impingement may relate to work, arson or unintentional incidents. When designing protection of façade openings even these exposure scenarios should be considered.

## 3. DEVELOPMENT OF NEW CODES AND STANDARDS

Until recently it was not considered practical to protect ventilation openings in facade by fire dampers or by automatically activated closure of fire resistance rated louvres over openings. New passive designs have been developed however, to block out fire and embers from windows, vents and gaps. Simple multiple-layer glazing provides just the required resistiveness to wildfire attack. Mesh and labyrinth type steel vents block embers and flames during this typical 3–6 min exposure.

Concerns were common globally and justified that direct flame impingement may spread during the open state of vents that seal up during first minutes of heat exposure. Work is completed in US and ongoing in Europe to develop test standards to addresses this issue.

### 3.1 Europe

National requirements within European countries vary widely. Some countries emphasize protection against wildfire, while countries at risk from "urban" type of exterior fire put resources on protection of buildings against other buildings on fire.

Work groups at CEN are currently working on two test standards for cavity barriers and air transfer grilles in Europe. The technical committee for these work groups, TC127, has instructed that testing towards sudden fire exposure should be included in both test standards.

The test standard for cavity barriers, prEN 1364-6 [2] is a continuance of the EOTA TR31 for closed cavity barriers into one single CEN test standard for both closed and open/ventilating cavity barriers. The proposed test standard is currently being prepared for enquiry.

The pass/fail criteria of the method should comply with the general requirements of EN 1363-1 [3] and the results will be valid for indoor or façade applications. The method is a small scale furnace test and functions well as a way of comparing different cavity barriers ability to prevent fire spread in building cavities without needing to do a considerably more expensive full scale test. For assessment of entire ventilated façade systems, however, full scale testing is necessary. There is a variety of such façade tests available [4]. Aspects on these tests will be given later.

In line with the instructions from the TC127 committee, the test standard has been prepared with an annex describing an optional test method for ventilating cavity barriers that assesses the vents ability to resist penetration of flames in the open initial state. This additional test involves a removable shield that exposes the test specimen to immediate heat exposure which simulate breaking of glass, panel delamination, cavity flashover, arson liquid fire and similar abrupt transitions to flame.

A similar optional test has been suggested in the test standard for air transfer grilles (prEN 1364-5) which may include wall vents in facades.

### 3.2 USA and Australia

A parallel to the work in Europe, test standard ASTM E2912-13 [5] was developed in USA. This is a test method for non-mechanical fire dampers in vented construction and provides an evaluation of the test specimen ability to resist passage of flames radiation and hot gases caused by sudden direct flame impingement. In short, the method functions as follows: a specimen is exposed to sudden flame impact by quickly sliding it across a test bench so that it is positioned directly above a gas burner flame. The pass criteria of integrity in ASTM E119 (standard for fire resistance [6]) is $< 4$ s of sustainable flame. This basically apply to the open state test as well, although the E2912 records time of failure only and do not set any criteria. The E2912-13 is an additional test for products that comply with the E119 furnace test to verify that the product meets the test criteria in open initial state as well.

SPFR explored an ad hoc test method with similar set-up for assessing open state fire resistance of exterior wall vents, air transfer grilles and linear fire barriers for concealed space and perimeter fire barrier applications [7]. Figure 1 shows thermal image and temperature diagram for such a test of a ventilating cavity barrier.

ASTM E1912-13 is by now referred to as the flame blocking test of ASTM E2886 [8] for eave vents and drafted for inclusion in test standards for more exterior fire protection product.

California State Fire Marshal are revising building codes. It has informed it will reference the new set of ASTM standards on wildfire protection in order to comply with a code requirement that opening protection shall resist the intrusion of flame and embers. The current code already require that mesh openings of vents shall be within 1.6 mm and 3.2 mm [9]. This means products and systems of limited ability to resist flame or embers will no longer be acceptable.

NFPA Standard 1144 [10] is favored by US states for protection of structures in wildfire prone areas. A draft revision appears to go the way of CSFM, i.e. referencing the new set of ASTM standards.

Australia employs much of the same fire protection methodology as the US. Standards are stricter and protection requirements given for five risk levels of exterior fire exposure, urban and wildfire types.

The Australian national science agency, CSIRO have recently led to stricter pass/fail criteria and test methods for opening protectives against flame duration, ember size and radiation levels addresses the open state before vents seal up. In general, the new criteria tend towards 2 mm mesh openings for protection against embers (previously 6.4 mm) and no flame or $< 4$ s flame duration. Radiation criteria are less changed although the critical levels and effects of radiation is still debated. The same trend applies for USA by NIST and IBHS research.

### 3.3 Baltic-Nordic countries [11]

A Baltic-Nordic edited handbook "Brandsäkra Trähus Edition 3" (in Swedish) on the design of fire protection of wooden buildings. Codes refer to the handbook which describe methods to protect openings, gaps and joints by use of fire resistance rated facades solutions, rainscreen cladding and

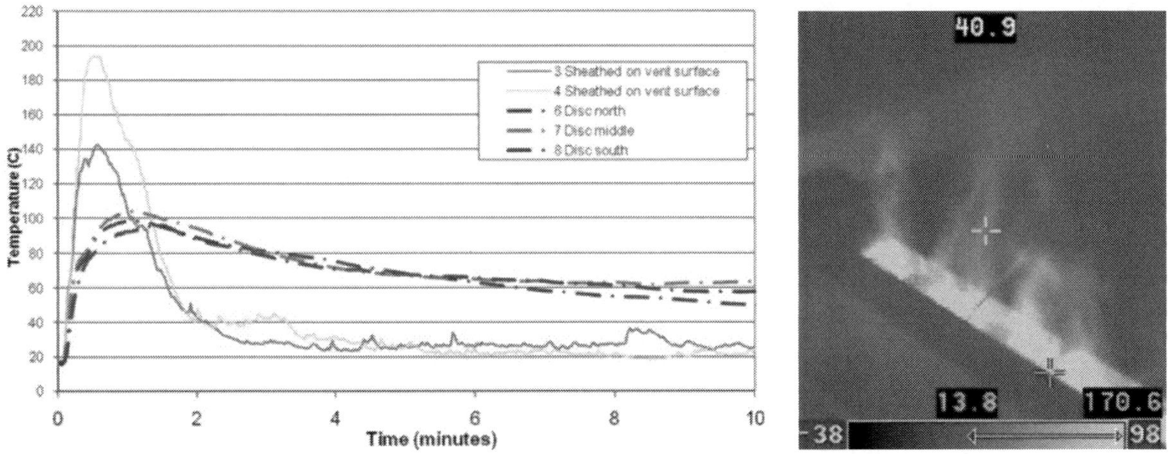


**Figure 1.** A cavity barrier type previously passing EN 1366-4 for EI 60 rating was used in a pilot test during development of sudden flame impingement test ASTM E2912-13. Diagram show readings of disc thermocouples to EN 1363-1 nos. 6, 7 and 8 on unexposed side. They all peaks around 100 °C at 60 s after direct flame exposure. For research, nondamped (sheathed) thermocouples nos. 3 and 4 were applied right above specimen and peak higher while still clearly below flame temperature during the entire open state from time of zero to about 60 s of flame exposure to specimen from below. Thermal images conform with readings of nondamped rather than the standard thermocouples. Note: thermal images show temperature by a color scale. Max temperature 170.6°C, i.e. no flame (SPFR).

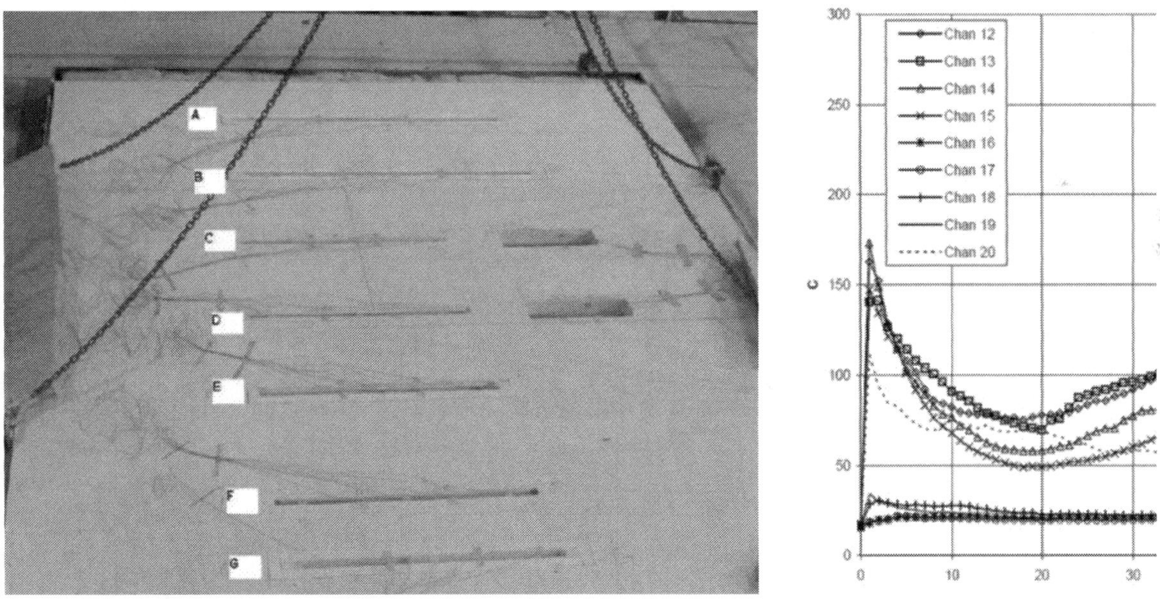


**Figure 2.** EI 30 cavity barrier in a EN 1366-4 horisontal furnace test of multiple specimens (SPFR)[13].

vents. The book describe fire protection of air gaps of façade cladding and points out the importance of protecting ventilated facades with cavity barriers that withstand the relevant exposures.

Such cavity vents were tested by ad hoc sudden flame impingement tests as a supplement to EN 1366-4 [12] in order to determine if acceptance criteria of EN 1363-1 are met in the open state. Figures 1 and 2 below shows test set up and time-temperature curves for these tests. These test results were considered by the working groups of ASTM and CEN along with other test evidence of UK and US types of ventilating cavity barriers.

### 3.4 Full scale façade tests

Several methods for full-scale façade tests exists: LEPIR II, SP FIRE 105, BS 8414-1, DIN E4102-20 [14–17]and more. The advantage of such tests is that all aspects of a complex wall can be assessed from one single test.

Ad hoc façade tests demonstrate that without a fire barrier flames can spread rapidly in the cavity behind cladding regardless of the materials used in the façade and they can extend 5–10 times of original flame height as they seek oxygen for combustion [18].

Proprietary tests of façade systems to SP Fire 105 full-scale façade test proves that ventilating rainscreen cladding including wind-breaker and battens may all be combustible while still preventing fire spread. This is enabled by cavity barriers plus fire retardant paint/chemicals on the outer surface only.

One drawback with the full scale façade standards is that they do not simulate the sudden flame exposure scenarios, such as breaking of window or arson liquid fire. In other words, those tests cannot be used to fully assess the fire resistant properties of ventilating fire barriers [1]. A solution to this issue could be to add a smaller scale test standard, such as ASTM E2912-13 or CEN prEN 1364-6.

## 4. CAVITY BARRIERS

As stated above, without proper fire barrier a fire can propagate unseen in a ventilated cavity system.

The challenge with many cavity barriers is that the must serve conflicting requirements, i.e. allow air to vent the cavity and prevent flames to spread in the event of a fire [18]. Most fire stop measures available today may prevent fire spread but do not provide venting. Or, the other way around: Some ventilating cavity fire barriers provide venting but do not block penetration of all flaming scenarios. Some designs apply to specific fire exposures only. So far there is no test standard that distinguishes type of fire exposure versus type of fire barrier.

To illustrate this we looked further into a range of cavity barriers available and which fire exposures they are aimed at. This is summarized in Table 1. It should be noted that the same exposures will apply to other facade elements as well, such as wall vents or eaves vents, see the paragraph on opening protection in facades further down.

To summarize, in order to maintain adequate ventilation as well as prevent all flame penetration when exposed to fire, cavity barriers should be designed to first block flames from entering. Secondly, stop fire spread to other cavity spaces should a flashover occur in one cavity. Then, once fully sealed the cavity barriers prevent hot smoke from spreading fire to other cavity spaces during the fire resistance time of the barrier or cladding. Finally, even when the cladding itself or the barrier are burnt away this does not mean fire have spread to the next cavity space, any a remaining cavity fire may then but start exposing the next cavity barrier and its associated cladding. The latter development presumes that the fire resistive outer surface of cladding prevents fire from bypassing cavity barriers.

It should be recognized that the rationale above is deducted from cited sources only.

## 5. OPENING PROTECTION IN FACADES

While ventilating cavity barriers are part of facade construction systems, most facades incorporate some form of vents or openings for venting as well. These are not specifically addressed in this paper. These vents are however, exposed to the same exterior fires although vertically oriented. In order to protect a building by a fire safe building envelope it must resist fire uniformly, so the same or similar standards as for fire resistance rated cavity barriers exists or are being developed to apply to vent openings in facades.

Many products are developed and applied. Most are either not tested for fire resistance at all or tested in furnaces to standards for enclosure fires. Furnace testing to standard exposure however does not

**Table 1.** Summary of different cavity barriers and applicable fire exposures and test methods.

| CAVITY BARRIER | FIRE EXPOSURES | TEST METHOD |
|---|---|---|
| **Intumescent strip**<br> | Room fire (enclosure fire, i.e. slow gradient in time, no direct flame)<br><br>Façade fire | Standard furnace test and fire resistance criteria. EU:<br><br>Comply 5 to 30/60/120 min of test. prEN 1364-6<br>or:<br>Comply full scale façade fire test, SP Fire 105 or other national test.<br><br>**Typical fire response profile**<br>• EI 60 min (5-60)<br>• No open state fire resistance |
| **Perforated metal barrier or metal sheet labyrinth**<br>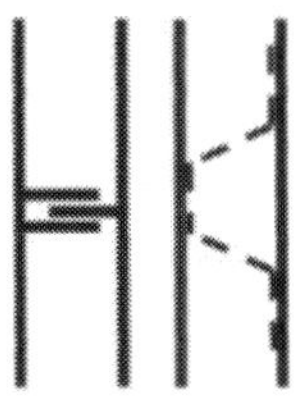 | Wildfire ember and flame attack.<br><br>Ad hoc rainscreen cavity fire tests. | EU: Ad hoc full scale tests done.<br>US: ASTM E2886 Test of ember and flame blocking<br><br>**Typical fire response profile**<br>• Labyrinth types pass E2886<br>• Perforated,large holes,report-ed fire spread delay 4-6 min |
| **Flame quenching mesh and intumescent element**<br> | Ember attack<br><br>Sudden direct flame exposure<br><br>Room fire (enclosure fire, i.e. slow gradient in time, no direct flame)<br><br>Façade fire | Standard furnace test and fire resistance criteria.<br>EU: Comply 0 to 30/60/120 min of test. prEN 1364-6$_{Flame}$ (method B)<br>or:<br>Comply full scale façade fire test, SP Fire 105 or other national test.<br>US: ASTM E2912-13 + ASTM E119<br><br>**Typical fire response profile**<br>• EI 60 min (0-60) (method B)<br>• Open state fire resistance<br>• Ember resistant |
| **SOLID BARRIER** (Non-ventilating)<br>Wood/mineral wool/calcium silicate/gypsum/steel sheet<br> | Room fire (enclosure fire, i.e. slow gradient in time, no direct flame)<br><br>Façade fire | EU:<br>prEN 1364-6 (cavity barrier test)<br><br>**Typical fire response profile**<br>• EI 30 min (expected)<br>Assumed to resist ember attacks and direct flame without testing. |

require failure of integrity and insulation to be recorded during first 5 min and the exposure in this period simulate a slow and uniform build-up of a smoke layer. The new standards for exterior vents addresses this by recording failures from time of 0 to 5 min under realistically simulated exterior exposures and include the open state before the vents are sealed or closed. The standards require a standard furnace enclosure fire test to be passed as well, for rest of the fire rated time.

Using for example ASTM E2886 with ASTM E2912 and ASTM E119 one can have fire rated vents listed for exterior openings in facades of type gable vents, foundation vents, eave vents, wall vents. Such listing verify resistance against embers, flaming and sustained fire from time of zero to 10 min, 30 min, 60 min or to 2 h or more.

## 6. CONCLUSION

New standards are available to verify fire resistance performance of vents for protection of openings or ventilated wall cavities in facades and others are being developed. The small scale test methods were evaluated to be trustworthy with respect to replicating credible exterior fire exposures. Exterior fires are different from enclosure fires by not establishing a hot smoke layer or ventilation controlled fire while presenting several unique fire ignition and spread modes. Wildfires and structures fires as well as various incidental fires scenarios are covered.

The published set of ASTM standards involves tests of ember blocking and direct flame blocking as well as of integrity and insulation at sustained fire. The pass/fail criteria of the tests apply to the open state before vents become fully sealed as well as sustained fire during the sealed state. In the open state the exposures are characterized primarily by embers and flaming.

At CEN two standards are being developed; for cavity barriers and air transfer grilles respectively. As at ASTM, CEN aims at covering vents to protect exterior and interior openings in vented construction. The test methods however, are set up differently. ASTM apply fuel-controlled exposures while CEN apply standard furnace tests. Depending on expected service of the test specimen the CEN task group propose to alter the installation of cavity barrier specimens to simulate a sudden flaming condition. Thereby it shall meet the instruction to record any failure of fire penetration during the open state. The CEN tests are not planned to incorporate ember blocking test for exterior vents.

Not all cavity fire barriers in facades are likely to be exposed to all identified credible scenarios. It is found purposeful that no single test should verify all scenarios. This is in accordance with published ASTM standards and intention of CEN. It allows for designs and products to fit specific applications and the exposure to comply with the relevant national regulation.

### References

[1] Jensen, G., "*Fire spread modes and performance of fire stops in vented façade constructions – overview and standardization of test methods*", *MATEC Web of Conferences,* **9**, 02002, 1st International Seminar for Fire Safety of Facades. Paris, France, November 14–15, 2013.

[2] prEN 1364-6, *Fire resistance tests for non-loadbearing elements – Part 6: Cavity Barriers.*

[3] EN 1363-1:1999, *Fire resistance tests. General requirements.*

[4] Smolka, Miroslav et al, "*Semi-natural test methods to evaluate fire safety of wall claddings*". *MATEC Web of Conferences,* **9**, 02002, 1st International Seminar for Fire Safety of Facades, Paris, France, November 14–15, 2013.

[5] ASTM E2912-13 *Standard Test Method for Fire Test of Non-Mechanical Fire Dampers Used in Vented Construction.*

[6] ASTM E119. *Standard Test Methods for Fire Tests of Building Construction and Materials.*

[7] Test report NBL A10109 dated 2010-10-21 from SINTEF NBL. *Ad hoc test of fire resistance during the open state of exposure to fire.*

[8] ASTM E2886. *Standard Test Method for Evaluating the Ability of Exterior Vents to Resist the Entry of Embers and Direct Flame Impingement.*

[9] 2010 California Building Code, Title 24, Part 2 (First Printing), Includes Errata/Supplement through July 1, 2012. *Chapter 7A – Materials and Construction Methods for Exterior Wildfire Exposure*, Section 706A Vents.

[10] NFPA 1144: Standard for Reducing Structure Ignition Hazards from Wildland Fire.

[11] Östman, B. et al, "*Brandsäkra trähus. Version 3*" p. 125-130. AB Stjärntryck, 2012.

[12] EN 1366-4:2006+A1:2010, *Fire Resistance Tests for Service Installations. Linear Joint Seals.*

[13] Test report 103011.11 dated 2009-03-27 from SINTEF NBL as according to NS-EN 1366-4:2006.

[14] LEPIR II. *Test method for fire spread*, 1970.

[15] SP FIRE 105 Issue 5. *Large scale testing of façade systems.* SP Boras, Sweden, 1994.

[16] BS 8414-1 *Fire performance of external cladding systems. Test methods for non-loadbearing external cladding systems applied to the face of building*. BSI, 2002.

[17] DIN E 4102-20 *Brand verhalten von Baustoffen und Buteilen, Teil 20: Besonderer Nachweis für das Brandverhalten von Aussenwandbekleidungen*. DIN 2009.

[18] Colwell, S. and Baker, T. "*Fire exposure to external thermal insulation for walls of multistory buildings. Third edition*" p. 5–6 and 14–15, BRE Trust, Garston, Watford, 2013.

MATEC Web of Conferences **46**, 01005 (2016)
DOI: 10.1051/matecconf/20164601005


# Development of full-scale façade tests in ISO TC92

Patrick van Hees

Department of Fire Safety Engineering, Lund University, Sweden

**Abstract.** During the 90ies, ISO TC92 SC1 started the development of full-scale tests for facades based upon an international review. The work resulted in two standards namely ISO 13785 part 1 and part 2. The work was performed in WG7 and a substantial input from international experts was given. The paper gives a short historical update on the development but also includes a short overview of the major type of hazards or risks, which can be identified with respect to façades and how they relate to the façade test standards developed. Finally the two methods are summarised and reference to recent work is given.

## 1. BACKGROUND

Within ISO TC92 SC1 fire development, WG7 deals mainly with full and intermediate reaction to fire tests. After completion of the ISO 9705 room corner test standard [1] in the early 90ies, ISO TC92 SC1 WG7 forwarded a questionnaire towards national member bodies in order to investigate what type of large-scale test needed to be developed after the room corner test. Two areas for further development were clearly indicated: sandwich panels and façades. The WG7 started up a task group to see the possibilities for development of standards in this area. For façades it was decided to develop both a screening method at small level and a large-scale test.

After a first screening of available large-scale tests, it was decided to look to three different national test methods being originally from Sweden, Germany and Canada [2–4]. The Swedish method SP 105 [4] was developed in the 80ies primarily by Lund University [5–7] and later incorporated and further developed in a SP method [8]. The German proposal was mainly based on the research conducted at Leipzig [9], which resulted in a draft German standard available during the 90ies [3]. Also other research was conducted in Germany [10] at that moment. This draft standard [3] has now been further developed in a newer version [11]. It is important to point out that the versions do have different thermal attacks, which is important when comparing the different test standards. The Canadian based on the UL method [2] was also available at that moment. Even that method has been further developed in more recent versions [12]. The major differences in the three methods were items such as fuel for the thermal attack (gas, wood or combustible liquid), level of thermal attack, dimensions and configurations (re-entrant corner or not), possibility for observations, etc. Although that intensive tests and documents were distributed in the WG7 only few information was published in peer review journals or at scientific conferences [13, 14]. Finally a more performance-based standard [15] was written having propane as the main ignition source but allowing other fuels such as wood and liquids as long as it could be shown that the requirement of the total heat flux in the test standard could be fulfilled. The final level of total heat flux was established by international consensus. The standard set-up is explained more in detail in one of the next paragraphs.

Simultaneously a screening method was developed in the same WG7. This action was made as it was felt that large-scale tests are very expensive and there is need from industry to have cheaper product development tools. The standard was based on the results of a number of research project related to wooden façades [16, 17]. The work was also incorporated in a PhD dissertation [17] showing the link between the intermediate scale test and a large-scale test. The work resulted in an intermediate scale test [18]. However it was not checked if the test also correlated with the part 2.

During the development also European actions were undertaken. These actions resulted in a report dealing with the risk related to façades, an overview of a number of national standards and their link to the different risks [19]. The risk of façades will also be taken up in this paper in the next paragraph. After the publication of the ISO 13784 part 1 and 2 standards, a number of research publication were published showing comparison of different tests methods and different façade systems [20–23]. More recently work has been done in EOTA with respect to large-scale tests [24]. All these studies show the considerable risk associated to façades and the diversity of national and regional standards worldwide [25–33].

## 2. HAZARD AND RISK DETERMINATIONS

When considering façades and curtain walling systems, different types of hazards/risks can be considered. The risks and hazards considered in this paper are certainly not all types but they include the most important ones. Recent fires and research publications [20, 34] have shown that fires involving the exterior of a building should not be neglected. First there is the type of thermal attack or fire source. This thermal attack can be divided in to the following four main types

1. Fire inside the room
2. Fire in a room and outside the window
3. Fire at the exterior of the building close the façade e.g. container fire
4. Fire from another building.

Together with the type of fire source, different cases can be distinguished each addressing a specific hazard or risk. A schematic overview is given of four different hazards/risks below. Each of them has to be covered by a different type of testing set up or regulations and such attempts were made in earlier standardisation work [19]. Such work should always be done before introduction of new test methods in regulations. Regulations worldwide can namely address fire safety of façades in different ways. Examples are given in [35] for the Nordic regulation or in [28, 29] for insurance applications. Even routes of performance based design [36] and fire safety engineering [37] can be used. However this is considered outside the scope of this article.

### 2.1 Case 1

In this case the fire risk is penetration of the fire through the facade system from one level to another level trough e.g. the joint systems of the façade systems when a fire occurs inside a room. It can be considered that this type of hazard/risk is more linked to the evaluation of the fire resistance of a specific system. Specific attention should be given to the complexity of such system and the possible requirements the fire barrier border should or should not go throughout the façade system. This type of risk can also include spread of a fire from an attic down into the façade system [38].

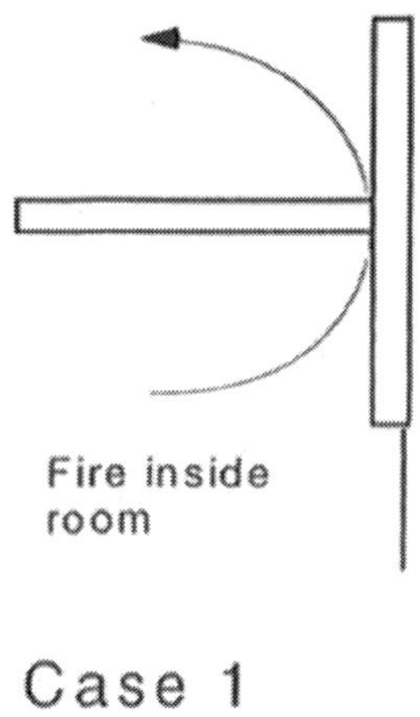


**Figure 1.** Case 1 of typical hazard/risk involving a façade. (Reproduced from [19].)

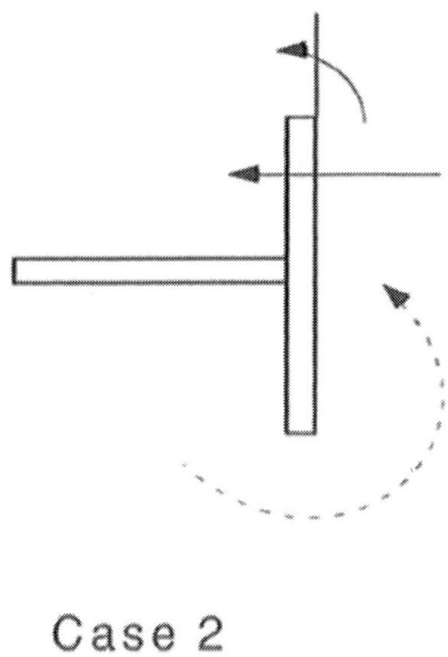


**Figure 2.** Case 2 of typical hazard/risk involving a façade. (Reproduced from [19].)

### 2.2 Case 2

When a fire occurs outside a room through the window there is a risk of penetration of the fire through the façade system or through the glazing to the next floor. This can occur even when appropriate joints are covering the risk in case 1. Also here most of the risk can be covered by a fire resistance test. Else regulation of safety distance can cover this risk (see case 2b, Fig. 3). An example is given in [39].

### 2.3 Case 3

When a fire occurs outside the window or at the exterior of the building (container fire) a considerable risk of fire spread on the façade surface or through the void between façade and building exists. Here the reaction to fire properties of the material used for the façade should be considered and is the main determining factor. However constructional details of windows and voids should not be neglected as well as means how to check fire stoppers inside the construction.

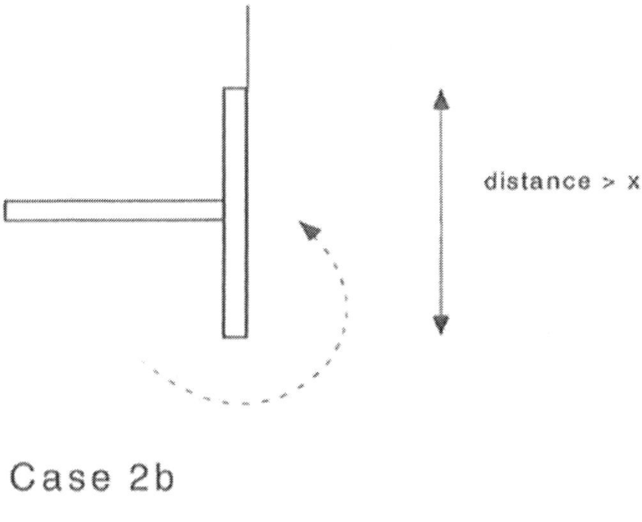


**Figure 3.** Case 2b of typical hazard/risk involving a façade. (Reproduced from [19].)

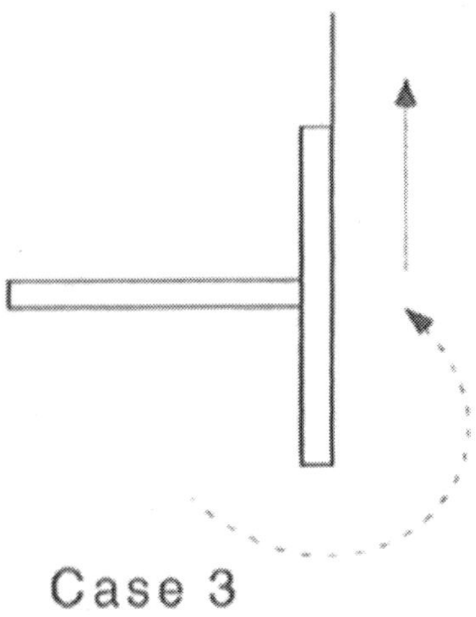


**Figure 4.** Case 3 of typical hazard/risk involving a façade. (Reproduced from [19].)

### 2.4 Case 4 Falling parts or debris

When a fire occurs through a window or al the exterior of the building nearby the façade it is important that no parts are falling clown from the façade. This would endanger the work of the rescue teams and also cause problems for evacuation of the building.

Constructional details from the façades are in this case extremely important. Glazing can be included if necessary but can be covered by other tests.

## 3. INTERMEDIATE SCALE FAÇADES TEST ISO 13785 PART 1

### 3.1 Need of screening test

Full-scale tests are expensive and there is need for the industry and for certification of products to reduce the scale of the tests. However it is necessary to explain the difficulties when scaling down the size of full-scale tests. Problems can occur if thermal exposure is not sufficient or not well scaled down. Another problem can be that the size of the specimens is too small so that mechanical behaviour of systems is

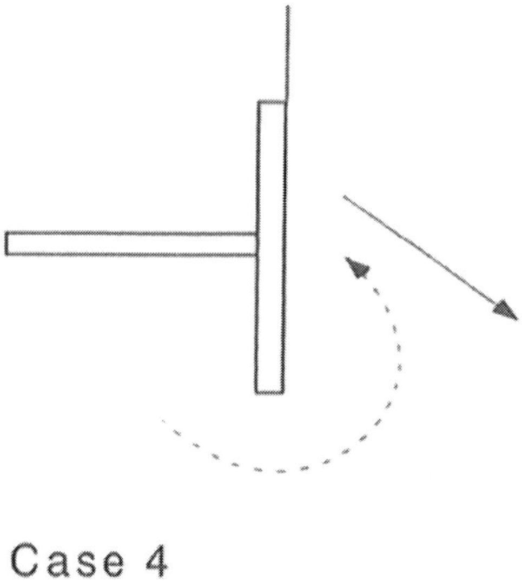


**Figure 5.** Case 4 of typical hazard/risk involving a façade. (Reproduced from [19].)

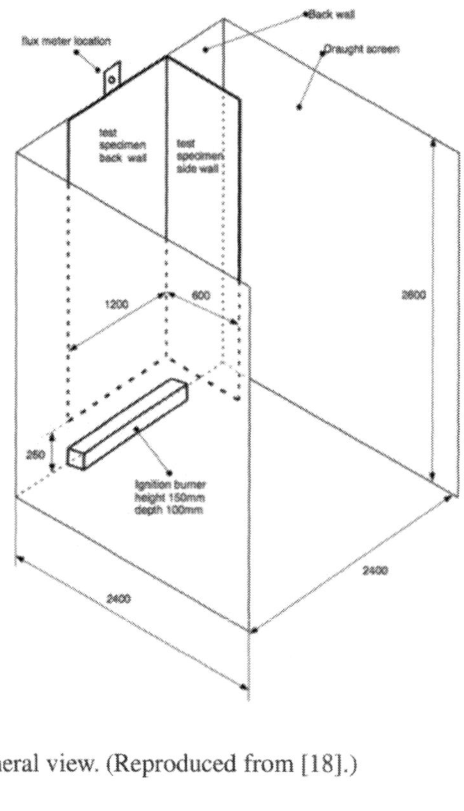


**Figure 6.** ISO 13785 part 1 – general view. (Reproduced from [18].)

not covered by the screening test. Screening tests can be used in combination with modelling but here it is important to investigate, verify and validate the models considerably [40].

## 3.2 Description of test method

Part 1 of the ISO standard consist of a test specimen of 1.2 × 2.9 m which is exposed at the bottom by a line burner with a heat release rate of 100 kW. Two draught screens are position perpendicularly to the

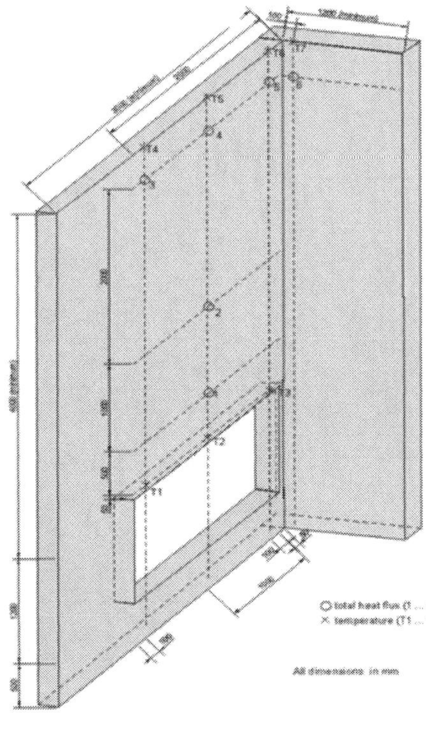

**Figure 7.** ISO 13785 part 2 – General view and measurement locations. (Reproduced from [15].)

wall to avoid unwanted draft and unstable flames. Measurement of temperature, total heat flux, flame spread and ignition are performed.

### 3.3 Recent developments

After the research performed with mainly wooden façades no real further investigations have been done on this method. Each systematic review of the standard in ISO resulted in confirmation of the standard.

## 4. LARGE SCALE FAÇADES TEST ISO 13785 PART 2

### 4.1 Description of test method

The method consists of a large-scale set-up built as an L-shaped configuration with two walls with dimensions of minimum 3 × 4 m and 1.2 × 4 m (width × height). The exposure to the façade construction is by flames emerging from a window with dimensions 2 × 1.2 m. The combustion chamber shall be of a regular shape with an internal volume not less than 20 m$^3$ and not more than 100 m$^3$. The recommended heat source is propane but other alternatives are given (wooden cribs and combustible liquids). Measurement of flame spread, temperature on surface and inside the construction as well as total heat fluxes at the surface are performed. An overview of the test set-up is given in Fig. 7 and different characteristics of the test method are listed in Table 1.

### 4.2 Recent developments

After development of the test standard there was little data available on the test method. However during the last 5 years a number national and regional projects were conducted which investigated the suitability of the method and compared it to other façade test standards [41–43]. During the latest systematic review

**Table 1.** Summary of the test method.

| Characteristic of the test method | Result | Characteristic of the test method | Result |
|---|---|---|---|
| Risks covered in the actual test method | Case 3 | Height of rig (m) | Min. 4 m above the opening |
| Use in legislation | No | Width of rig (m) | Min. 3 m + min 1.2 m. |
| Main set up | Fire room with comer shaped façade | Opening front wall (Width × Height) | 2 × 1 m |
| Exposure levels | 55 ± 5 kW/m$^2$ at 0.5 m<br>35 ± 5 kW/m$^2$ at 1.5 m<br>Both values for at least 15 minutes | Ventilation opening back wall | Dimensions subject to calibrations |
| Configuration | Corner | Ventilation | Natural or forced through ventilation opening |
| Windows and constructions details included | No | Size of Fire room | Between 20 and 100 m$^3$ |
| Measurements | Temperature, Heat Flux | Heat source | Propane as standard eat source (wood, heptane as alternatives) max. 120 g/s (approx. 5–6 MW) |
| Mounted area | Both walls | | |

in ISO a request was launched for revision of the standard. At this moment ISO TC 92 SC1 is waiting for a formal request to start revision of the method.

## 5. CONCLUSIONS

With the increase of high rise buildings and higher demands on thermal insulation, fire safety of external façades has become recently more important. Several hazards/risk can be identified with respect to façades and in order to tackle the flame spread risk on the surface, ISO TC92 SC 1 WG7 developed a large-scale test during the 90ies, which was published in 2001. As façades construction are very complex and are complete systems and not single materials or products, a full assessment by means of full-scale test is necessary. However there is a need for intermediate or screening to reduce testing and developing costs. Therefore an intermediate screening test was also developed by ISO and published in 2001. At the moment a myriad of national and regional test standards are available but the actual ISO test standard provides an internationally accepted alternative. Before introducing test standards into regulations, regulators should identify which risk they want to reduce and choose the appropriate performance criteria or safety levels. With this information, a suitable test method can be chosen or a suitable performance based solution based on fire engineering can be proposed.

### References


[1] ISO 9705, "Fire Tests-Full-scale Room Test for Surface Products", ISO, 1993.

[2] DIN E 4102-20 (draft), "Brandverhalten von Baustoffen und Bauteilen, Teil 20: Besonderer Nachweis für das Brandverhalten von Außenwandbekleidungen", DIN 2009.

[3] CAN/ULC-S134, "Standard Method of Fire Test of Exterior Wall Assemblies (CAN/ULC- S134-92)", Underwriters Laboratories of Canada, Scarborough, Ontario, 1992.

[4] SP FIRE 105, Issue 5, "Large scale testing of facade systems", SP Boras Sweden, Dnr 171-79-360, 1994.

[5] Ondrus, J., "Fire Hazards of Facades with Externally Applied Additional Thermal Insulation", Report LUTVDG/TVBB-3021 Lund Institute of Technology, Sweden. 1985.

[6] Ondrus, J., and Pettersson, O., "Fire Hazards of Facades with Externally Applied Additional Thermal Insulation Full Scale Experiments", Report LUTVDG/TVBB-3025 Lund Institute of Technology, Sweden, 1986.
[7] Ondrus, J., and Pettersson, O., "Fire Hazards of Window-Frames of Plastics, Aluminium and Wood Full Scale Experiments – a comparison", Report LUTVDG/TVBB-3037 Lund Institute of Technology, Sweden, 1987.
[8] Hermodsson, T. and Månsson, L., "Facades: Fire Testing of Materials and Constructions – A First Proposal for a Test Method (SP AR 1992:64)", Swedish National Testing and Research Institute, 1992.
[9] Hildebrand, C., "Façade Fire Tests," Workshop report, Institut für Baustoffe, Leipzig, Germany, 1988.
[10] G. M. F. Jeffs, H. G. Klingelhöfer, F. H. Prager, and H. Rosteck, "Fire-performance of a Ventilated Facade Insulated with a B2–classified Rigid Polyurethane Foam", *Fire and Materials*, **10**, Issue 2, 79–89, June 1986.
[11] DIN E 4102-20:2016-03 (draft): "Brandverhalten von Baustoffen und Bauteilen - Teil 20: Besonderer Nachweis für das Brandverhalten von Außenwandbekleidungen", DIN 2016.
[12] CAN/ULC-S134, "Standard Method of Fire Test of Exterior Wall Assemblies", Underwriters Laboratories of Canada, 2013.
[13] Sumathipala, K. and Kotthoff, I., "The Development of an International Fire Test Standard for Facades", Poster, Proceedings of the Fifth International Symposium Fire Safety Science. Melbourne, 1997.
[14] Babrauskas, V., "Façade Fire Tests: Towards an International Test Standard", *Fire Technology*, Third Quarter, pp. 219–230, 1996.
[15] ISO 13785 part 2, "Reaction to fire tests for façades — Part 2: Large scale test", ISO, 2001.
[16] Hakkarainen, T. and Oksanen, T., "Fire safety assessment of wooden facades", *Fire and Materials*, **26**: pp. 7–27. doi: 10.1002/fam.780, 2002.
[17] Hakkarainen, T., "Studies on fire safety assessment of construction products", Doctoral thesis, VTT publication nr. 459, Espoo 2002.
[18] ISO 13785 part 1, "Reaction to fire tests for façades — Part 1: Intermediate-scale test", ISO, 2001.
[19] Van Hees P., "Semi Natural Fire test for façades and curtain wall systems", Nordtest report 1433-99, SP AR 2000:39, Borås 2000.
[20] White N., Delichatsios, M., "Fire Hazards of Exterior Wall Assemblies Containing Combustible Components", CSIRO report EP142293, CSIRO, June 2014.
[21] Wade, C.A. and Clampett, J.C., "Fire Performance of Exterior Claddings", BRANZ Report FCR 1. Building Research Association of New Zealand. Judgeford, New Zealand, 2000.
[22] Wade, C.A., "Fire Performance of External Wall Claddings Under a Performance-based Building Code", *Fire and Materials*, **19**, pp 127–132, 1995.
[23] Smolkaa, M., Messerschmidt, B., Scott, J., le Madec, B., "Semi-natural test methods to evaluate fire safety of wall claddings", 1st International Seminar for Fire Safety of Facades, Paris, MATEC Web of Conferences Volume 9, MATEC Web of Conference, DOI: 10.1051/matecconf/20130902012, 2013.
[24] Kotthoff, I., Riemsch-Speer, I., "Mechanism of fire spread on facades and the new Technical Report of EOTA Large-scale fire performance testing of external wall cladding systems", 1st International Seminar for Fire Safety of Facades, Paris, MATEC Web of Conferences **9**, DOI: 10.1051/matecconf/20130902010, 2013.
[25] Önorm B 3800-5 (draft), "Fire behaviour of building materials and components – Part 5: Fire behaviour of facades – Requirements, tests and evaluations", Austrian Standards Institute, 2003.
[26] BS 8414-1, "Fire performance of external cladding systems. Part 1: Test method for non-loadbearing external cladding systems applied to the masonry face of a building", BSI, 2015.

[27] BS 8414-2, "Fire performance of external cladding systems, Part 2: Test method for non-loadbearing external cladding systems fixed to and supported by a structural steel frame", BSI, 2015.

[28] LPS 158: Issue 2.0:2010, "Requirements and tests for LPCB approval of non-loadbearing external cladding systems applied to the masonry face of a building, 2010.

[29] LPS 158: Issue 1.0:2010, Requirements and tests for LPCB approval of non-loadbearing external cladding systems fixed to and supported by a structural steel frame, 2010.

[30] L.E.P.I.R. 2 (Local expérimental pour incendie réel à deux niveaux), Arrêté du 10 septembre 1970 relatif à la classification des façades vitrées par rapport au danger d'incendie, Version consolidée au 16 février 2016, French test method for fire spread, downloaded from www.legifrance.gouv.fr on 20151220, 1970.

[31] NFPA 285, "Standard Fire Test Method for Evaluation of Fire Propagation Characteristics of Exterior Non-Load-Bearing Wall Assemblies Containing Combustible Components", NFPA, 2012.

[32] NFPA 268, "Standard Test Method for Determining Ignitibility of Exterior Wall Assemblies Using a Radiant Heat Energy Source", NFPA, 2012.

[33] BR 135, "Fire performance of external thermal insulation for walls of multi-storey buildings, second edition", BRE report, 2003.

[34] Peng, L., Ni, Z., Huang, X., "Review on the fire safety of exterior wall claddings in high-rise buildings in China", 9th Asia-Oceania Symposium on Fire Science and Technology, Procedia Engineering **62** (2013) pp. 663–670, Tokyo, 2015.

[35] Strömgren, M., Albrektsson, J., Johansson, A., Almgren, E., "Comparative analysis of façade regulations in the Nordic countries, 1st International Seminar for Fire Safety of Facades, MATEC Web of Conferences **9**, DOI: 10.1051/matecconf/20130901003, 2013.

[36] Mikkola, E., Hakkarainen, T., and Matala, A., "Fire safety of EPS ETICS in residential multi-storey buildings", VTT report VTT-R-04632-13, Espoo, 2013.

[37] Didieux, F., "Facade fire – fire safety engineering methodology", 1st International Seminar for Fire Safety of Facades, Paris, MATEC Web of Conferences **9**, MATEC Web of Conference, DOI: 10.1051/matecconf/20130903010, 2013.

[38] Johansson, N., "Fallstudie av konstruktionsbränder (in Swedish)", LTH report 3191, ISSN: 1402-3504 ISRN: LUTVDG/TVBB–3191—SE, Lund, 2015.

[39] Règlement de sécurité contre l'incendie relatif aux établissements recevant du public Livre II: Dispositions applicables aux établissements des quatre premières catégories Titre premier: Dispositions generals, Chapitre II: Construction Instruction technique n > 249 Relative aux façades, http://www.sitesecurite.com/ERPIT/IT24900.htm#0, Downloaded 20151220, 2015.

[40] Van Hees, P., "Validation and Verification of Fire Models for Fire Safety Engineering", Procedia Engineering, **62**, 2013, Pages 154–168, 9th Asia-Oceania Symposium on Fire Science and Technology, DOI:10.1016/j.proeng.2013.08.052, 2013.

[41] Yoshioka, H., Ohmiya, Y., Masaki Noaki; M. And Yoshida, M., "Large-scale Facade Fire Tests Conducted Based on ISO 13785-2 with Noncombustible Facade Specimens", *Fire Science and Technology* **31**, No. 1, 1–22, 2012.

[42] Macdonald, M., "A comparison of BS 8414-1 & -2, draft DIN 4102-20, ISO 13785-1 & -2", EN 13823 and EN ISO 11925-2", Report number CC275194 issue 2. BRE Global Ltd. 2012.

[43] Messerschmidt, B., Fellman, J., "Fire Testing of ETICS a comparative study". Proceedings of International Conference Interflam 2013, Windsor, UK. Interscience Communications, 2013.

MATEC Web of Conferences **46**, 02001 (2016)
DOI: 10.1051/matecconf/20164602001


# Experimental investigations of the fire behaviour of facades with EPS exposed to different fire loads

Christian Northe, Olaf Riese and Jochen Zehfuß

Technische Universität Braunschweig, Institute of Building Materials, Concrete Construction and Fire Protection (iBMB) Braunschweig, Germany

**Abstract.** In context of recent fire damages to façades and the related discussion about the safety of façade systems in case of fire, there is a need for discussion about the boundary conditions and the phenomena of façade fires based on fundamental investigation. On behalf of the European Institute for Fire Protection (EIFP) four major fire tests were carried out on a flat facade with external thermal insulation composite systems (ETICS) based on polystyrene with fire loads at the ground in front of the façade [1]. The tests were conducted at the Materialprüfanstalt Dresden based in Freiberg. The tests differ to the fire tests of the German building ministers in case of different fire loads and the corner-situation of the studied façade. The four fire tests are a step in systematic investigations to study examination criteria of the fire phenomena at façades. Further experimental and numerical investigations will follow.

## 1. FIRE TESTS AT IBMB

To evaluate different fire scenarios at the base of the façade fire pre-tests were carried out at iBMB. Wood cribs are selected and tested at a façade without ETICS. These fire loads are examples of waste containers or rubbish on the street. In addition, pool fires as a liquid fire sources are used to represent for example melted EPS on a job site scenario or a burning car to represent a fire with a higher heat release rate as well. A schematic measurement plan of the test façade is shown in Fig. 1. With 40 thermocouples, the temperature distribution on the façade is measured and the heat flux density (HF1 – HF5) is measured with four Gardon Gauge probes from 0.75 m to 2.75 m height. For the pool fire tests the plume temperature (M70 – M73) is measured at a height of 1 m, 2 m, 3 m and 4 m above the pool surface and moved 0.10 m left from the central axis. The measuring points of plume temperature are 0.70 m in front of the façade and the probes for heat flux density measurement are installed plane with the façade surface. The mass and the mass loss are recorded with a weighing system (MV).

### 1.1 Wood cribs

As a reference for solid fire loads wood cribs are used with a weight of 100 kg, 150 kg and 200 kg. The spruce wood was prepared at humidity of 50% and an air temperature of 23 °C. The wood cribs are

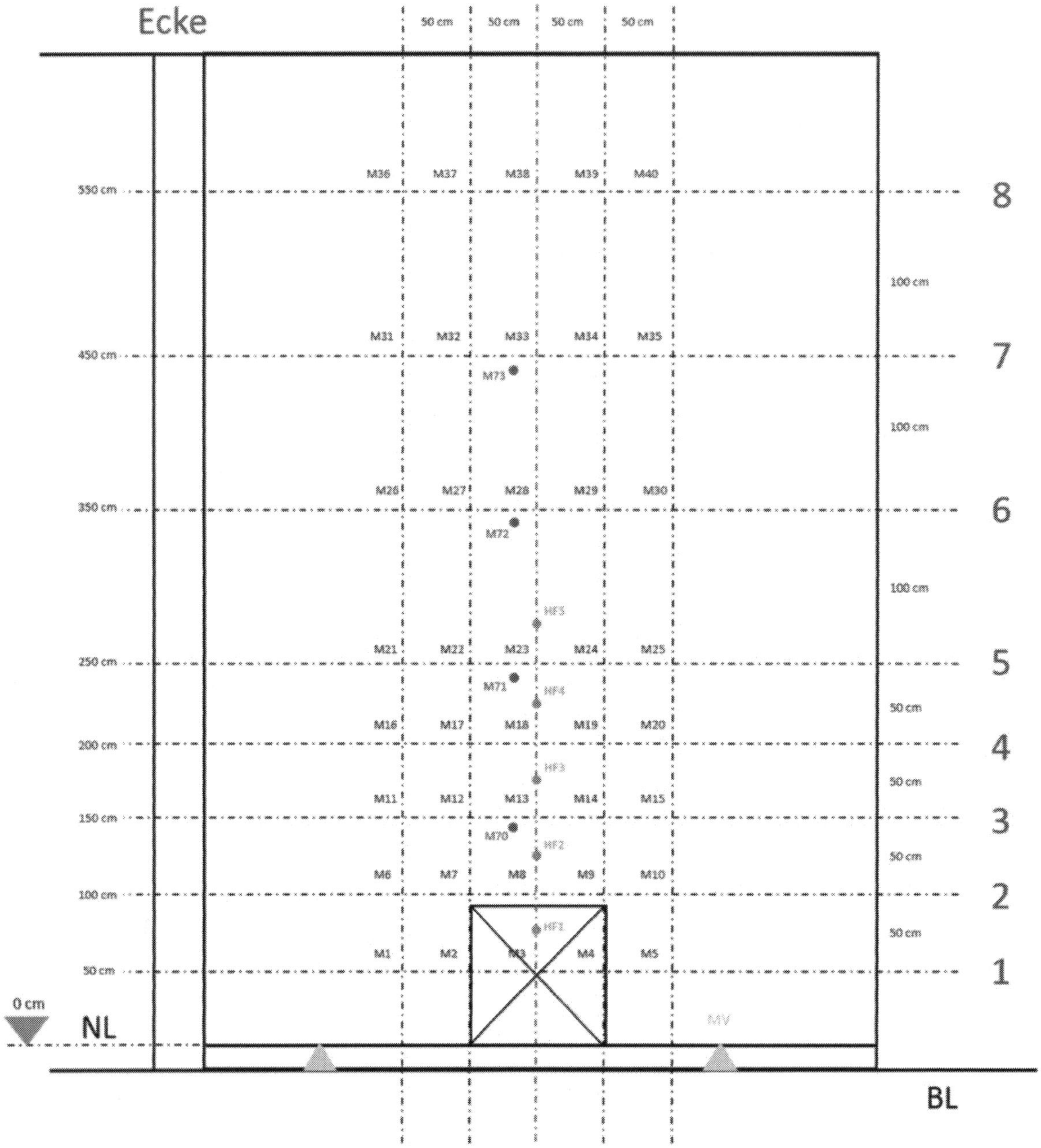


**Figure 1.** Experimental setup of the fire tests using the example of a wood crib.

made of single wooden sticks with an edge length of 39 (±2) mm × 39 (±2) mm and 1.00 m length. The density after preconditioning is 450 kg/m$^3$. The area of each crib is 1.0 × 1.0 m$^2$ and the height depends on the weight and the number of used wooden sticks. To ignite the cribs 3 × 200 ml methanol are used.

The heat release rate (HRR) is calculated by the mass loss rate $\dot{m}''$, the caloric value $H_u$ and the combustion efficiency $\chi$ according to Eq. (1) [2].

$$HRR = \dot{m}'' \cdot H_u \cdot \chi. \tag{1}$$

Figure 2 shows the HRR of the three tested wood cribs. The combustion efficiency was set to 0.9 and the caloric value for wood was set to 17.28 MJ/kg [3]. The maximum values are time shifted and reaches

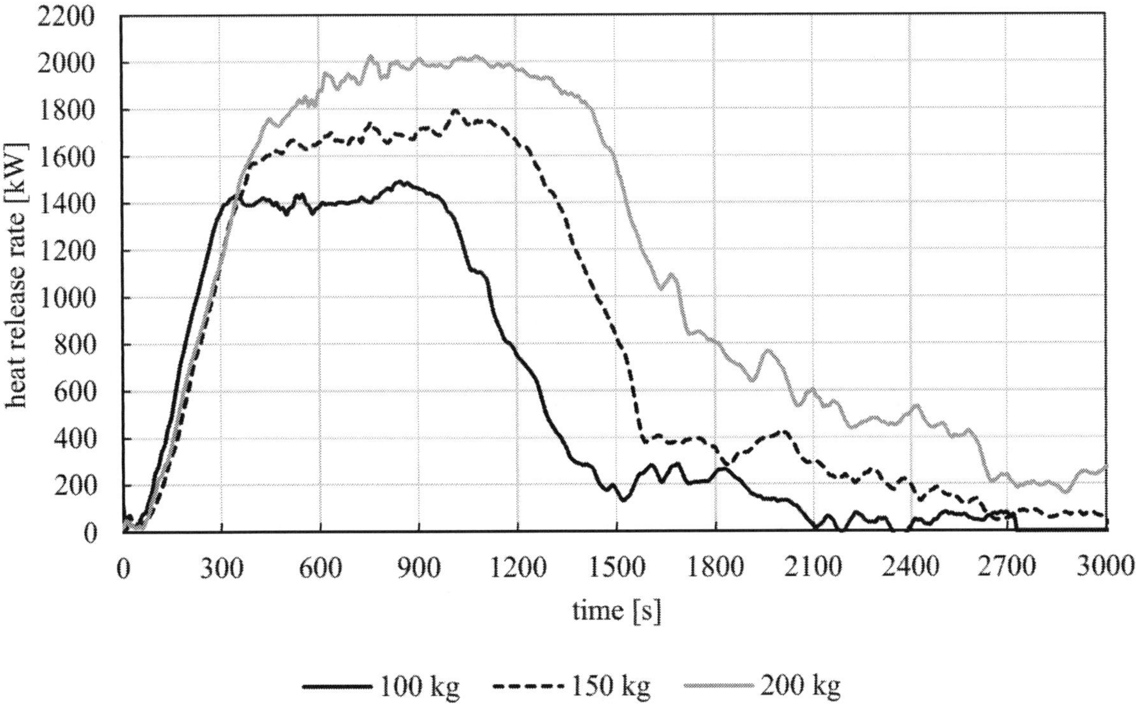


**Figure 2.** Heat release rate of wood cribs in the fire tests, 100 kg (solid black line), 150 kg (dashed black line), 200 kg (solid grey line).

from 1.4 MW for the 100 kg wood crib to 2.0 MW for the 200 kg wood crib. The three HRR-time-curves show a similar course including the fire development, the constant combustion and the combustion abatement. According to that, Figs. 3 and 4 show a similar course of the temperature-time- and heat flux density-time-curves at the centerline of the façade surface. The temperature maximum reaches from 700 °C for the 100 kg wood crib to 900 °C for the 200 kg wood crib. The measured heat flux density reaches time depending values of about 40 kW/m$^2$ at 1.25 m height above the zero line (NL).

### *1.1.1 Pool fires*

Comparing to the solid fuels two liquid fuels were tested as a fire source. Both liquids, methanol and isopropanol, were tested with a volume of 200 l in a steel pan with an area of 1.0 × 2.0 m$^3$ within a cooling pan with a size of 1.30 m × 2.80 m × 0.31 m, made of steel as well. According to the wood crib fire tests heat flux densities and temperatures at different heights and the mass loss were measured. For ignition a burning fuse was used. The caloric value of methanol is 19.44 MJ/kg and 27.00 MJ/kg for isopropanol [3]. The HRR-time-curve of isopropanol and methanol is shown in Fig. 6. Over a period of 75 minutes the methanol reaches a HRR of about 1 MW. The isopropanol pool fire reaches a maximum HRR of about 3.5 MW and was burned out after 28 minutes. In [4] is shown that a maximum value like this is similar to the HRR of passenger cars. The combustion efficiency was set to 1.0 for liquids [3]. In the centre axis of the façade surface the temperature reaches 900 °C for isopropanol and 500 °C for methanol at a height of 1.0 m, see Fig. 7. The measurement point M8 is 0.50 m above the measurement point M3 located and used because at this point a temperature maximum was recorded for the pool fires. The heat flux density reaches a value of 125 kW/m$^2$ for isopropanol and 30 kW/m$^2$ for methanol at a height of 1.25 m. The plume temperature reaches 850 °C for isopropanol and 350 °C for methanol at a height of 1.0 m above the surface of the liquid.

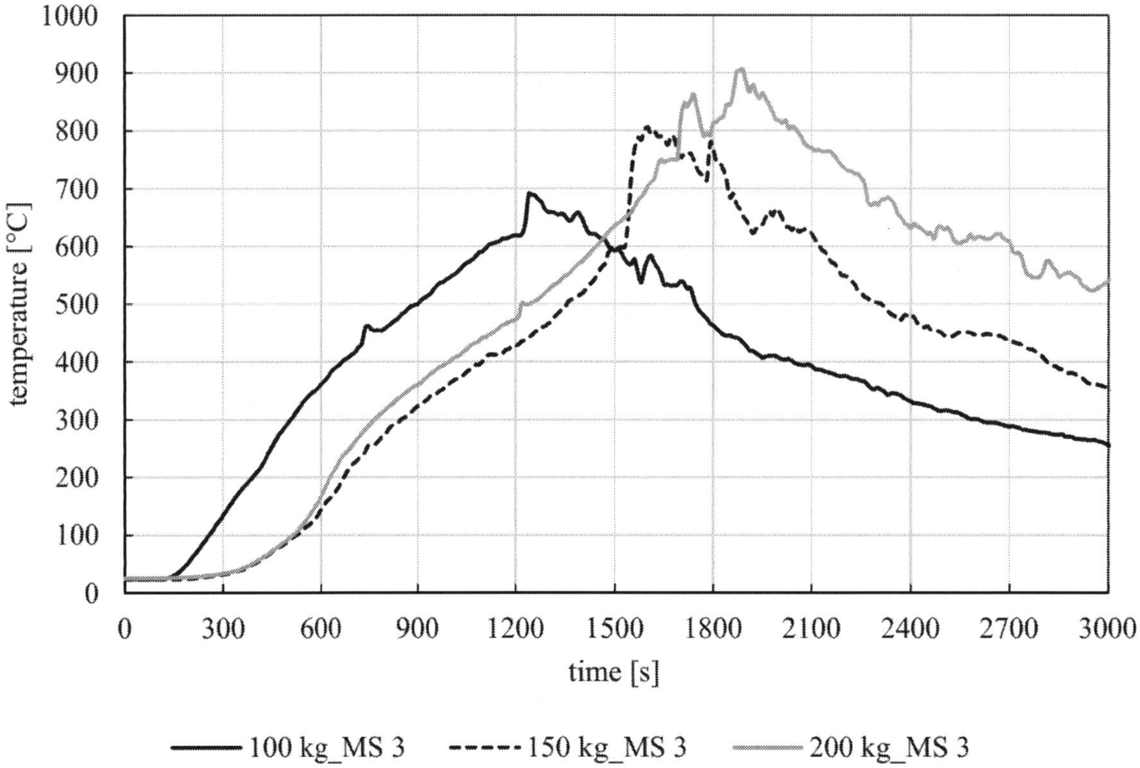


**Figure 3.** Comparison of the temperature-time-curves in the center axis of the façade for the fire tests at measurement point M3 at 0,50 m height, 100 kg (solid black line), 150 kg (dashed black line), 200 kg (solid grey line).

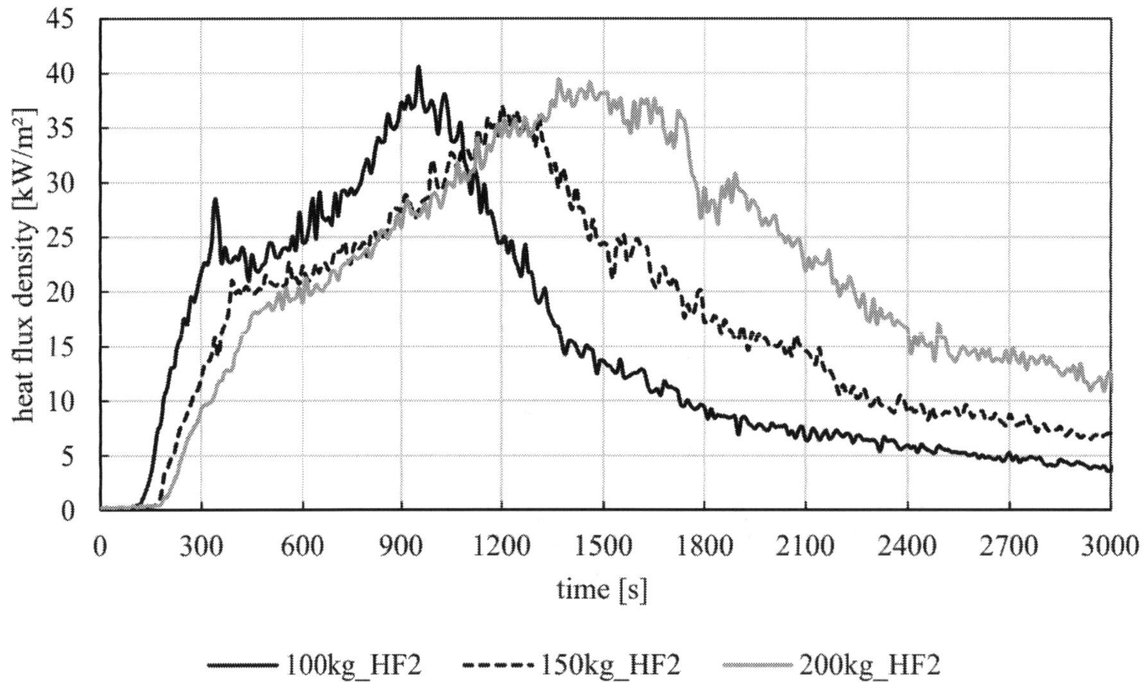


**Figure 4.** Comparison of heat flux density-time-curves on the façade surface for the fire tests at measurement point HF2 at 1.25 m height, 100 kg (solid black line), 150 kg (dashed black line), 200 kg (solid grey line).

**Figure 5.** Fire tests with pool fires; left methanol and right isopropanol.

## 2. LARGE-SCALE FIRE TESTS AT MPA DRESDEN

The large-scale (LS) fire tests were carried out at the Materialprüfanstalt Dresden based in Freiberg. The testing façade has an area of 6 m × 8 m (w × h) for the first and the second LS fire test and an area of 6 m × 9 m for the third and fourth test, no corner and a flat surface. For the fire tests an ETICS based on XPS at the base of the façade and EPS for the remaining façade was used. Altogether four LS fire tests had taken place in the period from August 2014 to September 2015. The used EPS/XPS has a thickness of 300 mm on which a layer of ground plaster, a reinforcing mesh and an upper layer of plaster were applied. An overview about the main features of the four LS-fire test is shown in Table 1. A schematic overview about the measuring points is shown in Fig. 10. The measurement points for temperature are at a distance of 1 m × 1 m and about 10 mm in front of the façade surface, number 1 to 24, type NiCr-Ni and at a depth of 150 mm inside the EPS insulation, number 25 to 45. According to the temperature measurement the wind speed and wind direction were recorded as well as the heat flux density at 4 m distance in front of the fire load. The first test is used as a calibration-test and the second and third tests as job site/ construction site tests. The fourth test was chosen to show the influence of holes or openings in the façade surface. The experimental conditions were not global laid down before all tests. They had always been redefined according to the latest test results. Table 2 gives an overview about all LS fire test, with comparative pictures at the same time of each test and with information's about the ETICS components after the four tests.

The first LS fire test was carried out in August 2014. The fire load was a 200 kg wood crib like it was used in the fire tests at iBMB. It was a flat façade with a surface area of about 8 × 10 m$^2$ without

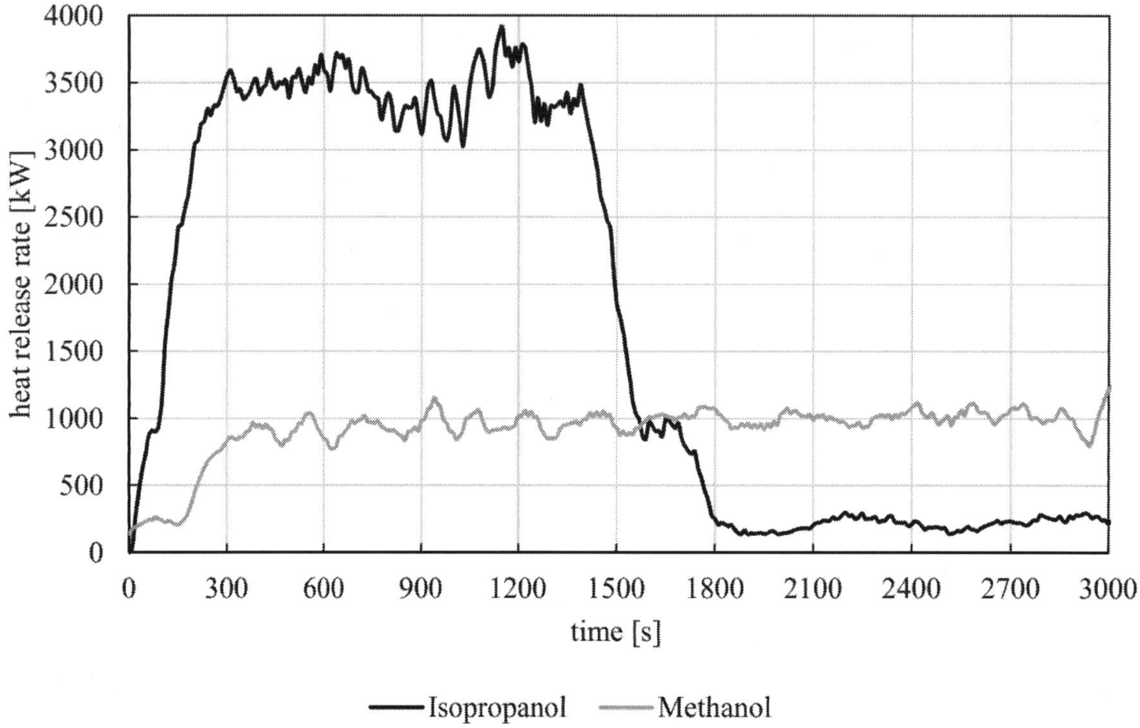


**Figure 6.** Comparison of heat release rate (HRR) of isopropanol (black line) and methanol (grey line).

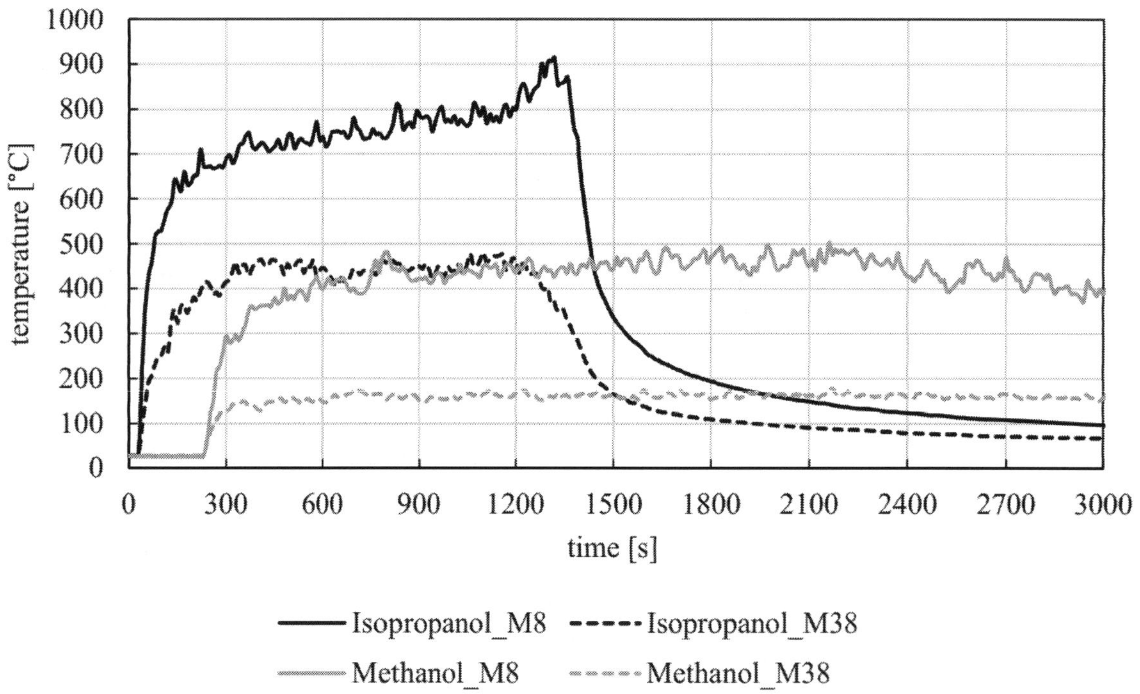


**Figure 7.** Comparison of temperature-time-curves at a height of 1.0 m (M8, bold line), 5.5 m (M38, dashed line) in the middle axis of façade surface for isopropanol (black lines) and Methanol (grey lines).

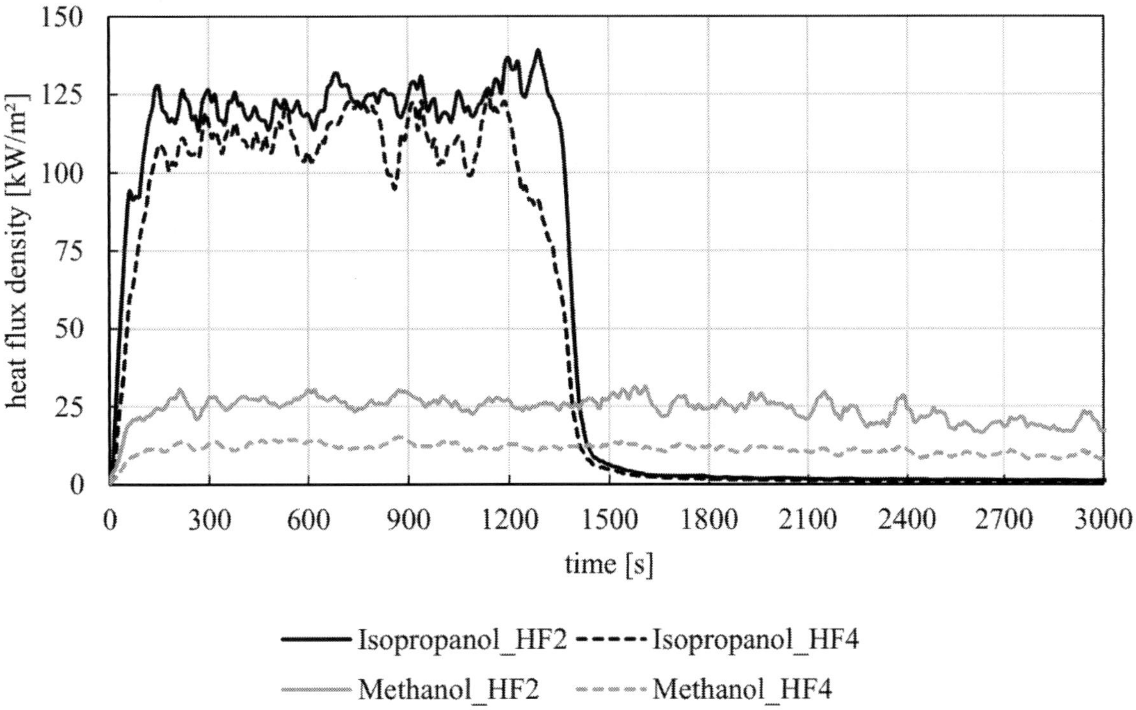


**Figure 8.** Comparison of heat flux density-time-curves (HFD) on the façade surface for the fire tests at measurement point HF2 at 1.25 m height and HF4 at 2.25 m height for isopropanol (black lines) and methanol (grey lines).

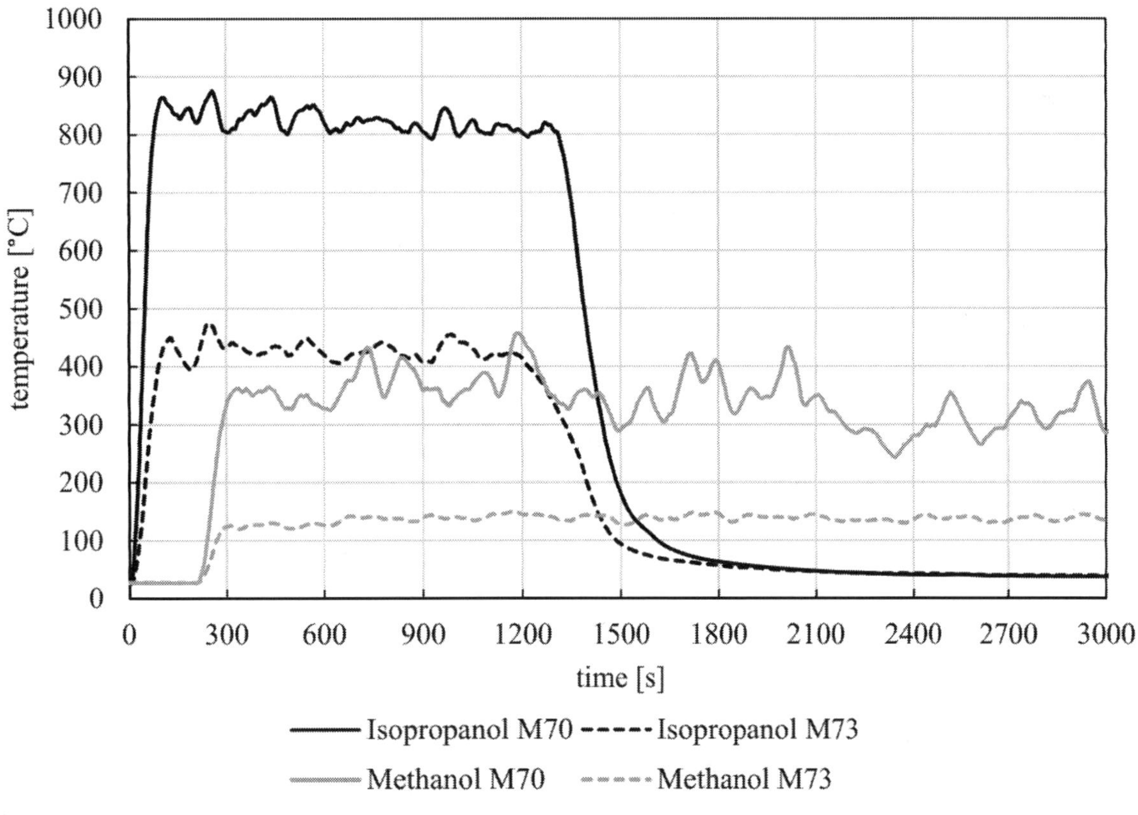


**Figure 9.** Comparison of plume temperature-time-curves at 1.5 m height (M70) and 4.5 m height (M73) for isopropanol (black lines) and methanol (grey lines).

**Table 1.** Overview about the main features of the ETICS that changed in progress.

| Comment | LS Fire Test 1 | LS Fire Test 2 | LS Fire Test 3 | LS Fire Test 4 |
|---|---|---|---|---|
| Fire load | 200 kg wood crib | 200 l isopropanol | 200 l isopropanol | 200 l isopropanol |
| HRR [kW] | 2000 ± 200 | 3500 ± 350 | 3500 ± 350 | 3500 ± 350 |
| HFD [kW/m$^2$] | 32 | 125 | 125 | 125 |
| Fire barrier | 1 | 1 | 3 | 3 |
| Openings | no | no | no | 3 (1.35 m × 1.01 m) |

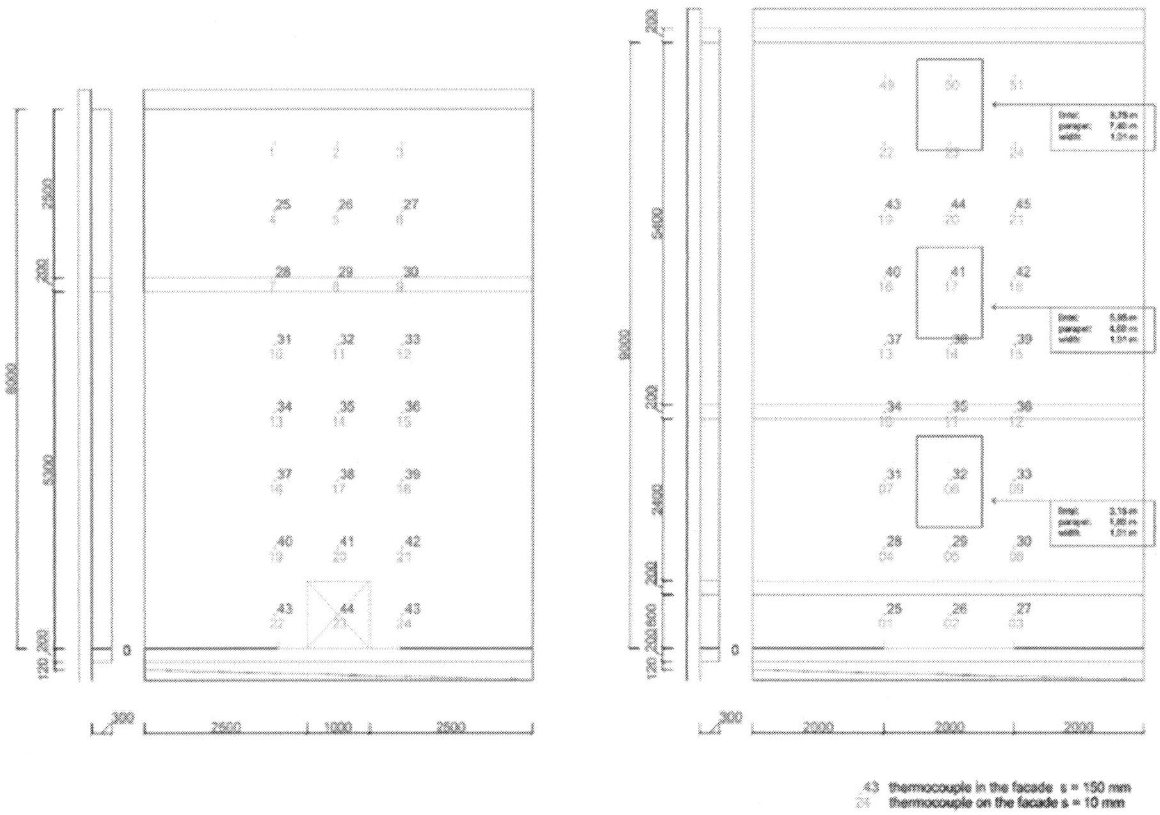

**Figure 10.** Schematic measurement plan of the first large-scale fire test (left) and the fourth large-scale fire test (right); bright: temperature measurement in front of the façade (10 mm), dark: temperature measurement at half depth of EPS (150 mm).

holes and corner configuration tested. A fire barrier was placed at a height of 5.30 m. A schematic measurement plan is shown in Fig. 10. The façade surface has slight traces of fire but no significant damage. The façade did not open. The temperature maximum on the façade surface reached a value of approximately 825 °C. Inside the ETICS (at 150 mm depth of the EPS) the temperature maximum reached about 335 °C.

The second fire test was carried out in November 2014 and had the same façade structure as the first, only the fire load was changed to a 200 l isopropanol pool fire. The isopropanol pool is used to represent a job site scenario with burning and melting polystyrene and also a fire with a higher HRR see [5], for example a car. The ETICS opened in the twelfth minute and burned down completely shown in Fig. 11. A temperature was measured of above 1000 °C. A precise position of the thermocouple could not be assigned because of the destruction of the measurement grid.

After the second LS fire test an instruction paper by DIBt (German approval authority for a uniform fulfilment of technical tasks in the field of public law) which regulates the construction of the façade

**Table 2.** Overview about the large-scale fire test of the "Pilotprojekt EIFP" [1].

| Comment | LS Fire Test 1 | LS Fire Test 2 | LS Fire Test 3 | LS Fire Test 4 |
|---|---|---|---|---|
| Date | 20.08.2014 | 03.11.2014 | 19.05.2015 | 01.09.2015 |
| 1 min |  |  |  |  |
| 5 min |  |  |  |  |
| 10 min |  |  |  |  |
| 15 min |  |  |  |  |
| 20 min |  |  |  | Cancelled within 12th Minute |
| 25 min |  |  |  | - |
| 30 min |  |  |  | - |
| 35 min |  |  |  | - |
| Condition of the plaster layer | In the area of primary flame decomposed and burned | Destroyed | Destroyed in flame area/ decomposed and burned | Destroyed |
| Condition of the ETICS | Polystyrene in flaming area decomposed | Almost destroyed | Damaged in main components | Façade burning until cancel through fire brigade |
| Condition of fire barrier | Not damaged | Not damaged | Not damaged | Not damaged |
| Comment | No fire propagation | Fire propagation to the upper end of façade | No fire propagation about middle fire barrier until 30 minutes | Fire propagation to the upper end of façade |

**Figure 11.** Second large-scale fire test, left: complete detachment of the plaster layer, middle: opening of the ETICS, right: burning fragments of the plaster layer and of melted EPS are thrown away.

design was issued [6]. The third test was planned considering [6]: one fire barrier at the transition from the base of the façade (made of XPS) to the main façade (made of EPS) at 0.80 m. A second fire barrier at the height of the first floor ceiling at 3.40 m and a third at the height of the third floor ceiling at 9.00 m, shown in Fig. 10 like it was made for the fourth test. The fire barriers were also secured with dowels. Within the third test carried out in May 2015, the façade opens first at about 30 minutes until the second fire barrier. The second opening happened during the 40$^{th}$ minute with the result of a completely damaged façade. Compared with the second test the measures suggested in [6] seem to delay the vertical fire spread significantly. Until the 30$^{th}$ minute temperature peaks of about 950 °C on the façade surface and about 450 °C inside the insulation were measured.

To show the unknown influence of openings in the façade surface LS fire test four was carried out in September 2015. The façade structure was the same like fire test three but three additional openings were made. The window lintel of the first opening was at a height of 3.15 m, the second at a height of 5.95 m and the third at 8.75 m height. All openings has a size of 1.35 m × 1.01 m (height by width). The influence of the three openings was obvious and the fourth fire test was cancelled within the twelfth minute by the fire brigade. In Fig. 12 problems with the lintel in the openings are illustrated. The temperature maximum on the façade surface reached a value of approximately 1300 °C. Inside the ETICS, the temperature maximum reached about 1050 °C until the opening of the façade. After the test it can be shown that the fire barriers are not damaged. The strong fire behaviour seems to be attributable to the early fail of the construction of the edges of the window lintels. The structural design of the edges was performed according to the manufacturer's instructions and the edges/ brackets were made of plastic.

## 3. LESSONS LEARNED AND PERSPECTIVES

The fire tests show that the fire load at the base of a façade is a decisive influence factor for the fire behaviour of the façade. With job-site scenarios or special circumstances, pool-fires could be realistic fire-loads for example melted and burning EPS. An additional use of non-combustible fire barriers in

**Figure 12.** Fourth large-scale fire test, left: burning of the edge profiles of the lintel, middle: exit from burning EPS, right: burning of the entire window opening.

comparative setups promotes an increased level of safety. The job-site tests are carried out by a higher fire-load (about 3.5 MW) than the wood crib (about 2 MW).

Based upon the third test with an equal façade setting and fire-load the fourth test differed through additional openings like windows and shows that the type of façade design or structure can have an influence on the burning behaviour.

More detailed information about the fire tests can be found in [5] and [7] and a complete and detailed description of each fire test is given by the part reports of each fire test and the final report of the research project [1].

## References

[1] Zehfuß, J.; Riese, O; Northe, C.; Küppers, J.; Pilotprojekt EIFP: Brandverhalten von Fassaden; January 2016 (previously unreleased).

[2] Babrauskas, V.; Grayson, S. J.; Heat Release in Fires, Elsevier Applied Science, 1992.

[3] DIN 18230-3:2002-08. Baulicher Brandschutz im Industriebau – Teil 3: Rechenwerte.

[4] Babrauskas, V; Heat Release Rates; In: Hurley, M. (Editor), SFPE Handbook of Fire Protection Engineering; Fifth Edition; Springer New York, 2016.

[5] Zehfuß, J.; Riese, O.; Northe, C.; Küppers, J.; Experimentelle und numerische Erkenntnisse zum Brandverhalten von WDVS-Fassaden auf Polystyrol-Basis; In: Bauingenieur (90) 2015; Edition 12/2015; Springer VDI Verlag.

[6] DIBt, Hinweis WDVS mit EPS-Dämmstoff, 16. Dezember 2014.

[7] Northe, C.; Riese, O.; Experimentelle Untersuchungen von Bränden an Fassaden mit unterschiedlichen Brandlasten; In: Braunschweiger Brandschutz-Tage 2015; September 15.–16.2015; Heft 227.

MATEC Web of Conferences **46**, 02002 (2016)
DOI: 10.1051/matecconf/20164602002


# Fire propagation over combustible exterior facades exposed to intensified flame in Japan

Yuhei Nishio[1], Hideki Yoshioka[2,a] Takafumi Noguchi[1], Manabu Kanematsu [3], Tatsuo Ando[1], Masamichi Tamura[1] and Yoshihiro Hase[4]

[1] The University of Tokyo, Japan
[2] National Institute for Land and Infrastructure Management (NILIM), Japan
[3] Tokyo University of Science, Japan
[4] Mitsubishi Plastics, Inc., Japan

**Abstract.** With regard to fire safety for exterior walls of a building, fire-resistance performance is considered, according to the current Building Standard Law of Japan. And it was revealed that the fire safety is not specifically regulated from the viewpoint of reaction-to-fire performance, such as fire propagation caused by combustible materials or products installed on the exterior side of fire-resistant load-bearing walls. Actual fire incidents in the world have shown that massive façade fire could occur at the exterior side of building wall even when the wall itself is fire resistant. In previous studies of the authors, a test method of façade fire was proposed for evaluating the vertical fire propagation over an external wall within the same building [1, 2]. Based on these studies, new domestic standard test method was established in Japan as JIS A 1310: 2015, "Test method for fire propagation over building façades" at the end of January 2015 [3]. But there was the argument that heat output of burner inside the combustion chamber was not sufficiently high in the previous study. In this paper, results of fire tests on combustible façades are discussed from the viewpoints of different strength of flame exposing facade. In this research, it was clearly found that JIS A 1310 with heat output of 900kW could be applicable for evaluating fire propagation behaviour over various types of combustible exterior façades.

## 1. INTRODUCTION

According to the current Building Standard Law of Japan, it is not restricted to install combustible materials over their exterior walls as long as fireproof buildings retain their required fire resistance performances. However, installation of combustible material over the exterior wall could cause large-scale fire propagation. In previous studies of authors, façade test method for evaluation of fire propagation over combustible cladding was proposed, using various types of combustible façade specimens [1, 2]. Heat output of gas burner in a combustion chamber was set 600 kW, because temperature in the chamber was about 1000 °C and temperature at lintel was about 900 °C, which are near at real fire condition.

---

[a] e-mail: yoshioka-h92te@nilim.go.jp

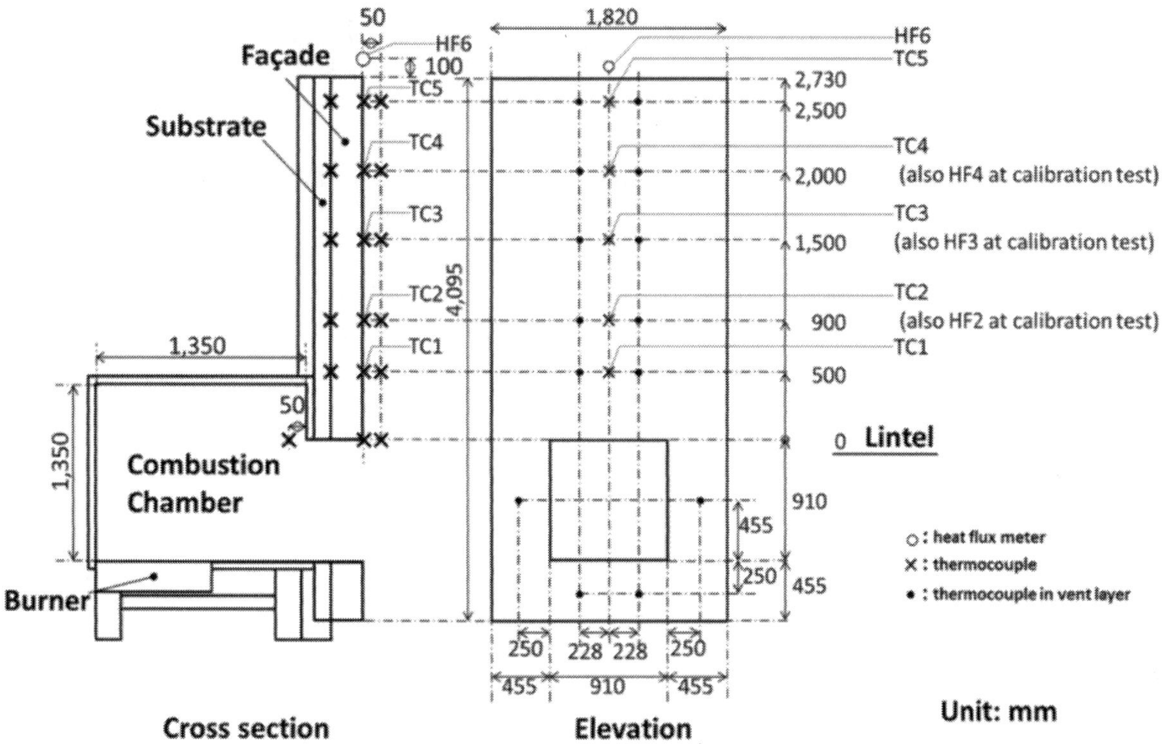


**Figure 1.** Cross section and elevation of test facility of JIS A 1310 including positons of thermocouples and heat flux meters.

However, some severe fire incidents occurred at high rise building covered with aluminium composite panel in Korea and Dubai, and a fire propagation also occurred at a building coated with resin paint in Japan even thought it was a small fire in scale. Compared with these combustible façade fires in actual cases, flame propagation at fire test with some combustible façade in previous study were smaller. It was considered that heat output in the previous study could not verify fire propagation over combustible façade properly.

The purpose of this paper is to investigate properties of fire propagation over various types of combustible façade specimens which could be applied in actual buildings with higher heat output from combustion chamber, compared with the previous study and to verify the proper heat intensity on JIS A 1310 as the test method for evaluating properties of fire propagation over combustible exterior façades as the result of the investigation.

## 2. TEST METHOD

Tests were performed according to JIS A 1310, which is vertical and façade-shaped simulating ejected flame from opening after flash over. Test facility of JIS A 1310 is shown in Fig. 1. As for test procedures, a propane burner is ignited when the test is initiated, and continues to supply a certain amount of propane gas constantly for 20 minutes by using a flow meter. For setting flow rate of propane gas, the heat output of propane burner was controlled by the oxygen consumption method in advance. A video camera was located diagonally opposite to the specimen's surface to visually observe fire propagation behaviour over combustible façade surface.

Unlike ISO 13785-2[4], specimen's size is smaller (from lintel to specimen's top is about 2.7m) and opening size is not rectangular but square (about 0.9m square). And specimen's shape is flat surface without wing-wall unlike ISO 13785-2.

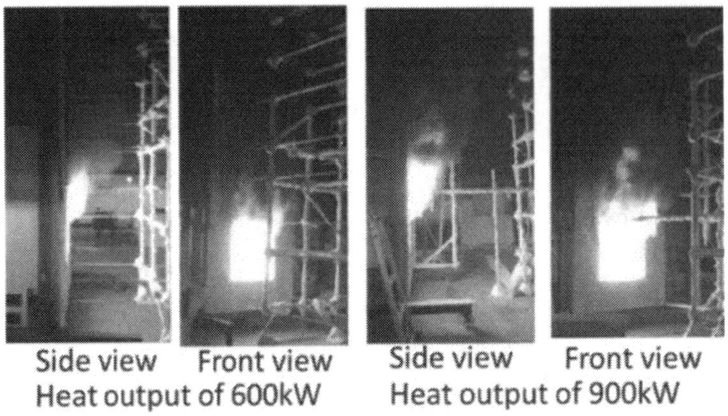


**Figure 2.** Photograph of flame venting through the opening at a calibration test.

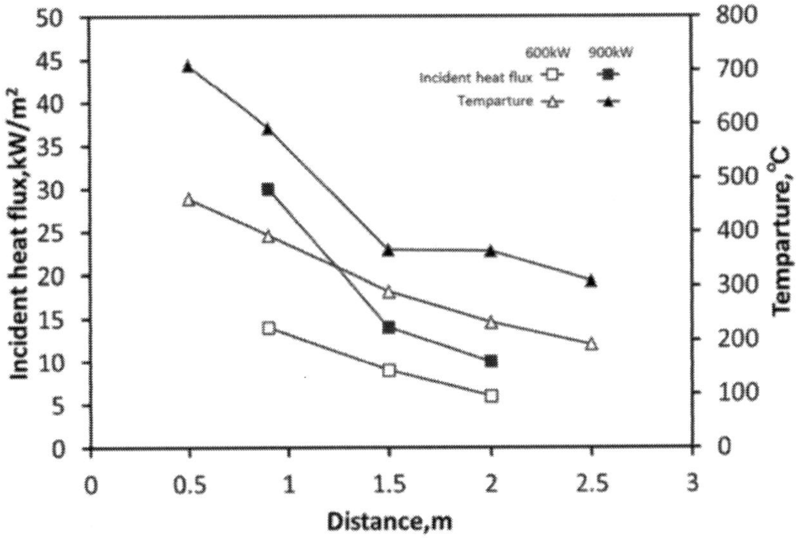


**Figure 3.** Calibration test result about relationship between distance above the opening and façade surface temperatures and façade surface incident heat fluxes.

## 3. FLAME PROPERTIES WITH NONCOMBUSTIBLE FACADE

As a calibration test, a fire test was performed with a non-combustible façade which surface was covered with 25 mm thick ceramic fibre blanket. Heat flux meters were installed at HF2, HF3 and HF4, and thermocouples were installed at the position indicated in Fig. 1 (TC1 to TC5).

The author conducted experiments according to JIS A 1310 with a heat output of 600 kW in the previous study. In this research, fire tests were conducted with a higher heat output than the previous study (about 900 kW).

Figure 2 shows appearance of venting flame in stable condition at a calibration test. It seems that flame in the case of 900 kW was higher, thicker and along the façade surface longer than flame in the case of 600 kW. Figure 3 shows comparison between heat output of 600 kW and 900 kW about average façade surface temperatures and façade surface incident heat fluxes in stable condition. At 0.5 m and 0.9 m upward from the lintel, temperature and incident heat flux on façade surface with 900 kW are much higher than that of 600 kW, especially at 0.9 m upward from the lintel, incident heat flux on façade surface was about twice as much of 600 kW. The difference between temperatures and incident

**Table 1.** Details of façade specimens.

| No. | Heat release Rate (HRR) | Details | Specimen | Category |
|---|---|---|---|---|
| 1-a | 600 kW | EPS(100 mm); Acrylic resin mortar (2 mm thick); | ETICS | Exterior |
| 1-b | 900 kW | Reinforcing net(single layer); Edge: back-wrapping; Finishing material: combustible | 100 mm | thermal insulation |
| 2-a | 600 kW | Same as No.1 except thickness of EPS(200 mm) | ETICS | |
| 2-b | 900 kW | | 200 mm | |
| 3-a | 600 kW | Horizontal planks of Japanese cedar on furring | W | |
| 3-b | 900 kW | | | Wooden |
| 4-a | 600 kW | Horizontal planks of Japanese cedar with chemical | W-FR | façade |
| 4-b | 900 kW | treatment on furring | | |
| 5-a | 600 kW | Core: polyethylene(3 mm thick); Surface: aluminium | ACP | Aluminium |
| 5-b | 900 kW | facing(0.5 mm thick) on both sides | | |
| 6-a | 600 kW | Core: $Al(OH)_3$+ polyethylene(3 mm thick); Surface: | ACP-FR | composite panel |
| 6-b | 900 kW | aluminium facing(0.5 mm thick) on both sides | | |
| 7-a | 600 kW | Water resistant exterior finish paint, | RP | Resinous paint |
| 7-b | 900 kW | (Dry film thickness: 2 to 3 mm) | | |

heat fluxes on façade surface of 600 kW and 900 kW is relatively smaller at 1.5 m or higher distance from the lintel.

## 4. SPECIMENS

Table 1 summarizes overall information on façade specimens. All the specimens are installed upon substrate, calcium silicate boards. Attached "-a" at end of specimen's number were conducted experiments at heat output of 600 kW, while attached "-b" at end of specimen's number were at 900 kW. As for an exterior thermal insulation system, which is composed of EPS boards, glass fibre mesh, adhesive (mortar based) and finish coat (cement-based), the authors conducted fire tests for two different thicknesses of EPS (100 mm and 200 mm). As for wooden façade, which have recently been used for decorative purposes at exterior façade, two types of specimens were used. The difference is mainly whether applied chemical treatment to enhance fire-retardant performance or not. Specimens Nos. 5 and 6 are thin aluminium composite panels. Both surfaces are covered by aluminium facings. The core material of No. 5 is polyethylene, while that of No. 6 is composed of 30% weight polyethylene and 70% weight aluminium hydroxide ($Al(OH)_3$) for fire-retardant purpose. The composite panel of No. 6 holds various fire-safety certifications around the world. No. 7 is coated with water-resistant exterior finish paint, coated over the substrate. Only specimen Nos. 5 and 6 have a vertical joint at the center line of the façade, in order to make a weak point which is possible even in an actual building.

## 5. FIRE PROPAGATION BEHAVIOUR IN COMBUSTIBLE FAÇADE

Observation of fire propagation behaviour is an essential element to understand fire performance of combustible façade. As follows, we show fire propagation behaviours in each specimen.

Figure 4 shows fire propagation behaviours in specimens of exterior thermal insulation system with 100 mm thickness of EPS. Although fire propagation at No. 1-a was slightly visible from 4 to 5 min after the start of the test, fire propagation at No. 1-b was easily visible from 1 to 5 min.

Figure 5 shows fire propagation behaviours in specimens with exterior thermal insulation system with 200 mm thickness of EPS. Different from the case of 100 mm thick EPS, fire propagations were observed at both specimens (No. 2-a and No. 2-b). Although fire propagation at No. 2-b was observed from 1.5 min after the start of the test, fire propagation at No. 2-a was not visible until 5 min.

**Figure 4.** Specimens of exterior thermal insulation system with 100mm thick EPS (No. 1-a and 1-b) 1, 1.5, 4, 5, 8 and 10 min after the start of the test.

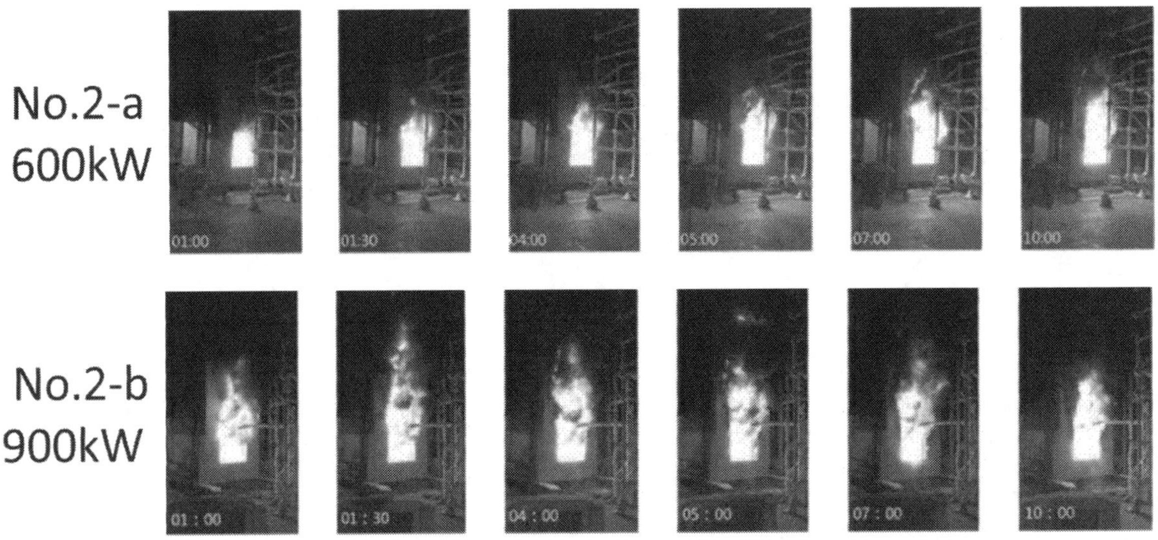


**Figure 5.** Specimens of exterior thermal insulation system with 200 mm thick EPS (No. 2-a and 2-b) 1, 1.5, 4, 5, 7 and 10 min after the start of the test.

Figure 6 shows fire propagation behaviours in specimens of wooden façade with no fire-retardant treatment. Although fire propagation at No. 3-a was occurred from 13.5 min after the start of the test, fire propagation at No. 3-b easily occurred from 1 min and continued until the end of the experiment.

Figure 7 shows fire propagation behaviours in specimens of wooden façade with chemical fire-retardant treatment. In spite of different flaming properties, remarkable fire propagations were not observed at both specimens (No. 4-a and No. 4-b).

Figure 8 shows fire propagation behaviours in specimens of aluminium composite panel with polyethylene core. At No. 5-a, fire propagation occurred 10 min after the start of the test and only the vertical joint of panels located on the centre line of the specimen were seriously damaged. On the other hand, at No. 5-b, fire propagation occurred 3 min after the start of the test and many parts of the specimen were burned out 8 min after the start of the test.

**Figure 6.** Specimens of wooden façade with no fire-retardant treatment (No. 3-a and 3-b) 1, 2, 9, 12, 13.5 and 19.5 min after the start of the test.

**Figure 7.** Specimens of wooden façade with chemical fire-retardant treatment (No. 4-a and 4-b) 0.5, 2, 5, 10 and 15 min after the start of the test.

Figure 9 shows fire propagation behaviours in specimens of aluminium composite panel with fire-retardant polyethylene core. Severe fire propagation did not occur with No. 6-a. On the other hand, with No. 6-b, although fire propagation occurred to some extent at 5 min after the start of the test, severe fire propagation, such as observed at No. 5-b, did not occur.

Figure 10 shows fire propagation behaviours of specimens coated with resinous paint. At No. 7-a, slight fire propagation occurred 3.5 min after the start of the test. On the other hand, at No. 7-b, fire propagated to the top of the specimens immediately and larger fire propagation occurred, but flames mostly disappeared just 5 min after the start of the test.

Compared with the experiments conducted by heat output of 600 kW, clearly different and more severe fire propagation behaviours were observed in some specimens, such as wooden façade without

**Figure 8.** Specimens of aluminium composite panel with polyethylene core (No. 5-a and 5-b) 2, 3, 5, 10, 15 and 19.5 min after the start of the test.

**Figure 9.** Specimens of aluminium composite panel with fire-retardant polyethylene core (No. 6-a and 6-b) 2, 3, 5, 8, 10 and 15 min after the start of the test.

fire-retardant treatment, aluminium composite panel with polyethylene core and rein paint. On the other hand, no severe fire propagation occurred even though intensified flame in specimens with fire-retardant treatment such as No. 4 and No. 6. It is concluded that fire propagation could occur in the fire test at heat output of 900 kW, while no severe flame propagation observed at the fire test with heat output of 600 kW.

## 6. COMOPARISION OF 600 KW AND 900 KW

Figures 11 and 12 show correlation of respectively, peak flame height and peak incident heat flux on the specimen's top (HF6 in Fig. 1) between test results with heat output of 600 kW and 900 kW. Flame height means the height from the lintel to the point which measured over 500 °C and was most distant

**Figure 10.** Specimens with resin paint (No. 7-a and 7-b) 1, 1.5, 2, 3, 3.5 and 5 min after the start of the test.

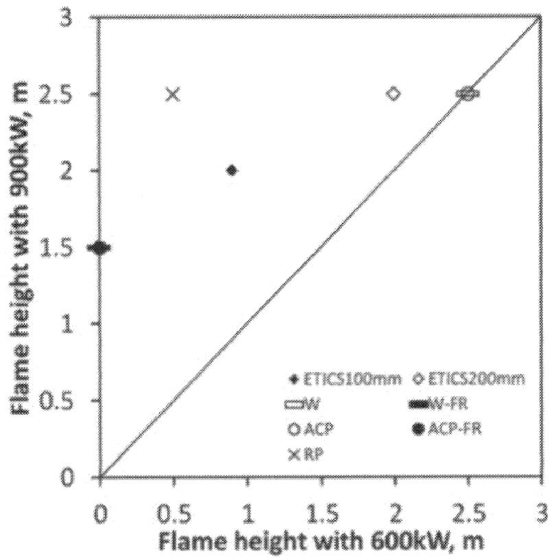


**Figure 11.** Peak flame height.

from the lintel (chosen among TC1 to TC5). Flame height was 0 m if the peak temperature at TC1 didn't exceed 500 °C.

With respect to flame height, flame heights with 900 kW were obviously higher because original flame height with no combustible façade was higher. For the distance from the lintel to specimen's top was about 2.7 m, maximum flame height were 2.5 m even though flame was likely to spread over the specimen's top, especially in the case of 900 kW. Difference between 600 kW and 900 kW about flame height was greater at specimens with resin paint.

With respect to heat flux at specimen's top (HF6), most of the specimens with heat output of 600 kW had about 10 kW/m$^2$ expect for aluminium composite panel with polyethylene core and wooden façade with no fire-retardant treatment. On the other hand, specimens with heat output of 900 kW had different values at HF6 in each specimen. Especially, at specimens of aluminium composite panel and resin paint, at which larger fire propagation were observed, relatively high heat fluxes were measured. Heat fluxes at specimen's top with heat output of 900 kW showed clear difference from specimens exposed of 600 kW.

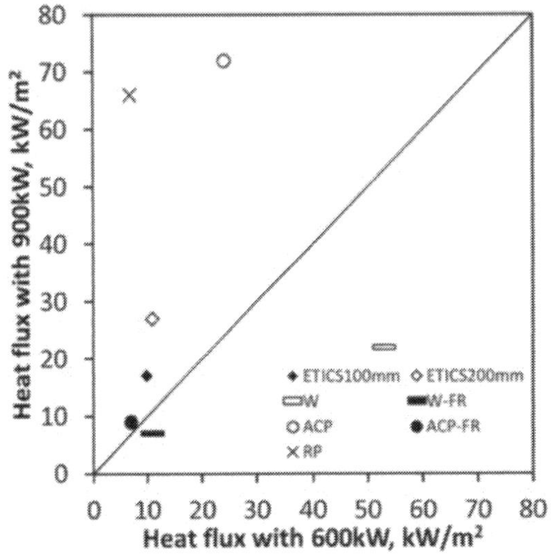


**Figure 12.** Peak incident heat flux on HF6.

## 7. CONCLUSION

As a result of conducting JIS A 1310 test with intensified flame, it becomes clearer to distinguish between combustible exterior façades with higher performance and those which propagates fire easily compared with the previous study.

Indeed JIS A 1310 with heat output of 600 kW could classify combustible exterior façades in which severe fire propagation occurred (specimens such as aluminium composite panel with polyethylene core, resin paint and wood with no treatment for fire retardancy) and those in which severe fire propagation didn't occur though ignition and combustion occurred by flame from the opening (specimens such as aluminium composite panel with fire-retardant polyethylene core and wood with treatment for fire retardancy), but there were combustible façade like rein paint at which flame propagation didn't occur with heat output of 600 kW. JIS A 1310 with heat output of 900 kW could make differences more remarkably in terms of flame propagation behaviour and heat flux measured at the specimen's top. In this research, it was clearly found that JIS A 1310 with heat output of 900 kW could relatively make fire performance clear of various types of combustible exterior façades.

Authors would like to acknowledge specimen suppliers, technicians and students for their contribution to façade fire tests in this paper. Especially, we wish to thank Yutaka Tanaike for great contribution to calculating heat release rate by oxygen consumption method.

### References


[1] H. Yoshioka et al.: "Study of Test Method for Evaluation of Fire Propagation along Façade Wall with Exterior Thermal Insulation," Fire Science and Technology, **30** (2011) No. 1, pp. 27–44.

[2] Hideki Yoshioka et al.: "Façade Tests on Fire Propagation along Combustible Exterior Wall Systems", Fire Science and Technology, **33**, No. 1, pp. 1~ 15, 2014.

[3] JIS A 1310: 2015, Test method for fire propagation over building façades.

[4] ISO 13785-1 & 2:2002, Reaction-to-fire tests for façades -Part 1: intermediate-scale test and Part 2: Large-scale test.

MATEC Web of Conferences **46**, 02003 (2016)
DOI: 10.1051/matecconf/20164602003


# New methods for testing fire resistance of wood façade systems

August Mårtensson

Brandkonsulten AB/Luleå University of Technology Stockholm, Sweden

**Abstract.** Arson in schools has been a huge problem in Sweden over the last fifteen years. The average amount of school arsons between 2000 and 2014 was 285 cases each year which corresponds to 50% of the total amount of reported fires in school buildings. This is a well-known problem and a lot of research has been done in this area. Investigations has been done about fire and heat detection systems, different technical factors significance in fire scenarios and how to prevent adolescents from starting fires.

Another part of the problem that partly been investigated is how the schools are constructed. Roughly 50% of the arsons are outside of the school building. In Sweden one and two storey buildings are allowed to be built with wooden façades in accordance with the building code, which is one of the reasons many schools are built with wooden façade systems.

The most critical part in a wood façade system from a fire safety perspective is concluded to be the eaves because of how they usually are built to let air pass through. Even though a wood façade isn't as well resistant to fire compared to a concrete façade, three versions of new test methods for combustible façades have been developed to make it possible to make sure in advance that a construction is resistant enough.

The new test methods are focused on specific details and parts of a façade system to provide a more informative and useful result compared to SP Fire 105. Observations and measurements of flame spread and temperature changes in the eave, over the window joints and in the air gap are made. With these parameters in consideration criteria's has been chosen for a critical temperature of 280 °C at a critical time of 20 minutes.

## 1. INTRODUCTION

In Sweden about 600 school fires occurs each and every year since 2000 and about half of them are caused by arson. It costs the society approximately 50 million euro's every year [1]. Studies and projects have been performed over the last fifteen years to solve this problem by investigating how these fires may be detected earlier and how to lower the occurrence of fires caused by arson.

One of the main reasons to the high costs is, except for the high number of fires, also that many fires ends in total damage and the buildings must be rebuilt from the ground [2]. If the fire is spreading into the building damages will be made from the smoke but also by the extinguishing water.

### 1.1 Question formulation and purpose

The following questions were formulated to be answered in the study:

- Is it possible to identify any patterns regarding specific weak parts in wooden façade constructions in school buildings?
- May façades consisting of wood be designed to prevent fire from spreading inside the wall and past separating structures?
- Is a new test method needed and are there any uses for such?
- How could a new test method be designed and customized for wooden school façades?

One purpose with this project has been to evaluate the possibility of finding a way to make it possible to reassure a wood façade system is moderately resistant against fire. In addition to that, also suggest how test methods could be designed to fulfil that purpose.

Another purpose was to investigate the need for special demands regarding specific elements in a wooden façade system and how these elements could be tested. A big part of this study is dealing with how this type of façade system can be tested and how a new test method for the same purpose could be designed.

### 1.2 Limitations

Developing a new test method has been determined to be too time consuming for this study. The test methods presented in this report are rough and must be carefully evaluated before they may be practiced.

## 2. METHOD

The information is based on statistics regarding the extent of the problem with school fires caused by arson. It is also based on and supported by statistics about how often buildings with wooden façades are burning. By processing this information it can be concluded which parts and details of a façade system that are less able to resist fire and be specific evaluated in new test methods.

The statistics comes mainly from the database IDA presented by the the Swedish Civil Contingencies Agency (MSB. In addition interviews have also been held with experts on the subjects of fire resistance and test methods for façade systems. The existing Swedish method for testing façades fire resistance, SP Fire 105, has been used as a base for the presented examples of new methods.

During this study other test methods have also been studied. Even though they have been established to be used for different scenarios they have been developed over a long time which is valuable. Experiences from other test methods can be of great use when creating new methods adjusted for the specific conditions described above.

## 3. STATISTICS AND THEORY

The statistics collected from the database IDA is based on reports written by the fire brigades in Sweden. These reports are filled out after every incident by one of the firemen and then sent to MSB. It could have been possible to collect statistic from other sources in this project, e.g. from the crime preventing counsel or from the insurance companies, but an assessment has been made that the statistics from MSB is sufficiently detailed for its purpose in this work, which is to make the reader understand the proportions of school fires caused by arson in Sweden. Deviations that might have been identified comparing different statistics is therefore concluded to be of less interest.

During the period 2000–2014 the fire brigade have been called out to school fires about 8 000 times. Of these fires almost 4 000 have been caused by arson, which is almost 50% [3]. Since 2000 the number

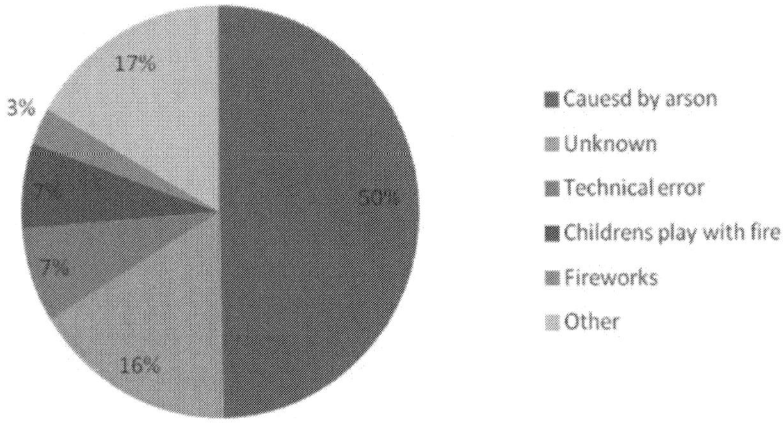


**Figure 1.** Percentage of contribution to fires in schools in Sweden during the years 2000–2014 sorted by cause.

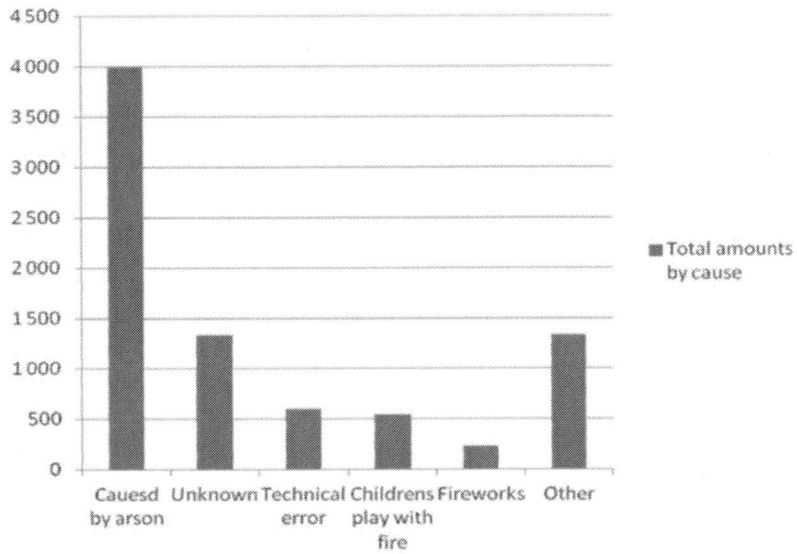


**Figure 2.** The amounts of different causes to fire in schools during 2000–2014.

of school fires caused by arson in Sweden is approximately 270 each year. 16% of the amount of 8000 is said to be by unidentified cause. That corresponds to 85 fires each year. The distribution and total amounts of different causes to school fires are presented below in Figs. 1 and 2.

School fires caused by arson is a well known problem in Sweden. A lot of studies and research has been done during the latest years, for instance in the project "*Anlagd brand*" ("*Arson*" in English) which was initiated in 2007. The final report of this project was presented in 2013 and it includes the results of various studies on the area [4]. One major part of the project was to find a way to lower the amounts of school fires caused by arsons. The purpose of this study, however, is to make it possible to build more fire safe constructions without raising the legal requirements for approved surface materials in façade systems, and thereby reduce the consequences when a fire occur.

## 3.1 Swedish laws and regulatory requirements

The law requirements of approved surface materials for a building are regulated by the use of the building and the number of storeys. It is also regulated by what kind of activities the building houses.

For instance it differs if it's a house with apartments or a house with offices. Mentioned factors divides buildings into different building categories according to Swedish building regulations [5]. A building in two stories used for school activities is assigned to building category Br 2. For buildings in category Br 2 or Br 3 it is permitted to use materials of class D-s2,d2 on the façade surface. This class corresponds to untreated wood panel. Since many newly built schools belong to building category Br 2 it is also common they are built with wood façades.

## 3.2 The significance of the fire brigade

When smoke is detected or a fire is identified in some way the fire brigade is called and arrives to the area of fire. When a building in category Br 2 is designed the main purpose from a fire safety perspective is to make sure no one will be harmed in case of fire. However, the fire brigade also have the possibility to constrain the damage made on the building. The fire brigade have different resources and ability in different areas in Sweden when it comes to staff, equipment and driving time to villages. That is important to have in mind when a new school is built. A more fireproof façade can prevent massive damage on a building, but only for a limited time.

## 3.3 Characteristic fire scenarios and existing test methods

There are different existing methods to test the properties of façade systems. The different methods are used in different parts of the world and are designed for different purposes. In Sweden it is one method commonly used and mentioned in the building regulations. The name of this method is *SP Fire 105* and is a full scale test where the set up consists of a façade construction in two stories, 6.71 m high and 4.0 m wide.

SP Fire 105 was designed during the nineties when buildings in Sweden in a great extent where additionally insulated with combustible insulation covered with protective plaster. Since then the method has been used more and more and by the time the regulations has sharpened, the method has been established as a part of the building regulations. As fuel in the test method 60 liters of heptane in a barrel is used. The method is intended to represent a realistic fire scenario where the flames have broken out through a window in a compartment after a flashover [6].

A international standard is also used to test the fire resistance of façades. The method is divided into two parts where part 2 is called *Large-scale test* [7]. It is similar to SP Fire 105 with a scenario of a fully developed fire spreading out through an opening from a compartment after flashover where the façade outside the compartment is tested.

*Part 1* is a smaller version of Part 2 and is used when a large test isn't required [8]. These two tests are exclusively designed to test the properties of the material on the façade surface. They do not consider variations or details in a façade. SP Fire 105 is better in that way since it is built with fictive windows and with an eave where temperatures are measured [6].

# 4. TECHNICAL FACTORS SIGNIFICANCE IN CASE OF FIRE

In a substantial report as part of the Arson-project mentioned above, the significance of different technical factors regarding fire is analyzed [9]. The results in the report are built on a survey answered by 37 Swedish fire investigators. This report has been an important source of information in this study and in designing a new test method. By focusing on these specific details in the test method and to have them carefully evaluated, the opportunity of limiting fire spread is made.

According to the statistic database *IDA* [3] school fires caused by arson started outside the buildings varies between 70–110 cases during 2004-2014. The most common places for arson outside a school is right next to the façade, next to the entrance, close under the roof or in the crawl space [9].

### 4.1 Eaves

The eave should be built with separating properties corresponding to EI 30 to stop fire spread to the attic. EI 30 corresponds to stopping heat and smoke during a fire for 30 minutes. Otherwise the fire might quickly spread to other parts of the building. However, when the attic belongs to the same fire compartment as the storey bellow, which is common in buildings of category Br 2, there are no regulations concerning the eaves [10].

The design of the eave in façade systems has a huge impact of the possibility to fire spread and is the foremost weakness in the façade systems of schools [9]. When a fire is started next to a façade it is mainly through the eave it spreads. Eaves are, because of this, important to include in a new test method.

### 4.2 Penetrations in the façade

A façade is not made by a single, flat wall. Often, if not always, there are details and installations integrated such as ventilation devices etc. When a penetration in the façade is made it is of great importance that the opening is well sealed. With this kind of details in the façade the risk of fire and smoke spread in the building increases [9].

### 4.3 Mounting of windows

The way windows are mounted in a façade is also of great importance. Like other penetrations for different installations the frame for a window must be carefully sealed to avoid fire and smoke spread in the façade system. If the construction holding the window is badly executed it is possible for fire and smoke to find its way into the façade system and spread inside the wall in the air gap [6].

### 4.4 Air gaps

To keep a wood façade ventilated and to avoid moisture damages it is common an air gap is part of the façade system. An air gap is also creating beneficial conditions for fire spread if a fire, somehow manage to get there. It is because of the openness of the cavity and the access to oxygen. The air gap, however, is not mentioned as a weakness of a façade system and has not been said to have been the reason of a serious school fire in Sweden [9]. It will still be included in the test method since it is considered to be a weakness of the façade system.

## 5. ARSON SCENARIOS AND SOURCE OF FIRE

In the new test method it is desirable that the source of fire represents an expected and realistic fire scenario for school fires caused by arson. In a report, four expected scenarios are presented and two of them are outside the school, next to its façade [11]. The two scenarios are:

1. Trash/combustible liquid placed by the façade.
2. Motorcycle/moped placed by the façade.

These two scenarios are described as common in school fires caused by arson. The heat release rate (HRR) varies between the different cases but do also depend on what kind of trash, what kind of liquid or what kind of vehicle that is used.

The category *Car or other vehicle* has been used as igniting source 20 times between the years 2004–2014 according to IDA. That is less than two times every year. However, many of the school fires with the most serious consequences have been started with a vehicle as igniting source. An experiment shows that a burning motorcycle can reach a maximum HRR of 1 300 kW and a heat radiation corresponding to

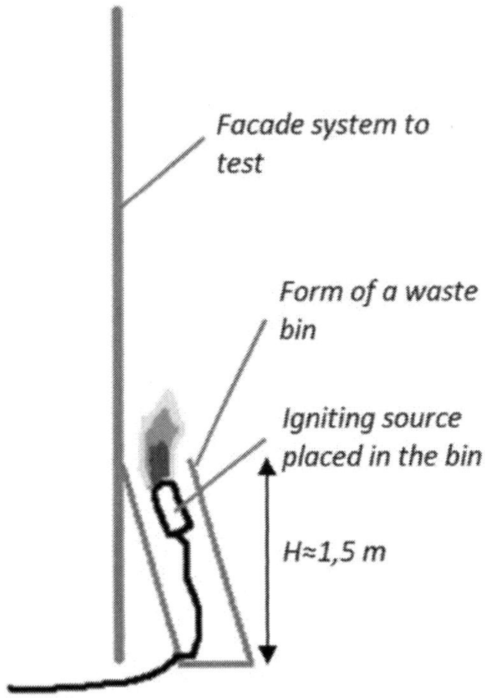


**Figure 3.** Igniting source corresponding to a waste bin.

30–70 kW/m$^2$, and temperatures higher than 300 °C 3 m above the vehicle. [11] In the new test method the source of fire and its HRR can be adjusted to correspond to scenario 1 or 2. It is a decision to make for the client ordering the test, why it is desirable that more than one HRR is possible.

A 125 l waste bin made of poly ethylene will reach a maximum HRR of 350 kW after about two minutes. The HRR also depends on the density and specific kinds of trash. A gas burner may be used as igniting source which can be placed leaning against the façade.

### 5.1 Design of burner

Two different designs of burners should be developed. One design for each scenario explained above.

The effect of the igniting source of each burner can be adjusted to a linear rising from 0 to 350 and 1 300 kW in two minutes and then kept steady over a specified time according to the clients demands. The design of the source of fire should look like an actual waste bin or an actual motorcycle. The one corresponding to a waste bin can be leaned against the façade as in Fig 3. The one corresponding to a motorcycle can have a similar design, but only a bit lower and wider.

## 6. RESULTS THE TEST WILL GENERATE

After a façade has been tested according to the new method the results should be presented describing how well it resisted the fire scenario. In this chapter the relevant information which should be obtained from each technical detail of the façade system is described.

### 6.1 Eaves

As mentioned earlier the eave is probably the most critical part of the façade system in a fire scenario. To generate a result regarding the fire resistance of the eave the temperature should be measured on

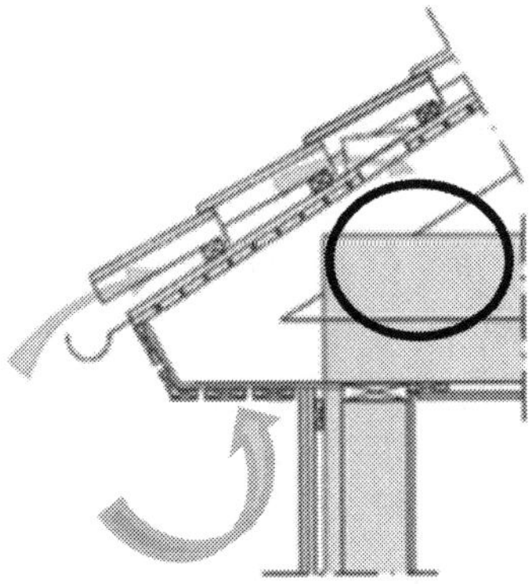

**Figure 4.** Eave with place to put a thermocouple and piece of cotton circled.

the inside of the eave. In addition to the temperature a piece of cotton may be placed inside the eave and observed during the test. It should then be noted at what time the piece of cotton ignites and the temperature curve shall be presented. The thermocouple and the cotton piece should be placed in the circled area in Fig. 4, but separated from each other.

## 6.2 Penetrations in the façade

It is desirable to create a standardized penetration in the façade and install a detail that will be used in the real façade system, within the hole. It should then be sealed accordingly to descriptions and how it is meant to be performed at the building place. A thermocouple may then be placed inside the façade construction above the hole so temperature varieties can be measured. Sudden temperature raises will indicate if the flames have spread into the wall. The temperature curve and moment of spread into the wall shall be documented.

## 6.3 Mounting of windows

By building a façade system with fictive windows like described in SP Fire 105 the frames which are to be used may be installed in the set up and tested. By mounting the window frame which shall be used in the building and sealing the construction with a noncombustible board behind the frame to keep the fire from the back side, the mounting can be tested. In the same way as above a thermocouple can be placed inside the façade system over the window frame. Sudden temperature raises will indicate if the flames have spread into the wall. The temperature curve and moment of spread into the wall shall be documented.

## 6.4 Air gaps

To test the air gap a thermocouple may be placed inside it. If the cavity shall be divided by fire stops it shall also be so in the test. Then thermocouples shall be placed on each side of the separating detail. Sudden temperature raises will indicate if the flames have spread into the air gap. The temperature curve and moment of spread into the wall shall be documented.

**Table 1.** Symbols and explanations of the different parameters obtained as result from the test method.

| Symbol | Explanation |
|---|---|
| $A_x yy$ | Size of fire and test time. |
| Bz | Measure of fire spread inside and on the façade. |
| C | Tested with fire stop by the eave. |
| D | Tested with sealed penetration in the façade. |
| $O_n$ | Other details included in the test which should be specified. |

# 7. METHODS OF CLASSIFICATION

In this following chapter three different methods of classifications will be described. They are formed differently to make it possible to compare them to each other and make it possible to develop one or more into a good and reliable test method.

## 7.1 Requirements the façade must endure

Three factors that should be possible to measure are presented. They are described in what way they may be accepted. The following factors shall be considered during a test.

- *Fire spread on the surface.* That the surface of the façade is burning is not necessary a problem. However, if it is clear that the spread is so aggressive the flames reach the eave, the eave should be protected with some kind of fire stop. If the fire spread is limited and the flames are low the façade system may be built without extra requirements regarding the eave.
- *Fire spread inside the façade.* Depending on the construction of the façade system a fire might find its way into the wall. In a ventilated façade the fire can quickly spread through the air gap and move up the building. Technical details can also be part of fire spread inside the wall. After a test the fire spread inside the wall should be evaluated. If the spread is considered to be too serious, fire stops may have to be installed in the cavity.
- *Fire spread through the eave into the attic.* If the fire spread through the eave and into the attic (which is supposed to be behind the eave) measures must be taken. If the fire has spread into the attic it can quickly move to other parts of the building. That's why this kind of spread must be prevented. This can be helped by some kind of fire stop.

## 7.2 Method 1

The following method of classification has been developed to look like the European standard *EN 13501-1:2007* that is used to determine materials reaction to fire. The classification is defined letters and numbers to explain what's been tested and which results that are achieved. The same method of classification is used in method 1, although the letters and numbers have a different meaning. Below in Table 1 the different parameters are listed and explained.

With definitions according to table 1 and the letters x, y, z and n corresponding to numbers a façade could be classified as in example 1 and 2 bellow where parameters are combined after achieved result.

*Example 1:* Tested with a fire load of category 2 for 15 minutes with a fire spread corresponding to the requirements fulfilling criteria 1, tested with fire stop by the eave.

$$\mathbf{A_2 15, B_1, C}$$

*Example 2:* Tested with a fire load of category 1 for 30 minutes with a fire spread corresponding to the requirements fulfilling criteria 3, tested with a specific detail corresponding to $O_1$.

$$\mathbf{A_1 30, B_3, O_1}$$

**Table 2.** Grading criteria's for measurements and observations from the test method.

| Method 2 | 0 | 1 | 2 | 3 | 4 |
|---|---|---|---|---|---|
| **Eave (points*2) Temperature** | > 280 °C in 10 min | > 280 °C after 10 min | < 280 °C after 10 min | > 280 °C after 20 min | < 280 °C after 20 min |
| **Piece of cotton** | ignition < 5 min | ignition < 10 min | ignition < 15 min | ignition < 20 min | ignition > 20 min |
| **Penetration Temperature** | > 280 °C in 10 min | > 280 °C after 10 min | < 280 °C after 10 min | > 280 °C after 20 min | < 280 °C after 20 min |
| **Mounting of window Temperature** | > 280 °C 10 min | > 280 °C after 10 min | < 280 °C after 10 min | > 280 °C after 20 min | < 280 °C after 20 min |
| **Air gap Temperature** | > 280 °C in 10 min | > 280 °C after 10 min | < 280 °C after 10 min | > 280 °C after 20 min | < 280 °C after 20 min |

**Table 3.** Grading criteria's for measurements and observations from the test method.

| Points | Grade | Explanation of grade |
|---|---|---|
| **24-28** | A | Achieved points indicate **strong** protection against spread of fire in and through the façade system. |
| **20-24** | B | Achieved points indicate **quite strong** protection against spread of fire in and through the façade system. |
| **14-20** | C | Achieved points indicate **quite week** protection against spread of fire in and through the façade system. |
| **0-14** | D | Achieved points indicate **week** protection against spread of fire in and through the façade system. |

Furthermore the parameters and categories need be defined and adjusted after desirable regulation level which may vary between tests.

## 7.3 Method 2

A grading system has been developed for technical factors presented in the article as weaknesses in a façade system. Possible points per measure/observation vary between 0–4 where 4 is the highest point to be achieved for each detail. However, since the eave is concluded to be the most critical detail in the façade system its points are doubled. Possible amount of points is between 0–28. Below in Table 2 the criteria's for each measurement/observation is described.

Table 3 that follows is used to classify the façade system after achieved points are summarized. The highest grade is A and the lowest D. The grades may be adjusted.

## 7.4 Method 3

For technical factors in the façade system described as weaknesses in this article, five criteria have been developed. In method 3 all criteria must be fulfilled for the façade system to be approved as fireproof. The same criteria are included in method 2 but in method 3 they are presented as requirements instead of scoring factors. The requirements that need to be fulfilled are presented bellow in Table 4. In this example 20 minutes is used as a critical time, but it is possible to adjust critical time and temperature after a clients demands.

**Table 4.** Grading criteria's for measurements and observations from the test method.

| **Method 3** | |
|---|---|
| | **Eave** |
| **Temperature** | < 280 °C after 20 min |
| **Piece of cotton** | Ignition > 20 min |
| | **Penetration** |
| **Temperature** | < 280 °C after 20 min |
| | **Mounting of window** |
| **Temperature** | < 280 °C after 20 min |
| | **Air gap** |
| **Temperature** | < 280 °C after 20 min |

# 8. ANALYSIS AND CONCLUSIONS

## 8.1 The need and use of new test methods

The conclusion is that existing methods many times are too tough for the purpose of testing against school fire caused by arson. Furthermore existing methods are not adjusted to take the weaknesses of a façade system into account. This is something a new test method should incorporate.

## 8.2 Identified patterns and technical factors significance in case of fire

Fire spread through the eave and up to the attic have been concluded to be the most common cause of fire spread from school fires caused by arson. If it is possible to make sure a fire doesn't spread through the eave a big amount of resources can be saved. It can lower the number of total damages of school buildings significantly which can save the society a lot of money. Other technical details mentioned in this article should also be included in a new test report. It won't be hard to integrate those parts and it may point out weaknesses that can be corrected before a school is built.

## 8.3 Critical time and temperature

In method 2 and 3 for classification a critical temperature is chosen to be 280 °C and a critical time is set to 20 minutes. What should also be said about this is that critical time and temperature may be adjusted. What is considered to be a reasonable critical time and temperature may however vary between different buildings and areas because of the fire brigades resources etc.

## References


[1] Höjer, H., Svenska skolor brinner mest [Swedish schools burn the most]. *Forskning & Framsteg. #7.* 14–17, 2012.

[2] Zetterström, H-E., Fire and risk engineer at the department of risk and damage techniques at Länsförsäkringar, interview, September 17$^{th}$ 2015.

[3] Swedish Civil Contingencies Agency (MSB), Statistics database – Fires in buildings, `http://ida.msb.se/ida2#page=a0109`, 20150901.

[4] Anlagd brand [Arson], Brandforsk särskild satsning om anlagd brand, `http://anlagdbrand.se/sv/Sidor/default.aspx`, 2015-09-21.

[5] Boverket, *Boverkets byggregler 22*, Karlskrona, Sweden, 2015.

[6] Boström, L., Department of fire resistance at SP Technical Research Institute of Sweden, Fire Research, Borås, Interview, September 11$^{th}$ 2015.

[7] ISO copyright office, International standard, ISO 13785-2, *Reaction-to-fire tests for façades — Part 2: Large-scale test*, Geneva, Switzerland, 2002.

[8] ISO copyright office, International standard, ISO 13785-1, *Reaction-to-fire tests for façades — Part 1: Intermediate-scale test*, Geneva, Switzerland, 2002.

[9] van Hees, P & and Johansson, N., "*Use Of Case Studies To Determine Technical Deficiencies With Respect To Fire Spread In School Buildings,*", Report: 3148), ISSN: 1402-3504. Lunds Faculty of engineering, Lund, Sweden, 2010.

[10] Björkman, B., Fallqvist, K., and Klippberg, A., *BBR: Brandskydd i boverkets byggregler*, ISBN: 978-91-7144-440-0, Brandskyddsföreningen, Stockholm, Sweden, 2014.

[11] Klason, L-G., Johansson, N. & P. Andersson, *Dimensionerande brand: anlagda skolbränder, [Design fires: arson in school buildings]*, report 2010:15, SP Technical Research Institute of Sweden, Borås, Sweden, 2010.

MATEC Web of Conferences **46**, 02004 (2016)
DOI: 10.1051/matecconf/20164602004


# A new fire performance test for cavity wall insulation

Kristin L.T. Jamison[1] and Daniel A. Boardman[2]

[1] FM Global, Research Division
[2] FM Approvals, LLC, Materials; 1151 Boston-Providence Turnpike, Norwood, MA 02062, USA

**Abstract.** Cavity walls containing combustible insulation present an increased risk for fire propagation in a confined, concealed space. Damage to the building resulting from ignition of combustible insulation can be extensive; especially so, in the absence of horizontal and vertical fire-stops. Current codes and standards assess fire performance of exterior wall assemblies subjected to external ignition sources. However, test methods do not address the potential fire hazard resulting from ignition of combustible insulation within the wall cavity. A fire performance test has been developed that evaluates the fire propagation behavior of combustible insulation in a configuration that is representative of the actual installation. The test utilizes a full scale cavity wall assembly and offers fire performance evaluation of insulation of any thickness for either a 51 mm or 102 mm wide air space. A propane sand burner was selected as the ignition source; in addition to being reliable and repeatable, its heat output was carefully calibrated to be representative of potential ignition scenarios that may occur within a cavity wall. During the 15 minute fire performance test, the sample is continuously subjected to the propane sand burner exposure fire. An acceptable sample will produce a peak chemical heat release rate less than 100 kW and a maximum visible flame height less than 1.8 m. This fire performance test method is being incorporated into FM Approvals Standard for Cavity Walls and Rainscreens, Class Number 4411 [1] and is suitable for incorporation into other codes and standards.

## 1. INTRODUCTION

Building trends are showing increased use of cavity walls and rain screens on exterior wall construction. Cavity walls typically consist of two walls separated by an air gap. The outer wall is typically of non-combustible construction (e.g., brick or masonry) while the inner wall can be of combustible or non-combustible construction. Cavity walls are primarily used to allow moisture that has penetrated the outer wall to drain down the cavity space where flashing directs the water to weep holes that remove the water from the cavity. These wall assemblies often contain insulation to increase thermal efficiency. Due to the reasons stated above, cavity walls are a desirable construction element. However, there is an increased risk of fire propagation and smoke development when combustible insulation, such as foamed plastics, is used within the cavity wall. In addition, other combustibles such as air or vapor retarders could be present. Such fires would be difficult to locate and extinguish due to both sides being covered, and could result in fire damage, as well as extensive smoke damage.

Foamed plastic insulation, such as polystyrene and polyurethane, became commercially available in the 1940's and 1950's [2]. It is a desirable insulation material due to its relatively high insulating value (i.e., R-value), its durability and relative ease of fabrication, it resists corrosion, water, and organisms, and its lightweight construction results in economy of shipping and reduced weight load in buildings [3]. However, many foamed plastics, if ignited, have the potential to create a rapid growing fire with high heat release. There are commercially available insulations that exhibit good fire performance, such as expanded glass, mineral wool and foil faced polyisocyanurate. However, combustible foam plastics such as extruded polystyrene, expanded polystyrene and polyurethane are often used in lieu of these good performers because of their economy and relatively high insulating value. Furthermore, they have not been precluded from use in construction in spite of adverse fire test experience. While current codes and standards assess the fire performance of exterior wall assemblies [4, 5] subject to an external ignition source, none to date address the potential for an ignition source inside a cavity wall assembly.

In response to the international energy crisis of the 1970's, the use of foam plastic insulation in building construction increased in an attempt to gain thermal efficiency; codes and standards of the time lacked appropriate guidance and test standards to address the potential fire hazard. The Federal Trade Commission (FTC), in response to a number of large fire losses, conducted an investigation into the manner in which foam insulation fire performance was being described. Based on this investigation, the FTC Consent Cease and Desist Order was issued on November 4, 1974 and instructed the foam plastic industry to stop using terms such as "non-burning", "self-extinguishing" or "non-combustible" when describing foam plastic products. The flammability terminology had been based on results from fire testing per ASTM D1692: Standard Method of Test for Flammability of Plastic Sheeting and Cellular Plastics [6] and ASTM E84: Standard Test Method for Surface Burning Characteristics of Building Materials [7]. Both tests are conducted with the test specimen in the horizontal position. ASTM D1692, which was first issued in 1959, subjected the test specimen to a horizontal Bunsen burner flame for 60 seconds. A successful specimen was labeled as "non-burning" or "self-extinguishing" if it passed the test, meaning that ignition did not occur, or the flame spread rate was low. The ASTM E84 test, introduced in 1941 and known as the Steiner tunnel, subjects a test specimen, which is installed along the top of the horizontal test tunnel, to an open flame ignition source. The rate of horizontal flame propagation is used to provide a relative ranking termed the flame spread index (FSI). Both tests highlighted that some foam plastics have different burning characteristics than conventional materials such as cellulosic materials; they can shrink away from the flame, melt, drip and intumesce to an extent that results of the tests can indicate artificially favorable fire characteristics [7–9].

The FTC determined that there was misuse of small-scale fire test data and issued a requirement that products which referenced a numerical flame spread rating would contain a disclaimer stating "this numerical flame spread rating is not intended to reflect hazards presented by this or any other material under actual fire conditions". The ASTM D1692 standard was withdrawn in 1976; however, ASTM E84 continues to be used today in lieu of its deficiencies for foam plastics. Research continues to show that "favorable" flame spread ratings of foam plastic materials (as determined by ASTM E84) do not guarantee a favorable fire performance in certain fire tests, including room tests and tests of flame spread within wall cavities [8, 10]. Current codes require that all commercially available foam plastic be tested to ASTM E84 and have a FSI less than 75 and a smoke development index (SDI) less than 450. When installed in a habitable space the foam plastic must be covered by a thermal barrier [4, 5, 11] unless the insulation has passed NFPA 286: Standard Methods of Fire Tests for Evaluating Contribution of Wall and Ceiling Interior Finish to Room Fire Growth [12]. This method measures flame spread in a room configuration, including fire spread along walls, ceilings, and combinations of both. It is more indicative of real world fire hazards, and is preferred over ASTM E84 and it's equivalent UL 723 [13], but is more expensive [14]. Foam plastics that perform well during ASTM E84 testing have resulted in spectacular failures when tested to NFPA 286. These conflicting results lead to questions concerning the technical validity of using ASTM E84 for some types of foam plastics.

As a result of the 1974 FTC Consent Order the plastics manufacturers were tasked with conducting research to determine a safe means of using foam plastics. The foam plastic industry paired with Underwriters Laboratories, LLC (UL) and the National Bureau of Standards (now the National Institute of Standards and Technology or NIST) to develop new fire tests for foam insulation [15]. These included material tests (individual components) and assembly tests (a specific combination of building components representative of the actual construction). NFPA 285: Standard Fire Test Method for Evaluation of Fire Propagation Characteristics of Exterior Non-Load-Bearing Wall Assemblies Containing Combustible Components [16] was developed as a result of that research and the test method has been incorporated into international codes such as the International Building Code (IBC) [4] and NFPA 5000: Building Construction and Safety Code [5]. Results of NFPA 285 are based on the specific assembly and changes in a single component may result in substantial changes in the assembly's fire performance. Additionally, a test assembly that successfully passes NFPA 285 provides assurance of good fire performance under fully constructed conditions with careful attention to the assembly integrity. However, rapid and intense fire development may occur if the combustible insulation is exposed to an ignition source via a penetration in the finished assembly or improper fire stopping, or by grinding, welding or cutting during the building construction phase.

Walls that require a fire-resistance rating are tested to NFPA 251: Standard Methods of Tests of Fire Endurance of Building Construction and Materials [17], or an equivalent standard [5, 18, 19], which subjects an assembly to a standard fire exposure and provides an hourly rating for fire-resistance. This method, as with NFPA 285, only indicates the fire performance of a specific assembly under fully constructed conditions whose exterior surface is subjected to a fire exposure. Another test method used to assess the fire performance of exterior walls is NFPA 268: Standard Test Method for Determining Ignitability of Exterior Wall Assemblies Using a Radiant Heat Energy Source [20]. This test method determines the propensity of ignition of exterior wall assemblies from exposure to a radiant heat source in the presence of a pilot flame. It is not intended to evaluate the fire endurance of wall assemblies, nor is it intended to evaluate the effect of fires originating within the building or within the exterior wall assemblies [20].

The 2015 edition of NFPA 5000, which has similar requirements to Chapter 26 of the IBC, specifies that exterior walls of buildings of Type I, II, III or IV construction containing foam plastic shall meet the following requirements: fire resistive rating, if applicable [17–19], contain a thermal barrier [11] covering the foam plastic insulation, the foam plastic must meet NFPA 259: Standard Test Method for Potential Heat of Building Materials [21] and ASTM E84 requirements, and the assembly must meet NFPA 285 and NFPA 268 requirements. There is no one test that can accurately determine the fire performance of a wall assembly under all fire conditions; therefore it is necessary, as is seen in NFPA 5000, to utilize material tests in conjunction with assembly tests to create a general indication of fire performance. The existing assembly tests do not directly expose the combustible insulation and do not address the potential of a small ignition source within the wall cavity that simulates exposure from hot work or electrical equipment. Additionally, ASTM E84 is the sole material test requirement that assesses flame spread. For these reasons, it was desirable to develop a test method that adequately assesses the fire performance of combustible insulation when installed in its final configuration within a wall cavity.

The purpose of this program was to recommend a test method capable of evaluating the flammability of insulations when installed in non-combustible cavity walls [22]. A full scale cavity wall panel test method was evaluated using sample insulation materials. A sand burner ignition source was developed to ensure the appropriateness of the intensity and duration of the exposure fire. The effect of the air cavity width was also considered as the potential fire severity might increase with increasing air cavity size (up to a certain threshold). Air cavity widths of 51 mm and 102 mm were evaluated as the width typically falls within range of these values [23].

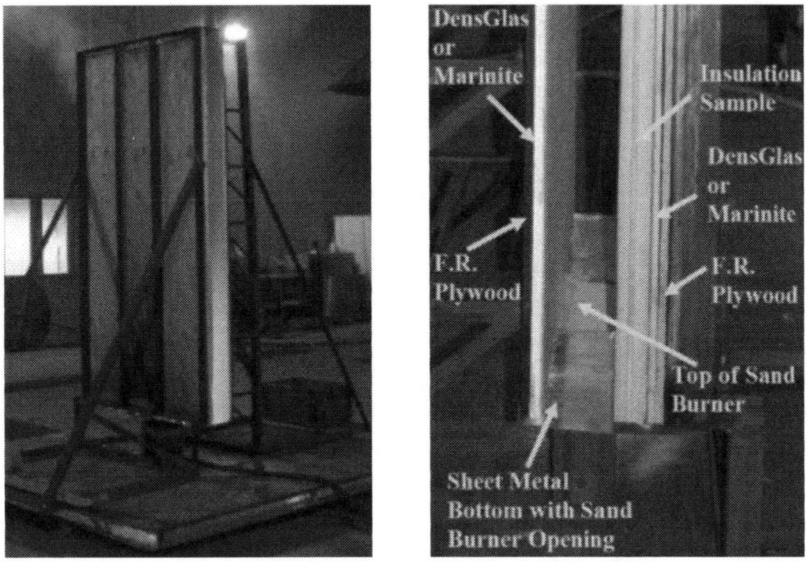


(a) (b)

**Figure 1.** Cavity wall panel test apparatus (a) and close-up view of the sand burner, insulation sample and test apparatus (b).

## 2. TEST APPROACH

Testing was conducted at the FM Global Research Campus in the Fire Technology Building, West Glocester, Rhode Island, USA. Rigid extruded polystyrene and sprayed polyurethane foam insulation were selected for testing. Selection of these insulations allowed for observations to be made of a sample that exhibited poor fire performance (rigid extruded polystyrene) and a sample that exhibited good fire performance (sprayed polyurethane).

Each sample material was installed in a test apparatus that replicated the geometry of a cavity wall and was subjected to a sand burner fire exposure. The test apparatus consisted of two 1.2 m wide by 2.4 m high steel frames. A single layer of 13 mm DensGlas Gold (glass faced gypsum board) or Marinite-I was fastened to each frame with dry-wall screws to simulate non-combustible sheathing or veneer. Each piece of DensGlas or Marinite was backed by a sheet of 13 mm thick fire retardant (FR) plywood. The test sample was then fastened to one frame. The two frame units were positioned parallel to each other with either a 51 mm or 102 mm air gap between the face of the sample and that of the non-combustible panel. Steel sheet metal flashing was installed along the top of the cavity wall to simulate a vertical fire-stop. Additionally, the bottom of the cavity wall was sealed with sheet metal in order to maintain realistic air flows within the cavity. A slot was placed in the bottom piece of steel sheet metal to accommodate the 305 mm long by 51 mm or 102 mm wide sand burners. The test configuration is shown in Fig. 1.

A total of eleven 1.2 m wide by 2.4 m high samples were tested under a 5-MW calorimeter. Heat release rate (HRR) measurements were taken throughout the tests. All tests were digitally recorded and visual observations of the fire growth, flame height and thermal damage were made.

### 2.1 Exposure fire

It was important to select an exposure fire that was representative of a potential accidental ignition source within a cavity wall (*i.e.*, welding and cutting torches). An oxy-acetylene torch, although it serves as a representative ignition source, is prone to be operator dependent and is not a consistent ignition source.

Therefore, to provide a steady and repeatable fire exposure and to be consistent with other standards [12, 24–26], two propane sand burners were selected for use in this study.

A 51 mm × 305 mm sand burner was selected for the 51 mm wide cavity tests and a 102 mm x 305 mm sand burner was selected for the 102 mm wide cavity tests. To be consistent with current FM Approvals test methods and to incorporate an exposure fire known to be representative of common accidental fires, it was decided that the new sand burners should have the same chemical HRR per unit surface area output as the 60 kW sand burner used in the FM Approvals Standard 4910 parallel panel test [25]. This 60 kW, 305 mm × 610 mm sand burner has a chemical HRR per unit surface area of 323 kW/m$^2$. Keeping the same HRR per unit area output resulted in a theoretical 5 kW exposure from the 51 mm × 305 mm sand burner and a 10 kW exposure from the 102 mm × 305 mm sand burner.

These target HRRs were assessed for appropriateness by comparison to those generated by an oxy-acetylene torch. For the purpose of this comparison, use of a torch with a number 4 tip is assumed, which operates at flow rates between 4.7 and 14 L/min [27]. The theoretical chemical HRR of the oxy-acetylene torch can be determined by considering the following expression which used the ideal gas law:

$$HRR = \frac{P}{(R/M)\,T}\dot{V}\Delta H_c. \tag{1}$$

Where $P$ is the pressure, $T$ is the temperature, $R$ is the universal gas constant (8.314472 J/mol/K), $M$ is the molar mass of the gas, $\dot{V}$ the volumetric flow rate and $\Delta H_c$ is the total heat of combustion of the gas. Assuming complete stoichiometric combustion at standard pressure and temperature conditions (*i.e.*, $P = 1$atm; $T = 298.15$ K) in which the working gas is acetylene ($\Delta H_c = 48.27$ MJ/kg, $M = 26.04$ g/mol), Eq. (1) yields a theoretical chemical HRR of 4.0 kW and 12.1 kW for the 4.7 and 14 L/min flow rates, respectively. The target chemical HRR outputs from the two sand burners are within the range of the theoretical chemical HRR output from an oxy-acetylene torch.

Prior to conducting fire testing of the samples, the two sand burners were calibrated in order to determine the propane flow rate required to achieve the desired 5 kW and 10 kW fire exposures. Equation (1) can be rearranged to calculate the volumetric flow rate that should be provided to a sand burner given a desired HRR. For example, for the 51 mm × 305 mm sand burner with a target $HRR = 5.0$ kW, assuming standard conditions (i.e., $P = 1$atm; $T = 298.15$ K) in which the working gas is propane ($\Delta H_c = 46.35$ MJ/kg, $M = 44.0956$ g/mol) one obtains a volumetric flow rate of 3.6 L/min. Likewise, for the 102 mm × 305 mm sand burner with a target $HRR = 10$ kW, one obtains a volumetric flow rate of 7.2 L/min.

## 2.2 Heat flux measurements

It was important to ensure that the incident heat flux to the sample surface from both burners was reasonable and similar in magnitude. Measurements of the incident heat flux were recorded during the calibration of the sand burners. A heat flux gage 0–100 kW/m$^2$ was placed along the centerline of a Marinite wall panel (in place of the insulation sample) at heights of 152 mm and 305 mm above the sand burner surface. Figure 2 shows the incident heat flux as a function of burner HRR.

The incident heat flux measured 305 mm above both sand burners at the target HRRs was 20 kW/m$^2$. However, at these target HRRs, the heat flux measured 152 mm above the burner was 30 kW/m$^2$ and 50 kW/m$^2$ from the from the 51 mm and 102 mm wide sand burners, respectively. This would result in significantly less severe thermal shock to the test samples in the 51 mm wide cavity wall test than in the 102 mm wide cavity wall test. Since it was desirable to maintain a similar maximum heat flux to the samples in both tests, new target HRRs of 5.8 kW and 9.5 kW were selected for the 51 mm and 102 mm wide burner, respectively. The corresponding propane volumetric flow rates were 4.2 L/min and

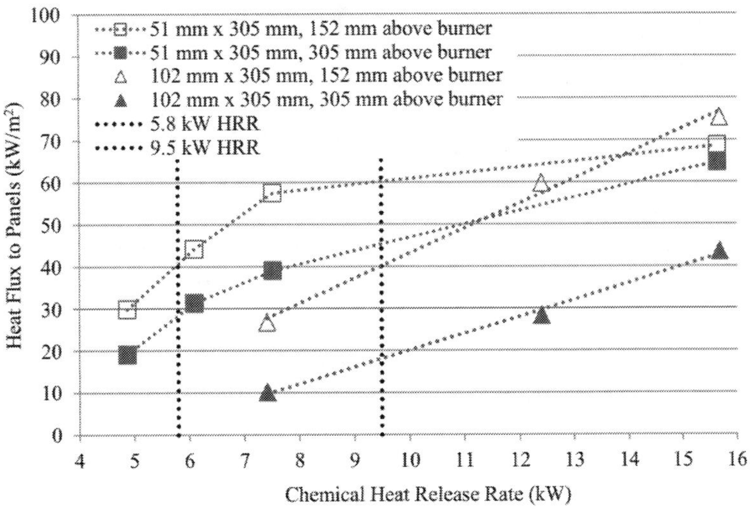


**Figure 2.** Incident heat flux to panels resulting from varying chemical heat release rates.

6.8 L/min, respectively. At these new HRRs, the heat flux to the sample surface 152 mm above the propane burners was 40 kW/m$^2$.

### 2.3 Cavity wall insulations samples

Six tests were conducted with rigid extruded polystyrene. For each test, two pieces of rigid extruded polystyrene boards, 0.6 m wide by 2.4 m long by 51 mm thick, were secured to the non-combustible sheathing, with the seam running vertically up the sample, to form the 1.2 m wide by 2.4 m high samples as shown in Fig. 3(a).

The remaining five tests were conducted with sprayed polyurethane foam. The polyurethane foam was applied to the 1.2 m by 2.4 m non-combustible sheathing at a nominal thickness of 51 mm as shown in Fig. 3(b). All five sprayed polyurethane samples were of the same size and construction. It is important to note that, unlike the rigid extruded polystyrene with a uniform 51 mm thickness, the sprayed polyurethane samples had non-uniform cross-sectional thickness due to the spray application technique. While a nominal 51 mm thickness was maintained, certain areas of the samples were thicker or thinner by up to 25 mm.

## 3. TEST PROCEDURE

Tests were conducted under the FM Global 5-MW fire products collector. The combustion products drawn by the collector are transported to an instrument station where gas species concentrations, temperature and pressure are measured. The gas analysis is capable of providing realtime chemical HRR information based on the total gas mass flow rate and measured concentrations of CO and $CO_2$.

Samples were constructed with either a 51 mm or 102 mm air gap between the exposed surfaces of the parallel panels. In the case of the uneven sprayed polyurethane samples, this gap was measured from the average surface plane of the samples. Depending on the gap between the face of the sample and that of the opposing non-combustible panel, test samples were either subjected to a 51 mm or a 102 mm wide sand burner fire exposure. Rigid extruded polystyrene and sprayed polyurethane samples are shown in Fig. 3(c) and 3(d), respectively, for the 102 mm wide test configuration. A summary of the samples and the test conditions is given in Table 1.

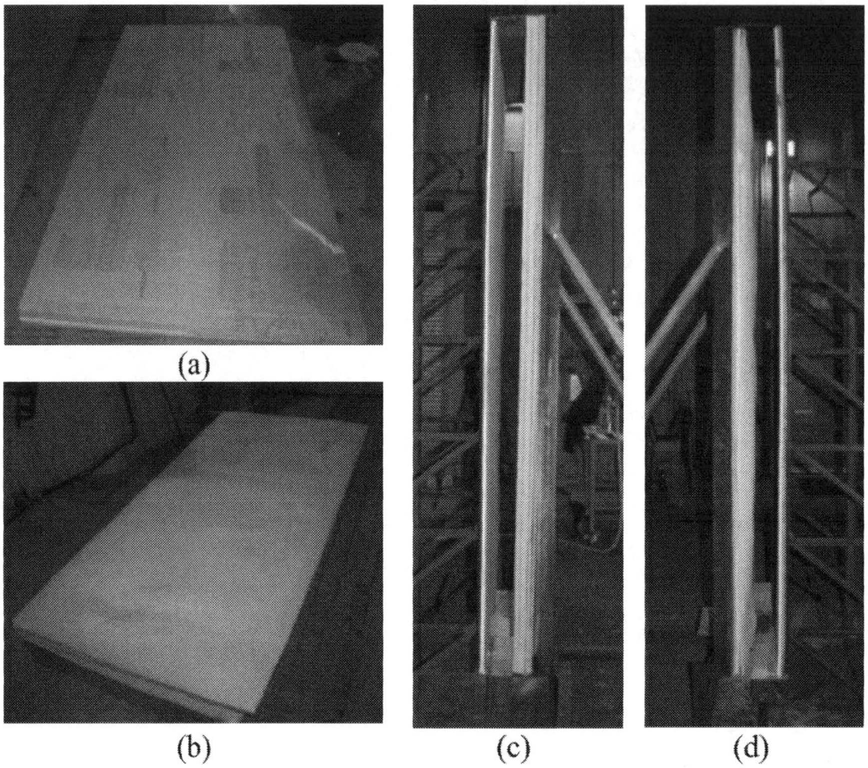

**Figure 3.** Rigid extruded polystyrene sample (a), sprayed polyurethane sample (b), rigid extruded polystyrene sample with 102 mm wide cavity (c) and sprayed polyurethane sample with 102 mm wide cavity (d).

The first six tests were conducted with rigid extruded polystyrene samples. Demonstration tests, conducted in a previous study indicated that poor fire performance could be expected from the rigid extruded polystyrene samples [28]. Tests 1 through 6 were conducted to verify this performance in the proposed test configurations and to establish the repeatability of the tests. The fire exposure period for all six tests was 15 minutes. Tests 1-3 were conducted with a 51 mm air gap (i.e., 5.8 kW exposure) and Tests 4-6 were conducted with a 102 mm air gap (i.e., 9.5 kW exposure).

Tests 7 through 11 were conducted with the sprayed polyurethane foam samples which were expected to exhibit good fire performance. Tests 7 through 9 were conducted to verify this good performance in comparison to Tests 1 though 6. The air gap was 51 mm in Test 7 and 102 mm in Tests 8 and 9; however, the exposure time was increased to 20 minutes to ensure that the 15 minute exposure period was long enough to adequately distinguish between good and poor performance. Tests 10 and 11 were conducted with a 102 mm air gap and an increased sand burner heat output in order to ensure that the previous tests were severe enough. The sand burner was adjusted to achieve a 15 kW heat release rate output with an initial exposure period of 15 minutes in Test 10 and an increased exposure period of 30 minutes in Test 11.

## 4. TEST RESULTS AND DISCUSSION

Results from the cavity wall panel tests are summarized in Table 1 below. The peak chemical heat release rate, maximum visible flame height and the height of the charred area of the samples were recorded.

**Table 1.** Summary of cavity wall test parameters and results.

| Test # | Insulation Type | Test Duration (min) | Cavity /Sand Burner Width (mm) | Sand Burner Fire Exposure (kW) | Peak Chemical Heat Release Rate† (kW) | Maximum Flame Height (m) | Height of Char (m) |
|---|---|---|---|---|---|---|---|
| 1 | Rigid extruded polystyrene | 15 | 51 | 5.8 | 835 | > 2.4 | 2.4 |
| 2 | | 15 | 51 | 5.8 | 1020 | > 2.4 | 2.4 |
| 3 | | 15 | 51 | 5.8 | 1100 | > 2.4 | 2.4 |
| 4 | | 15 | 102 | 9.5 | 1270 | > 2.4 | 2.4 |
| 5 | | 15 | 102 | 9.5 | 1630 | > 2.4 | 2.4 |
| 6 | | 15 | 102 | 9.5 | 1460 | > 2.4 | 2.4 |
| 7 | Sprayed polyurethane foam | 20 | 51 | 5.8 | <100 | 1.8 | 1.8 |
| 8 | | 20 | 102 | 9.5 | <100 | 1.2 | 1.2 |
| 9 | | 20 | 102 | 9.5 | <100 | 1.1 | 0.9 |
| 10 | | 15 | 102 | 15 | <100 | 2.3 | 2.1 |
| 11 | | 30 | 102 | 15 | <100 | 2.1 | 1.8 |

† The lowest chemical heat release rate which the 5-MW calorimeter is capable of accurately measuring is 100 kW [29]

## 4.1 Rigid extruded polystyrene test results

Figures 4(a) and 4(b) show representative images of the maximum flame height for Tests 1-3 and 4-6, respectively. The observed flame heights were above the 2.4 m height of the test panels during all six tests. It is noted that this specific material had a FSI of 5 as measured by the ASTM E84. The observed damage to the samples in Tests 1-3 was also very similar, with charred insulation extending the full length of the samples. The central portion of the insulation samples was entirely consumed during each of the three tests. A picture of the sample from Test 1 is shown in Fig. 4(c) which is representative of the observed damage to all three samples. The damage to the samples in Tests 4-6 was also very similar, with nearly all of the insulation consumed during the test. A picture of the sample from Test 5 is shown in Fig. 4(d) which is representative of the damage observed on all three samples.

In terms of the peak chemical heat release rate, the resulting fires in the 51 mm wide cavity tests (835 to 1,100 kW) were less severe than those in the 102 mm wide cavity tests (1,270 to 1,630 kW). The wider cavity space allows for increased air entrainment and more burning of the samples. This increases the wall temperatures. Also, the optical thickness of the flames increases, resulting in higher heat fluxes [30].

## 4.2 Sprayed polyurethane foam test results

As expected, the sprayed polyurethane samples performed better than the rigid extruded polystyrene samples. In terms of peak chemical heat release rate, the fire intensity observed in the rigid extruded polystyrene tests (835 to 1,630 kW) was much more severe than in the fires from the sprayed polyurethane tests (< 100 kW). Maximum visible flame heights were lower and observed damage and charring did not extend the full length of the test samples. Maximum flame heights achieved in Tests 7 and 9 are shown in Fig. 5(a) and 5(b), respectively, while post test damage and charring are shown in Fig. 5(c) and 5(d). Refer to Table 1 for a summary of results. Test 7, with a 51 mm wide cavity space, and Tests 8 and 9, with 102 mm wide cavity spaces, reached maximum flame heights within the first

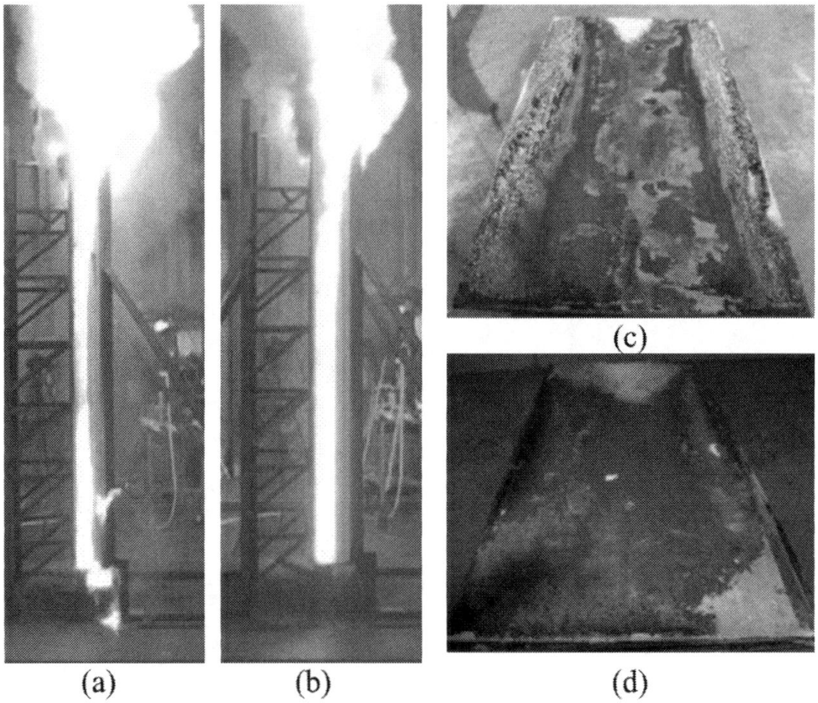

**Figure 4.** Observed maximum flame height Test 1 (a) and Test 5 (b), sample damage Test 1 (c) and Test 5 (d).

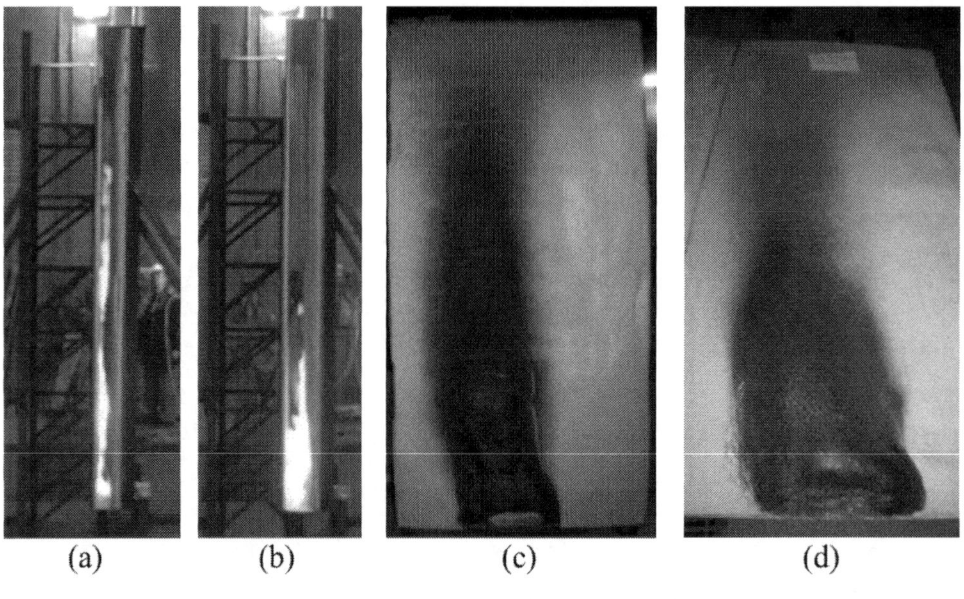

**Figure 5.** Observed maximum flame height Test 7 (a) and Test 9 (b), sample damage Test 7 (c) and Test 9 (d).

15 minutes of the tests. No additional upward flame spread was observed during the extended exposure period of these tests.

The final two tests (Tests 10 and 11) were conducted with an increased fire exposure of 15 kW. Both tests showed increased flame heights and damage to the samples when compared to Tests 7 through 9. However flame heights and charring still did not reach the full 2.4 m height of the test panels. The observed maximum sustained flame height in Test 11, with a 30 minute exposure was lower than that observed in Test 10 with a 15 minute exposure. This is further evidence that irregularities in the samples, rather than test conditions, contributed to unexpected variations in the flame heights. It is also important to note that the maximum flame height in Test 11 was observed during the initial 15 minutes of the test, further indicating that the 15 minute test duration is sufficient.

## 5. CONCLUSIONS

A test protocol for incorporation of cavity walls into FM Approvals Standard for Cavity Walls and Rainscreens, Class Number 4411 has been presented and validated. Test results indicate that the proposed cavity wall panel test method is appropriate for evaluating the flame propagation characteristics of cavity wall insulation materials. The rigid extruded polystyrene samples supported flame spread beyond the 2.4 m height of the test apparatus while the sprayed polyurethane samples supported flame spread to less than 1.8 m. Charring or complete consumption of rigid extruded polystyrene samples was visible over the full 2.4 m length of the samples while charring on sprayed polyurethane samples was limited to a maximum length of 1.8 m. The peak chemical heat release rates measured during the rigid extruded polystyrene tests were 835 to 1,630 kW while the peak chemical heat release rates measured during the sprayed polyurethane tests were below 100 kW. Tests conducted with sprayed polyurethane samples subjected to an increased exposure fire of 15 kW did not yield significantly different results, with the sprayed polyurethane still exhibiting good fire performance. The flame height was less than 2.4 m and charring was less than 2.1 m. Based on these test results, the exposure fires of 5.8 kW in the 51 mm wide cavity test and 9.5 kW in the 102 mm wide cavity test provide an appropriate thermal insult to the samples in order to make a judgment of acceptable performance. In all tests, maximum visible flame heights were observed before the tests reached 15 minutes. In those tests with a 30 minute exposure, no significant thermal damage to the samples was observed after 15 minutes. These results indicate that the 15 minute exposure period is appropriate for evaluating the performance of cavity wall insulation materials in the cavity wall panel test. Tests were conducted with a 51 mm wide and a 102 mm wide cavity between parallel surfaces. The results of six tests conducted with rigid extruded polystyrene, three in each test configuration, indicate that the 102 mm wide test is more severe than the 51 mm wide test. Higher flame heights and more intense fires (in terms of chemical heat release rate) were observed in the 102 mm wide tests. Similar observations and chemical heat release rate measurements were made in each of the three tests conducted under the same conditions, indicating that the test is capable of providing repeatable results.

## 6. RECOMMENDATIONS

The fire performance tests evaluated in this study should be used to quantify the heat release rate and maximum flame height of combustible components (insulation and air/vapor retarders) within a cavity wall construction. Even though the two test configurations, a 51 mm and a 102 mm wide cavity test, lead to differences in insulation fire severity that are relatively small, both may be adopted, if the added flexibility from the choice is deemed to be of value. The 51 mm × 305 mm and 102 mm × 305 mm propane sand burners are recommended for the 51 mm wide and 102 mm wide cavity tests, respectively. The 51 mm × 305 mm sand burner should be adjusted to a propane flow rate of 4.2 L/min to achieve a chemical heat release rate of 5.8 kW. The 102 mm × 305 mm sand burner should be adjusted to a propane flow rate of 6.8 L/min to achieve a chemical heat release rate of 9.5 kW.

A 15 minute test duration for both the 51 mm wide test and the 102 mm wide test is recommended. The propane sand burner output should be constant throughout the test duration. Propane to the sand burner should be shut off and the samples should be extinguished at the end of the 15 minute exposure period. Chemical heat release rate and visual flame height should be measured throughout the duration of the test. Based on the results of testing conducted in this study, acceptable samples shall exhibit a chemical heat release rate less than 100 kW and a maximum visual flame height equal to or less than 1.8 m.

## References


[1] "FM Approvals Standard for Cavity Walls and Rainscreens, Class Number 4411", FM Approvals, Norwood, MA, 2016.

[2] "Invention of STYROFOAM™"". Dow Chemical. Retrieved January 19, 2016.

[3] "Caution – Plastics in Construction," Record, v. 77, p. 6-11, Factory Mutual Insurance Corporation, 1997.

[4] IBC 2012, Chapter 26, International Code Council, 2012.

[5] NFPA 5000, "Building Construction and Safety Code," National Fire Protection Association, Quincy, MA. 2015.

[6] ASTM D1692, "Standard Method of Test for Flammability of Plastic Sheeting and Cellular Plastics," American Society for Testing and Materials, West Conshohocken, PA. 1968.

[7] ASTM E84, "Standard Test Method for Surface Burning Characteristics of Building Materials," ASTM International, West Conshohocken, PA.

[8] Babrauskas, V., Lucas, D., Eisenberg, D., Singla, V., Dedeo, M., & Blum, A., "Flame Retardants in Building Insulation: A Case for Re-evaluating Building Codes," Building Research and Information, **40**, 738-755, 2012.

[9] FM Global, "FM Global Data Sheet 1-57: Plastics in Construction," Norwood, MA, January 2009, Interim Revision July 2015.

[10] Choi, K.K. and Taylor, W. "Combustibility of Insulation in Cavity Walls," National Research Council Canada, Division of Building Research, 1984.

[11] NFPA 275, "Standard Method of Fire Tests for the Evaluation of Thermal Barriers," National Fire Protection Association, Quincy, MA. 2013.

[12] NFPA 286, "Standard Methods of Fire Tests for Evaluating Contribution of Wall and Ceiling Interior Finish to Room Fire Growth," National Fire Protection Association, Quincy, MA. 2015.

[13] UL 723, "Standard for Test for Surface Burning Characteristics of Building Materials", Underwriters Laboratories LLC, Northbrook, IL, 2008.

[14] NFPA 101, "Life Safety Code," National Fire Protection Association, Quincy, MA. 2015.

[15] "Fire Research on Cellular Plastics: The Final Report of the Products Research Committee," National Bureau of Standards, Washington, DC. 1980.

[16] NFPA 285, "Standard Fire Test Method for Evaluation of Fire Propagation Characteristics of Exterior Non-Load-Bearing Wall Assemblies Containing Combustible Components," National Fire Protection Association, Quincy, MA. 2012.

[17] NFPA 251, "Standard Methods of Tests of Fire Endurance of Building Construction and Materials," National Fire Protection Association, Quincy, MA. 2006.

[18] ASTM E119, "Standard Test Methods for Fire Tests of Building Construction and Materials," ASTM International, West Conshohocken, PA. 2015.

[19] ANSI/UL 263, "Standard for Fire Tests of Building Construction and Materials," American National Standards Institute/Underwriters Laboratories LLC, 2015.

[20] NFPA 268, "Standard Test Method for Determining Ignitability of Exterior Wall Assemblies Using a Radiant Heat Energy Source," National Fire Protection Association, Quincy, MA. 2012.

[21] NFPA 259, "Standard Test Method for Potential Heat of Building Materials," National Fire Protection Association, Quincy, MA. 2013.

[22] Jamison, K.L.T., Boardman, D.A., and Khan, M.M. "A New Fire Performance Test for Cavity Wall Insulation," Proceedings of Fire & Materials, San Francisco, CA, February 2015.

[23] "Cavity Walls - Design Guide for Taller Cavity Walls", Masonry Advisory Council, Park Ridge, IL, 2002.
[24] ISO 9705:1993, "Fire Tests – Full-scale Room Test for Surface Products," International Organization for Standardization, Geneva, Switzerland, 1993.
[25] "FM Approvals Standard for Cleanroom Materials Flammability Test Protocol, Class Number 4910", Norwood, MA, June 2009.
[26] "FM Approvals Standard for Wafer Carriers for Use in Cleanrooms, Class Number 4911", Norwood, MA, April 2010.
[27] "Victor Torch Handles, Cutting Attachments, and Heating Nozzles Safety and Operating Instructions (Form No. 0056-0138)", Thermadyne Industries Incorporated, St. Louis, MO, May 14, 2008.
[28] "FM Global Property Loss Prevention Data Sheet 1-12, Ceilings and Concealed Spaces", FM Global, Norwood, MA, May 2008.
[29] Heskestad, G., "A Fire Products Collector for Calorimetry into the MW Range", Technical Report J.I. 0C2E1.RA, Factory Mutual Research Corporation, Norwood, MA, June 1981 (Available upon request).
[30] Khan, M.M., Jamison, K.L.T. and de Ris, J.L., "Upward Fire Growth over Thermally Thin Corrugated Paperboard," Proceedings of the $10^{th}$ International Symposium on Fire Safety Science, 2011.

MATEC Web of Conferences **46**, 02005 (2016)
DOI: 10.1051/matecconf/20164602005


# Falling parts of external walls claddings in case of fire – ITB test method – results comparison

Jacek Kinowski, Bartłomiej Sędłak and Paweł Sulik

Fire Research Department of Building Research Institute, Ksawerów 21, 02-656 Warsaw, Poland

**Abstract.** Paper presents ITB methodology of testing and assessment of buildings with regard to falling of external walls claddings in case of fire. In the ITB methodology of impact on the facade (external claddings) was assumed in form of flames, generated from a specially designed propane, sand burner with a defined fuel flow rate and laminar airflow from the inside of the furnace resulting in determined temperature and height of the flame. Evaluated criterion is also discussed. Test results of several different types of claddings is analysed and presented in relation to the national Polish regulation along with some technical assessment regarding individual.

## 1. INTRODUCTION

In this paper methodology of testing and assessment of buildings with regard to falling of external walls claddings in case of fire is presented. Some of the test results are analysed and presented in relation to the national Polish regulation.

The necessity to ensure safety of people evacuating from the building and members of rescue teams in case of fire is one of the essential requirement concerning fire safety [7, 14, 15]. Therefore this requirement also shall apply to the risk associated with the possibility of parts falling from the exterior wall in case of fire. It is worth emphasizing, that the requirement applies not only to fire resistant parts of facades, but practically to the entire exterior wall of every construction work. For fire resistant curtain walls this problem is better recognized [5–7, 9–13] and the risk is arguably eliminated [4]. For other types of outer walls solution effective evaluation and influencing of limiting that risk is possible only with implementations of appropriate test method.

## 2. ITB TEST METHOD

ITB test method was developed and implemented in Fire Research Department of the Building Research Institute. After establishment of initial assumptions method was formed in the process of several calibration tests, Fig. 2. During the calibration tests, following conditions were taken into account:

- the height and temperature of the flame tip (based on EOTA Technical Report N073 [3] guidelines),
- the temperature on the lintel in correlation with standard fire curve,
- distribution and shape of the flames [8].

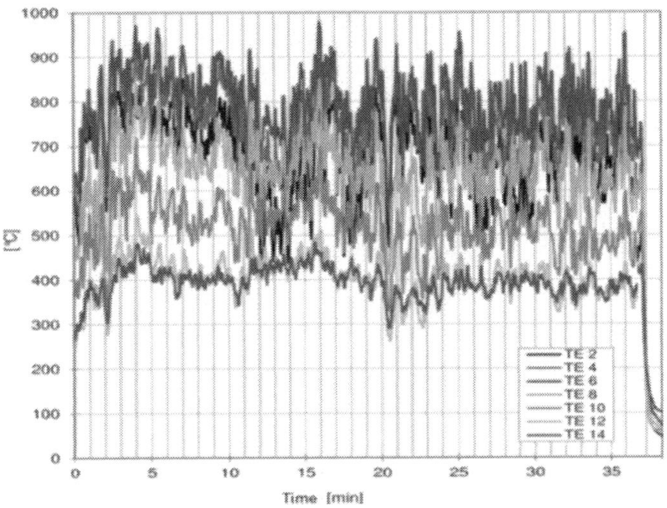


**Figure 1.** Calibration test, temperature measurements.

**Figure 2.** Calibration test.

Dimensions of the representative test specimens were determined, as follows:

- width: 2100–2400 mm,
- height: 2000–2500 mm.

The criterion evaluated in ITB test method for assessment of buildings with regard to falling of external walls claddings in case of fire was that the mass of falling parts cannot be greater than 5 kg.

## 3. INSTRUMENTATION

In the ITB methodology impact on the facade (external claddings) was assumed in form of flames, generated from a specially designed propane sand burner with a defined fuel flow rate and laminar airflow from the inside of the furnace resulting in determined temperature and height of the flame.

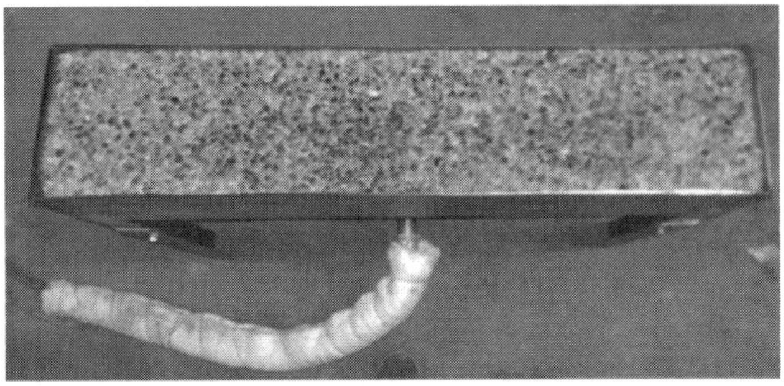

**Figure 3.** Propane burner.

**Figure 4.** Composite panel fixed mechanically.

The temperature rise on unexposed surface of the test specimens was measured during the calibration tests with standard plate thermocouples (type K). Results of temperature measurements on unexposed surface of the wall were used for determination of the flow rate of the propane burner designed based on EN 14390:2007 [2] standard, but of much larger size ($170 \times 800$ mm). For regular fire tests the only checked value was air flow and burner fuel flow rate (with high – precision mass flow controller).

## 4. TEST RESULTS

After establishing of strictly defined test conditions several fire tests were performed. Tests were carried out in compliance with the adopted procedure, that tested, inter alia, solutions with facade claddings in the forms of: composite panels (Fig. 4), ceramic claddings (Fig. 6) stone claddings (Fig. 5), fibre cement plates (Fig. 7) and sandwich panels (Fig. 8) and other solutions not included in this presentation. All six fire tests selected for the presentation were conducted in Fire Testing Laboratory of Building Research Institute in Pionki.

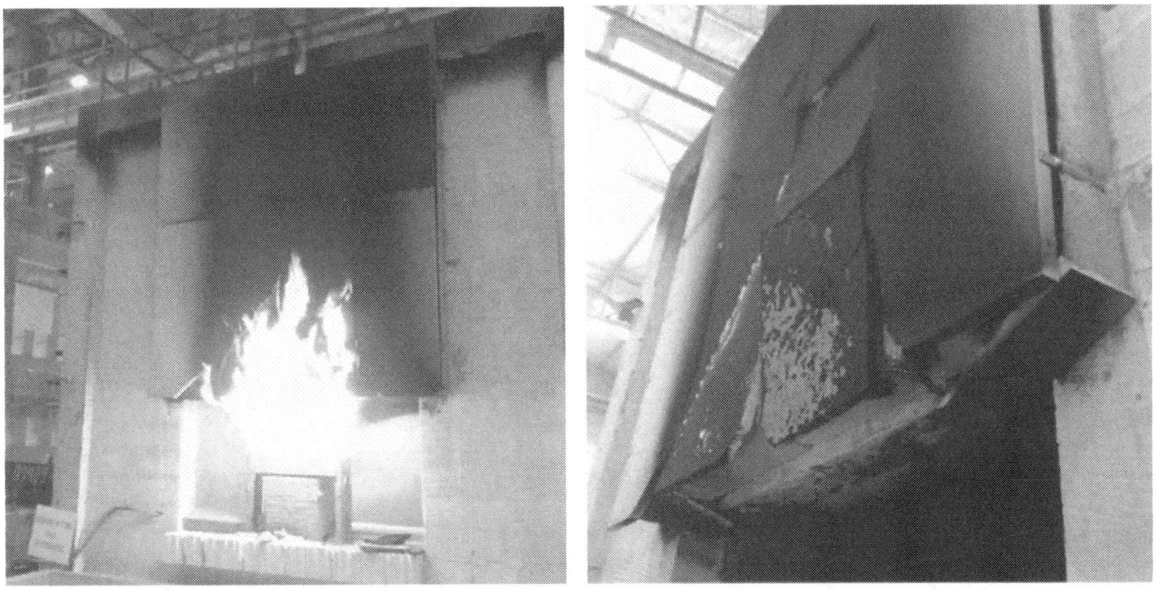

**Figure 5.** Stone claddings fixed mechanically.

**Figure 6.** Ceramic claddings fixed mechanically.

## 5. DISCUSSION

Analyzing the results of selected fire tests, performed in accordance with ITB test method, it was observed that much safer solutions are those wherein the claddings are fixed mechanically in opposition to solutions that used adhesives [15]. However, in some cases, this has no relevance to the outcome of the fire test due to the behavior of some materials when heated, where ceramic claddings are perfect example (Fig. 5).

Other interesting observations is that after 30 minutes since the start of the fire test, test specimen degrades very little or nothing at all. In Polish National Regulations [1, 14–16], in terms of time, requirement applies to 60 or even 120 minutes periods in witch spandrel of external walls should

**Figure 7.** Fibre cement plates fixed by adhesive.

**Figure 8.** Sandwich panels fixed mechanically, type 1.

maintain its integrity and the fixing of the external wall claddings shall prevents its parts from falling. Regularity observed in fire test leads to the conclusion, that this parameter in Polish National Regulations seems excessive.

Established criterion (5 kg for single falling element) appears to be reasonable for buildings of height below 25 m. In case of higher buildings [16] even this, relatively low, mass seems to be too high to ignore the potential threat, as the energy of 5 kg element after a period of free-fall could be dangerous. For such buildings, careful planning of escape routes should be required to reduce the danger for people evacuating and rescue teams.

Statistically in Poland injuries sustained by people evacuating from the building in case of fire and members of rescue teams, caused by falling parts of external walls, are extremely rare. And that is despite

**Figure 9.** Sandwich panels fixed mechanically, type 2.

the fact, that many of the building works facades does not meet discussed requirements. Nevertheless it's very important to have a test method – engineering tool that allows reliable evaluation of different technical solutions using explicit criterion.

## References


[1] Decree of Ministry of Infrastructure dated 12.04.2002 on the technical requirements for buildings and their location (*Polish national regulation*).

[2] EN 14390:2007 Fire test. Large-scale room reference test for surface products.

[3] EOTA Technical Report N073: Large Scale Fire Performance Testing of External Wall Cladding Systems.

[4] Kinowski J., "Bezpieczeństwo pożarowe przeszklonych ścian osłonowych (kurtynowych)", *Świat Szkła*, 20(5), 8–14, 2015 *(in polish)*.

[5] Kinowski J., Sędłak B., Sulik P., Izydorczyk D., "Fire resistance glazed constructions classification, changes in field of application", *Applications of Structural Fire Engineering, 15–16 October 2015, Dubrovnik, Croatia*, 460–465, 2015. DOI: dx.doi.org/10.14311/asfe.2015.073.

[6] Kinowski J., Sędłak B., Sulik P., "Izolacyjność ogniowa aluminiowo – szklanych ścian osłonowych w zależności od sposobu wypełnienia profili szkieletu konstrukcyjnego", *Izolacje*, 20(2), 48–53, 2015 *(in polish)*.

[7] Kinowski J., Sulik P., "Bezpieczeństwo użytkowania elewacji", *Materiały Budowlane*, 9, 38–39, 2014 *(in polish)*.

[8] Schriftenreihe des Bundesministers fur: Raumordnung, Bauwesen und Stadtebau, Bau- und Wohnforschung, 1978 *(in german)*.

[9] Sędłak B., Kinowski J., Borowy A., "Fire resistance tests of large glazed aluminium curtain wall test specimens–results comparison", *MATEC Web of Conferences*, **9**, p. 02009, EDP Sciences, 2013, DOI: 10.1051/matecconf/20130902009.

[10] Sędłak B., Kinowski J., "Badania odporności ogniowej ścian osłonowych – przyrosty temperatury na szybach", *Świat Szkła*, **18**(11), 20–25, 2013 *(in polish)*.

[11] Sędłak B., Sulik P., "Odporność ogniowa pionowych elementów przeszklonych", *Szkło i Ceramika*, **R.66** (nr. 5), 8–10, 2015 *(in polish)*.

[12] Sędłak B., "Systemy przegród aluminiowo szklanych o określonej klasie odporności ogniowej", *Świat Szkła*, **18**(10), 30–33, 41, 2013 *(in polish)*.

[13] Sulik P., Kinowski J., Sędłak B., "Fire resistance of aluminium glazed curtain walls, test results comparison depending on the side of fire exposure", *Applications of Structural Fire Engineering, 15–16 October 2015, Dubrovnik, Croatia*, 478–483, 2015. DOI: dx.doi.org/10.14311/asfe.2015.076.

[14] Sulik P., Sędłak B., "Bezpieczeństwo pożarowe przeszklonych elewacji", *Materiały Budowlane*, (nr 9), 18–20, 2015. DOI: 10.15199/33.2015.09.04.

[15] Sulik P., Sędłak B., Kinowski J., "Bezpieczeństwo pożarowe ścian zewnętrznych – Część 2 – Mocowanie okładzin elewacyjnych", *Ochrona przeciwpożarowa*, **1**(51), 9–12, 2015 *(in polish)*.

[16] Sulik P., Sędłak B., Turkowski P., Węgrzyński W., "Bezpieczeństwo pożarowe budynków wysokich i wysokościowych", *Budownictwo na obszarach zurbanizowanych – nauka, praktyka, perspektywy*, 14–19 September 2014, Krynica, Poland, 105–120, 2014 *(in polish)*.

MATEC Web of Conferences **46**, 02006 (2016)
DOI: 10.1051/matecconf/20164602006


# Study on critical places for maximum temperature rise on unexposed surface of curtain wall test specimens

Paweł Sulik, Bartłomiej Sędłak and Jacek Kinowski

Fire Research Department of Building Research Institute, Ksawerów 21, 02-656 Warsaw, Poland

**Abstract.** The paper discusses the main issues related to the fire resistance of glazed curtain walls including the tests methodology and way of classification of this type of building elements. Moreover, the paper presents an attempt to determine the weak points of aluminium glazed curtain wall test specimens regarding to the maximum temperature rise measurements, based on the fire resistance tests performed in recent years by Fire Research Department of Building Research Institute. The paper analyse the results of temperature rise on unexposed surface of 17 aluminium glazed curtain wall specimens tested for internal fire exposure in accordance with EN 1364-3:2006 [3] and EN 1364-3:2014 [4], which achieved the fire resistance class of min. EI 15.

## 1. INTRODUCTION

Curtain wall is a type of wall which usually consists of vertical and horizontal structural members, connected together and fixed to the floor supporting structure of the building to form a lightweight space enclosing continuous skin, which provides, by itself or in conjunction with the building construction all the normal functions of the external wall but doesn't take on any of the loadbearing characteristics of the building.

In case of buildings for a specific purpose, (e.g. hotels, schools, hospitals, malls) or those with significant height [31] non-loadbearing elements of building, such as curtain walls should be designed and constructed in a way that in case of fire it will limit the spread of fire inside the building, ensure the safety of rescue team and allow the evacuation of users. Those specific requirements are not usually considered individually (e.g. ensuring proper evacuation is connected with the structural load bearing capacity as well as fire and smoke propagation within the building), therefore individual elements of buildings can play several roles during a fire.

This refers also to the curtain walls which are usually required in terms of design and execution to ensure that in case of fire they shall for a specific period of time prevent its development between the adjacent floors of the building, between the adjacent objects, and in a specific situations also between the adjacent rooms in the building (connection of the curtain wall with inner wall of the building). Moreover, the fixing of the curtain walls shall be made in a way that prevents falling of the parts of element during the rescue team activities and evacuation of the users [8, 11, 27, 30].

---

e-mail: p.sulik@itb.pl, b.sedlak@itb.pl, j.kinowski@itb.pl

Glazed curtain walls with a specific fire resistance class are usually made as mullion – transom structures in which the areas between the aluminium or steel profiles are filled with fire resistant glass units. The loadbearing members in those structures are mullions, typically in the form of box section profiles (commonly with additional reinforcement – steel or aluminium stiffening profiles). *In order to improve thermal insulation, the profiles are also filled with special insulating inserts. Material of the inserts as well as the filling volume is dependent on the destination to use for the specific class of fire resistance of the curtain wall. The materials used for insulation inserts are usually gypsum plasterboards, silicate – cement or silicate – calcium.* [10]

*The glazing as well as solution of profile insulation is selected depending on the expected fire resistance class of the curtain wall. The most commonly used in practice are glass units comprising a fireproof glass (toughened glass panes divided by the special gel or foil expanding under the influence of high temperature) coupled with the outer glass pane. Glass units are usually fixed in the curtain wall by means of pressure plates screwed to the main aluminium profile.* [26] The solution of the glazing fixing method has a key role in achieving the appropriate fire resistance class especially in the case of increasingly common glass units of large dimensions [20, 21, 25].

*Important elements in the profile structure are intumescent gaskets placed on the perimeter of the glass panes. Intumescent gasket expands under the influence of high temperatures and closes the spaces through which the fire could get to the other side of the partition.* [18]

It is worth mentioning that solutions of this type are also used for roofing [12, 13] – glazed roofs are made of exactly the same profile as those used in curtain walls.

The fire resistance class of curtain wall cannot be calculated or assessed based on its project or other specification, therefore the only way to determine this characteristic is to perform the fire resistance test.

## 2. FIRE RESISTANCE TESTS AND CLASSIFICATION

The fire resistance test of glazed curtain wall in full configuration, according to the European Union provisions, shall be performed in accordance with EN 1364-3:2014 [5], following the procedures outlined in EN 1363-1:2012 [2] and where appropriate EN 1363-2:1999 [3].

The standard presents two different methods of heating and test specimen configuration depending on the side of fire exposure (external or internal). In case of internal fire exposure heating of the test specimen is held by the standard temperature-time curve, taken as an appropriate to reflect the fully developed, following the flashover fire inside the building (presented by the solid line in Fig. 1). For external fire exposure heating of the test specimen can be held by external temperature-time curve, which is relevant for the fire outside the building (presented by the dashed line in Fig. 1) or by the standard temperature-time curve. The influence of the heating conditions to the fire resistance of glazed curtain walls is presented in papers [1, 9, 22, 23].

*The testing structure for the internal test shall comply of testing specimen and supporting construction made of two floor slabs and two associated wall constructions. For external fire exposure the testing structure is composed only of the test specimen and testing frame (optionally).* [19] *The specimen for the test shall be fully representative of the curtain wall intended for use in practice (including any surface finishes and fittings which are essential and may influence its behavior in the test) or be designed to obtain the widest applicability of the test result to other similar constructions.* [16]

During the fire resistance test performance criteria of integrity, insulation and radiation can be verified.

Integrity is the ability of tested element to prevent the passage through it of flames and hot gases and to prevent the occurrence of flames on the unexposed side. In the test integrity is verified:

- by means of gap gauges (integrity fails if the 6 mm gap gauge can be passed through the crack in curtain wall and can be moved a distance of 150 mm along the gap, or if the 25 mm gap gauge can be passed through the crack in the specimen such that it projects into the furnace),

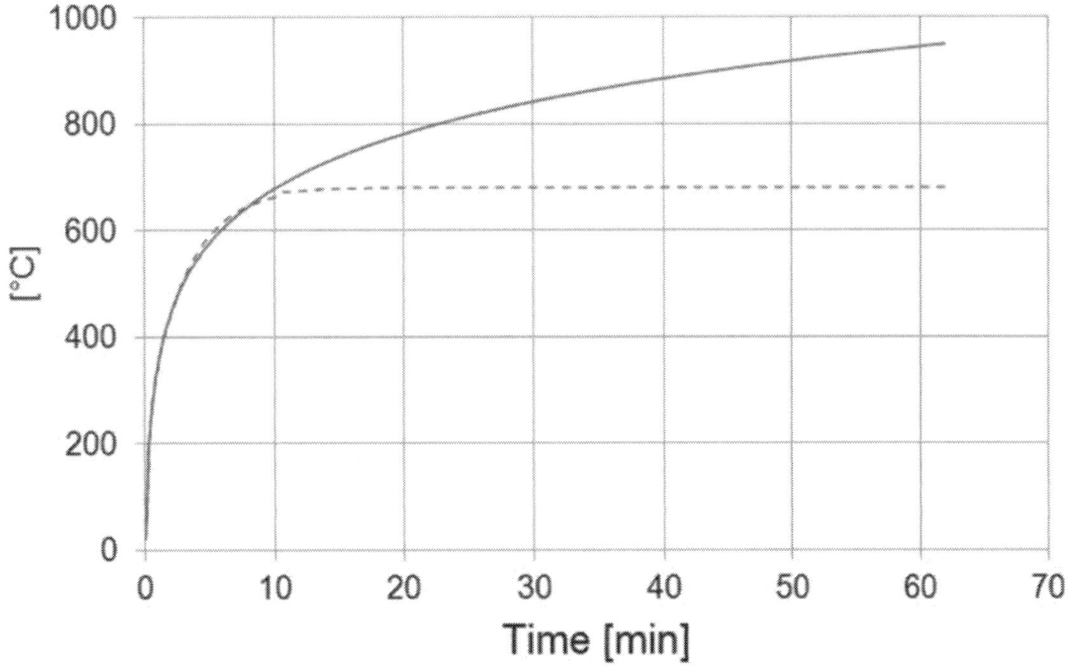


**Figure 1.** Temperature-time curves, solid line – standard curve, dashed line – external fire curve [20].

- visually (integrity fails when flaming longer than 10 s occurs on the unexposed surface of curtain wall),
- by means of cotton pad used on the unexposed surface of tested specimen, for max. of 30 s, or until ignition.

Insulation is the ability of glazed curtain wall to restrict the temperature rise of the unexposed face to below specified levels. The temperature rise on the unexposed surface of tested specimen is checked with thermocouples attached with use of heat resistant adhesive. Thermocouples are attached in the specific places given in the testing standard, arranged in the specific surfaces of the curtain wall, perimeter seal and vertical linear gap seal (if applicable). In order to maintain insulation – average temperature rise on the unexposed surface of test specimen cannot be greater than 140 K and the maximum temperature rise cannot be greater than 180 K.

Radiation is the ability of glazed curtain wall to prevent the passage of the fire, due to the transfer of significant heat through the element to the adjacent materials. Radiation in the test is measured with the radiometer placed in the 1 m distance from the geometric center of the unexposed surface of heated part of curtain wall. Radiation is evaluated on the basis of the time at which the maximum radiation measured does not exceed 15 kW/m$^2$.

Test methodology as well as detailed information regarding the selection of appropriate test specimens has been widely discussed in the existing papers [16, 17, 22, 29].

Classification of glazed curtain wall fire resistance is prepared in accordance with the requirements given in EN 13501-2:2009 [6], basing on the test results presented in test report. Fire resistance classes defined by the standard are presented in Table 1. Moreover each fire resistance class given in the Table 1 shall be supplemented by additional marking defining the side of fire exposure in the test.

## 3. TEST SPECIMENS

For the preparation of this paper 17 fire resistance tests of aluminium glazed curtain walls in full configuration (complete assembly) conducted in Fire Testing Laboratory of Building Research Institute

**Table 1.** Fire resistance classes [24].

| | | | | | | |
|---|---|---|---|---|---|---|
| E | 15 | | 30 | 60 | 90 | 120 |
| EI | 15 | | 30 | 60 | 90 | 120 |
| EW | | 20 | 30 | 60 | | |

(***E*** *– integrity,* ***I*** *– insulation,* ***W*** *– radiation)*

**Table 2.** Test specimens dimensions and characteristic.

| No. | Test date | Test duration duration [min] | System | Type of insulation inserts | Type of glazing | Width [mm] | Height [mm] | Span length [mm] |
|---|---|---|---|---|---|---|---|---|
| **1** | 2007-02-05 | 47 | 1 | 4 | 11 | 3822 | 4100 | 3536 |
| **2** | 2007-07-09 | 62 | 1 | 1 | 12 | 3822 | 4100 | 3536 |
| **3** | 2009-06-05 | 66 | 1 | 1 | 1 | 3822 | 4952 | 3536 |
| **4** | 2010-09-30 | 36 | 1 | 4 | 9 | 3822 | 4952 | 3536 |
| **5** | 2010-10-15 | 63 | 1 | 1 | 10 | 3822 | 4952 | 3536 |
| **6** | 2013-01-14 | 62 | 4 | 1 | 3 | 5825 | 5725 | 5000 |
| **7** | 2013-01-21 | 36 | 4 | 3 | 4 | 5825 | 5725 | 5000 |
| **8** | 2014-07-09 | 33 | 6 | 1 | 6 | 4522 | 5667 | 4700 |
| **9** | 2014-07-15 | 63 | 6 | 1 | 7 | 4522 | 5667 | 4700 |
| **10** | 2014-10-23 | 50 | 7 | 1 | 8 | 4730 | 5426 | 4700 |
| **11** | 2015-01-12 | 66 | 5 | 1 | 5 | 4122 | 5667 | 4700 |
| **12** | 2015-02-12 | 68 | 7 | 1 | 8 | 4730 | 5426 | 4700 |
| **13** | 2015-06-17 | 70 | 3 | 2 | 2 | 4520 | 4550 | 3900 |
| **14** | 2015-10-16 | 55 | 1 | 1 | 14 | 4122 | 5667 | 4700 |
| **15** | 2015-10-23 | 34 | 1 | 1 | 15 | 4122 | 5667 | 4700 |
| **16** | 2015-12-08 | 56 | 1 | 1 | 14 | 4122 | 5667 | 4700 |
| **17** | 2016-01-18 | 19 | 3 | no insulation | 13 | 4970 | 5350 | 4700 |

in Warsaw and Pionki in the last few years has been analyzed. All of the selected specimens were tested for internal fire exposure in accordance with EN 1364-3:2006 [4] or EN 1364-3:2014 [5] and achieved the fire resistance class of min. EI 15. The analysis has been prepared for 7 different systems of curtain walls with 4 types of insulation inserts and 15 types of glass units, with span length from 3536 mm up to 5000 mm. Tested specimens dimensions and characteristics are presented in Table 2.

## 4. FIRE RESISTANCE TESTS RESLULTS

For the purpose of attempt to identify the weak points of aluminium glazed curtain wall test specimens (regarding to the maximum temperature rise) there were selected 7 places, related with thermocouple arrangement presented in the EN 1364-3:2006 [4] and EN 1364-3:2014 [5]:

- T1 – thermocouples placed on the glass panes for average and maximum temperature measurements,
- T2 – thermocouple placed in the level of upper simulated floor soffit, in the middle width of the test specimen,
- T3 – thermocouples placed in the level of upper simulated floor soffit, in the line with the mullions,
- T4 – thermocouples placed at the junction of mullions and transoms,
- T5 – thermocouples placed on transoms at middle way between two mullions,
- T6 – thermocouples placed on the mullion at middle way between two transoms,
- T7 – thermocouples placed on the glass panes 20 mm from the mullion or transom edge (checked only for specimens 10–17 tested in accordance with EN 1364-3:2014 [5]).

**Table 3.** Temperature rises in 15th minute of the test.

| Test No. | T1 [K] (15 min) | T2 [K] (15 min) | T3 [K] (15 min) | T4 [K] (15 min) | T5 [K] (15 min) | T6 [K] (15 min) | T7 [K] (15 min) |
|---|---|---|---|---|---|---|---|
| 1 | 27.0 | 0 | 10.0 | 66.0 | 12.0 | 31.0 | x |
| 2 | 10.0 | 0 | 30.0 | 70.0 | 59.0 | 44.0 | x |
| 3 | 9.0 | 1.0 | 22.0 | 62.0 | 31.0 | 17.0 | x |
| 4 | 7.3 | 0.5 | 60.5 | 72.4 | 74.7 | 72.3 | x |
| 5 | 1.9 | 0 | 10.9 | 45.7 | 4.8 | 21.0 | x |
| 6 | 9.1 | 6.7 | 9.9 | 28.6 | 11.1 | 13.2 | x |
| 7 | 75.3 | 69.5 | 68.1 | 81.3 | 81.7 | 78.2 | x |
| 8 | 18.2 | 34.6 | 35.9 | 65.8 | 54.8 | 55.8 | x |
| 9 | 12.8 | 0.2 | 52.4. | 70 | 71.1 | 65.6 | x |
| 10 | 14.4 | 1.4 | 0.4 | 10.1 | 0.3 | 0.7 | 74.1 |
| 11 | 9.1 | 0.2 | 4.6 | 13.2 | 15 | 28.6 | 12.9 |
| 12 | 11.9 | 0.7 | 61.4 | 77.7 | 12.1 | 77.6 | 22.5 |
| 13 | 9.5 | 8.8 | 29.5 | 74.2 | 30.1 | 17.6 | 12.6 |
| 14 | 4.8 | 0.1 | 0.2 | 35 | 8.9 | 7.5 | 6.8 |
| 15 | 10.9 | 0.3 | 8.2 | 21.3 | 21.8 | 9 | 14.9 |
| 16 | 4.7 | 0.1 | 0.8 | 8 | 14.8 | 5 | 6.2 |
| 17 | 60 | 33.8 | 36 | 69.3 | 37.5 | 25.2 | 27.8 |

"x" – no data available.

**Table 4.** Temperature rises in 30th minute of the test.

| Test No. | T1 [K] (30 .min) | T2 [K] (30 min) | T3 [K] (30 min) | T4 [K] (30 min) | T5 [K] (30 min) | T6 [K] (30 min) | T7 [K] (30 min) |
|---|---|---|---|---|---|---|---|
| 1 | 65 | 3 | 75 | 99 | 76 | 93 | x |
| 2 | 30 | 2 | 73 | 71 | 74 | 72 | x |
| 3 | 26 | 5 | 69 | 78 | 83 | 56 | x |
| 4 | 28 | 4.7 | 62.3 | 80.5 | 80 | 83.2 | x |
| 5 | 14.9 | 2 | 73.5 | 74.8 | 72.2 | 74.7 | x |
| 6 | 25.5 | 21.2 | 37 | 41.6 | 69 | 36.9 | x |
| 7 | 78.7 | 74.7 | 92.5 | 137 | 127.9 | 146.1 | x |
| 8 | 46 | 56.1 | 66.6 | 68.1 | 66.8 | 81.2 | x |
| 9 | 35.3 | 5.3 | 69.7 | 70.6 | 74.7 | 74.3 | x |
| 10 | 60.8 | 8 | 1 | 62.4 | 1.4 | 32.1 | 73.6 |
| 11 | 25.5 | 2.6 | 17.7 | 34.6 | 29.3 | 38.1 | 36.9 |
| 12 | 33.7 | 65.9 | 70.8 | 79.3 | 67.5 | 79 | 58.1 |
| 13 | 25.9 | 35.4 | 81.3 | 110.2 | 79.2 | 80 | 48.8 |
| 14 | 23.8 | 8.2 | 13.6 | 43.4 | 31.5 | 33.8 | 30 |
| 15 | 33.2 | 4.5 | 26.3 | 42 | 46 | 22.1 | 40.4 |
| 16 | 24.7 | 9.1 | 19.4 | 32.6 | 35.6 | 31.3 | 29.3 |
| 17 | x | x | x | x | x | x | x |

"x" – no data available.

The results of temperature rise measurement in all of the locations listed above, in the 15th, 30th, 45th and 60th minute of the test are presented in tables 3–6, respectively. The temperature rise on unexposed surface of the test specimens was measured with surface thermocouples type K specified in PN-EN 1363-1:2012 [2].

## 5. DISCUSSION

Analysing the test results presented in Tables 3–6 it can be observed that the greatest temperature rise was recorded only in points T4–T7. The percentage of greatest temperature occurrence in specific place, in respect to all of the tests and available data at given time is presented in Table 7. Because in many

**Table 5.** Temperature rises in 45th minute of the test.

| Test No. | T1 [K] (45 min) | T2 [K] (45 min) | T3 [K] (45 min) | T4 [K] (45 min) | T5 [K] (45 min) | T6 [K] (45 min) | T7 [K] (45 min) |
|---|---|---|---|---|---|---|---|
| 1 | 79 | 6 | 78 | 175 | 96 | 166 | x |
| 2 | 45 | 4 | 74 | 81 | 86 | 87 | x |
| 3 | 36 | 10 | 73 | 96 | 101 | 80 | x |
| 4 | x | x | x | x | x | x | x |
| 5 | 31.5 | 6.2 | 76.9 | 100.9 | 81.3 | 81.6 | x |
| 6 | 33.9 | 29.2 | 36.4 | 47.6 | 70.2 | 54.7 | x |
| 7 | x | x | x | x | x | x | x |
| 8 | x | x | x | x | x | x | x |
| 9 | 58.9 | 15.1 | 70.5 | 87.1 | 80.9 | 109.6 | x |
| 10 | 62.3 | 22.5 | 1.9 | 60.1 | 6.2 | 54.1 | 73.4 |
| 11 | 33.9 | 9.8 | 27.5 | 50 | 42.1 | 43.4 | 54.7 |
| 12 | 54.9 | 67.5 | 71 | 98.7 | 69.7 | 72.8 | 68.1 |
| 13 | 36.9 | 51.1 | 103.5 | 127.5 | 72 | 84.7 | 88.5 |
| 14 | 36.4 | 24.2 | 29.6 | 64.4 | 57.7 | 54.1 | 44 |
| 15 | x | x | x | x | x | x | x |
| 16 | 38.3 | 26.6 | 30.8 | 69.3 | 62.7 | 72.6 | 45.6 |
| 17 | x | x | x | x | x | x | x |

"x" – no data available.

**Table 6.** Temperature rises in 60th minute of the test.

| Test No. | T1 [K] (60 min) | T2 [K] (60 min) | T3 [K] (60 min) | T4 [K] (60 min) | T5 [K] (60 min) | T6 [K] (60 min) | T7 [K] (60 min) |
|---|---|---|---|---|---|---|---|
| 1 | x | x | x | x | x | x | x |
| 2 | 56 | 7 | 74 | 104 | 99 | 121 | x |
| 3 | 41 | 16 | 77 | 127 | 119 | 103 | x |
| 4 | x | x | x | x | x | x | x |
| 5 | 42.7 | 12.1 | 78.1 | 117.1 | 93.8 | 104.3 | x |
| 6 | 45.1 | 47.4 | 64.5 | 62.1 | 71.9 | 63.4 | x |
| 7 | x | x | x | x | x | x | x |
| 8 | x | x | x | x | x | x | x |
| 9 | 52.9 | 24.3 | 81.7 | 122.1 | 116.4 | 146.2 | x |
| 10 | x | x | x | x | x | x | x |
| 11 | 45.1 | 18.9 | 31.3 | 64.7 | 51.1 | 62.1 | 65.5 |
| 12 | 68.4 | 68.8 | 73.7 | 102.8 | 70.3 | 93.5 | 83.1 |
| 13 | 44.6 | 58.9 | 125.7 | 136 | 92.8 | 90.3 | 122.9 |
| 14 | x | x | x | x | x | x | x |
| 15 | x | x | x | x | x | x | x |
| 16 | x | x | x | x | x | x | x |
| 17 | x | x | x | x | x | x | x |

"x" – no data available.

cases the recorded temperatures differed slightly further calculation was prepared on the basis of scores given for the temperature level (1 – for the lowest temperature up to 6–for the greatest) – this calculation was evaluated only for places T1–T6, which occurs in all of the tests. Example of the scoring for the 15th min of the test is presented in Table 8 and the percentage of greatest temperature occurrence in specific place based on the scoring is presented in Table 9.

As it can be observed in the Tables 7 and 9 the lowest temperatures were recorded in the places T1–T3. This phenomenon can be explained in a simple way. Thermocouples T1 were placed in the middle part of the glazing where the temperature rise is low due to the glass unit structure. The glass units are composed of the fire resistant glass pane as an internal layer coupled by means of the steel frame with the external glazing (usually toughened or laminated). The first part of glazing – fire resistant glass pane

**Table 7.** Percentage of greatest temperature occurrence in specific place.

| Time | T1 | T2 | T3 | T4 | T5 | T6 | T7 *) |
|---|---|---|---|---|---|---|---|
| **15 min** | 0.00% | 0.00% | 0.00% | 58.82% | 29.41% | 5.88% | 12.50% |
| **30 min** | 0.00% | 0.00% | 0.00% | 31.25% | 37.50% | 25.00% | 14.29% |
| **45 min** | 0.00% | 0.00% | 0.00% | 41.67% | 16.67% | 25.00% | 33.33% |
| **60 min** | 0.00% | 0.00% | 0.00% | 50.00% | 12.50% | 25.00% | 33.33% |

*) The percentage for the T7 was calculated only from the tests No. 10–17.

**Table 8.** Example of the scoring for 15th minute of the test.

| No. | T1 | T2 | T3 | T4 | T5 | T6 |
|---|---|---|---|---|---|---|
| **1** | 4 | 1 | 2 | 6 | 3 | 5 |
| **2** | 2 | 1 | 3 | 6 | 5 | 4 |
| **3** | 2 | 1 | 4 | 6 | 5 | 3 |
| **4** | 2 | 1 | 3 | 5 | 6 | 4 |
| **5** | 2 | 1 | 4 | 6 | 3 | 5 |
| **6** | 2 | 1 | 3 | 6 | 4 | 5 |
| **7** | 3 | 2 | 1 | 5 | 6 | 4 |
| **8** | 1 | 2 | 3 | 6 | 4 | 5 |
| **9** | 2 | 1 | 3 | 5 | 6 | 4 |
| **10** | 6 | 4 | 2 | 5 | 1 | 3 |
| **11** | 3 | 1 | 2 | 4 | 5 | 6 |
| **12** | 2 | 1 | 4 | 6 | 3 | 5 |
| **13** | 2 | 1 | 4 | 6 | 5 | 3 |
| **14** | 3 | 1 | 2 | 6 | 5 | 4 |
| **15** | 4 | 1 | 2 | 5 | 6 | 3 |
| **16** | 3 | 1 | 2 | 5 | 6 | 4 |
| **17** | 5 | 2 | 3 | 6 | 4 | 1 |

**Table 9.** Percentage of greatest temperature occurrence in specific place based on the scoring.

| Time | T1 | T2 | T3 | T4 | T5 | T6 |
|---|---|---|---|---|---|---|
| **15 min** | 13.45% | 6.44% | 13.17% | 26.33% | 21.57% | 19.05% |
| **30 min** | 10.98% | 6.23% | 15.43% | 24.63% | 21.07% | 21.66% |
| **45 min** | 11.90% | 6.35% | 13.10% | 25.40% | 19.84% | 23.41% |
| **60 min** | 8.33% | 6.55% | 16.67% | 25.60% | 20.24% | 22.62% |

were generally the same ones as used for the tests of doorsets [7, 28] or glazed partitions [13, 14]. Due to the specific of this elements testing the mentioned glass pane has to be prepared for stopping the increase of average temperature rise to a level of 140 K. This fire glass together with the additional air space and external glass panes creates a good insulating barrier and cause the low temperature rise in place T1. The temperature rises in points T2 and T3 were low, because those thermocouples were placed in the spandrel area, which is usually very accurately sealed and insulated.

In most of the examined cases the greatest temperature rise was recorded on the mullion transom connections (T4) and these places can be regarded as critical. The great temperature rise in those junctions can be explained by the large deformations of the facade specimens during the fire test. The deformation causes unplugging the transoms of the mullions which facilitates the flow of hot gases. Additionally, in those places often occur the special connectors, which limiting the space of insulation inserts.

An interesting phenomenon is fairly high temperature rise in place T7. Temperature measurements in this places were added in the new version of the standard (EN 1364-3:2014 [5]) and as it can be observed it was the correct move, because that also can be a places of the greatest temperature rise.

## References


[1] Borowy A., "Fire resistance testing of glazed building elements", *Požární ochrana 2014: Sborník přednášek XXIII. ročníku mezinárodní konference, Ostrava, 03–04.09.2014*, 15-17.

[2] EN 1363-1:2012 Fire resistance tests – Part 1 General requirements.

[3] EN 1363-2:1999 Fire resistance tests – Part 2 Alternative and additional procedures.

[4] EN 1364-3:2006 Fire resistance tests for non-loadbearing elements - Part. 3: Curtain walling - Full configuration (complete assembly).

[5] EN 1364-3:2014 Fire resistance tests for non-loadbearing elements - Part. 3: Curtain walling - Full configuration (complete assembly).

[6] EN 13501-2+A1:2009 Fire classification of construction products and buildings elements - Part 2: Classification using data from fire resistance tests, excluding ventilation services.

[7] Izydorczyk D., Sędłak B., Sulik P., "Thermal insulation of single leaf fire doors, test results comparison in standard temperature-time fire scenario for different types of doorsets" *Applications of Structural Fire Engineering,15–16 October 2015, Dubrovnik, Croatia*, 484–489, 2015. DOI: dx.doi.org/10.14311/asfe.2015.077.

[8] Kinowski J., "Bezpieczeństwo pożarowe przeszklonych ścian osłonowych (kurtynowych)", *Świat Szkła*, **20**(5), 8–14, 2015 *(in polish)*.

[9] Kinowski J., Sędłak B., Sulik P., Izydorczyk D., "Fire resistance glazed constructions classification, changes in field of application", *Applications of Structural Fire Engineering, 15–16 October 2015, Dubrovnik, Croatia*, 460–465, 2015. DOI: dx.doi.org/10.14311/asfe.2015.073.

[10] Kinowski J., Sędłak B., Sulik P., "Izolacyjność ogniowa aluminiowo – szklanych ścian osłonowych w zależności od sposobu wypełnienia profili szkieletu konstrukcyjnego", *Izolacje*, **20**(2), 48–53, 2015 *(in polish)*.

[11] Kinowski J., Sulik P., "Bezpieczeństwo użytkowania elewacji", *Materiały Budowlane*, **9**, 38–39, 2014 *(in polish)*.

[12] Roszkowski P., Sędłak B., "Metodyka badań odporności ogniowej dachów przeszklonych", *Świat Szkła*, **16**(6), 50–52, 2011 *(in polish)*.

[13] Roszkowski P., Sędłak B., "Badania odporności ogniowej poziomych elementów przeszklonych" *Świat Szkła*, **19**(12), 46–51, 2014 *(in polish)*.

[14] Sędłak B.,"Badania odporności ogniowej przeszklonych ścian działowych", *Świat Szkła*, R. **19** (nr. 2), 30–33, 2014 *(in polish)*.

[15] Sędłak B., "Bezpieczeństwo pożarowe przeszklonych ścian działowych", *Świat Szkła*, R. **20** (nr. 5), 34–40, 2015 *(in polish)*.

[16] Sędłak B., "Badania odporności ogniowej przeszklonych ścian osłonowych – Część 1", *Świat Szkła*, **17**(9), 52–54, 2012 *(in polish)*.

[17] Sędłak B., "Badania odporności ogniowej przeszklonych ścian osłonowych – Część 2", *Świat Szkła*, **17**(10), 53–58, 60, 2012 *(in polish)*.

[18] Sędłak B., "Systemy przegród aluminiowo szklanych o określonej klasie odporności ogniowej", *Świat Szkła*, **18**(10), 30–33, 41, 2013 *(in polish)*.

[19] Sędłak B., "Badania odporności ogniowej przeszklonych ścian osłonowych wg nowego wydania normy PN-EN 1364-3", *Świat Szkła*, **19**(7–8), 49–53, 2014 *(in polish)*.

[20] Sędłak B., "Odporność ogniowa ścian osłonowych z dużymi przeszkleniami – Część 1", *Świat Szkła*, **19**(3), 16–19, 25, 2014 *(in polish)*.

[21] Sędłak B., "Odporność ogniowa ścian osłonowych z dużymi przeszkleniami – Część 2", *Świat Szkła*, **19**(5), 28–31, 2014 *(in polish)*.

[22] Sędłak B., Kinowski J., "Badania odporności ogniowej ścian osłonowych – przyrosty temperatury na szybach", *Świat Szkła*, **18**(11), 20–25, 2013 *(in polish)*.

[23] Sędłak B., Kinowski J., Borowy A., "Fire resistance tests of large glazed aluminium curtain wall test specimens–results comparison", *MATEC Web of Conferences*, 9, p. 02009, EDP Sciences, 2013, DOI: 10.1051/matecconf/20130902009.

[24] Sędłak B., Sulik P., "Odporność ogniowa pionowych elementów przeszklonych", *Szkło i Ceramika*, R.**66** (nr. 5), 8–10, 2015 *(in polish)*.

[25] Sędłak B., Sulik P., "Odporność ogniowa wielkogabarytowych pionowych elementów przeszklonych", *Materiały Budowlane*, (nr. 7), 26–28, 2015. DOI: 10.15199/33.2015.07.06 *(in polish)*.

[26] Sulik P., Kinowski J., Sędłak B., "Fire resistance of aluminium glazed curtain walls, test results comparison depending on the side of fire exposure", *Applications of Structural Fire Engineering, 15–16 October 2015, Dubrovnik, Croatia*, 478–483, 2015. DOI: dx.doi.org/10.14311/asfe.2015.076.

[27] Sulik P., Sędłak B., "Bezpieczeństwo pożarowe przeszklonych elewacji", *Materiały Budowlane*, (nr. 9), 18–20, 2015. DOI: 10.15199/33.2015.09.04 *(in polish)*.

[28] Sulik P., Sędłak B., "Odporność ogniowa drzwi z dużymi przeszkleniami", *Świat Szkła*, R. **20** (nr. 3), 38–42, 2015 *(in polish)*.

[29] Sulik P., Sędłak B., Kinowski J., "Bezpieczeństwo pożarowe ścian zewnętrznych – Część 1 – Elewacje szklane: wymagania, badania, przykłady", *Ochrona przeciwpożarowa*, **4**(50), 10–16, 2014 *(in polish)*.

[30] Sulik P., Sędłak B., Kinowski J., "Bezpieczeństwo pożarowe ścian zewnętrznych – Część 2 – Mocowanie okładzin elewacyjnych", *Ochrona przeciwpożarowa*, **1**(51), 9–12, 2015 *(in polish)*.

[31] Sulik P., Sędłak B., Turkowski P., Węgrzyński W., "Bezpieczeństwo pożarowe budynków wysokich i wysokościowych", *Budownictwo na obszarach zurbanizowanych – nauka, praktyka, perspektywy*, 14–19 September 2014, Krynica, Poland, 105–120, 2014 *(in polish)*.

MATEC Web of Conferences **46**, 02007 (2016)
DOI: 10.1051/matecconf/20164602007


# Investigations of the performance of facades made of ETICS with polystyrene under external fire exposure and fire safety measures for their improvement

Ingolf Kotthoff[1], Sebastian Hauswaldt[2], Olaf Riese[3] and Jan Riemesch-Speer[4]

[1] Ingenieurbüro für Brandschutz bei Fassaden (IBF), Stadtlengsfeld, Germany
[2] MFPA Leipzig GmbH, Leipzig, Germany
[3] Technische Universität Braunschweig, Braunschweig, Germany
[4] Deutsches Institut für Bautechnik (DIBt), Berlin, Germany

## 1. INTRODUCTION

Exposure to flames of a fully developed room fire venting out of an external wall opening after flash-over is most commonly used as the reference fire scenario for assessing the fire performance of external wall cladding systems. This is reflected by various testing standards (e.g. ISO 13785-2 [1], E DIN 4102-20 [2]). The impact of other fire scenarios had been judged as less critical or most similar in the past. Hence, there was a widespread opinion that other scenarios were covered by these fire tests (cf. [3]). This assumption has also underlain assessment of ETICS fire performance.

Only in spring 2013 was there an fire scenario had been introduced in the discussion on fire performance assessment of external wall cladding systems in Germany – the exposure to fires outside near the base of the building (e.g. waste containers, small cars etc.), and this just for ETICS based on polystyrene insulation material.

## 2. DISCUSSONS IN GERMAN BUILDING REGULATORY BOARDS

After some particularly serious fires and extensive media coverage of these events, the Conference of the *Länder* ministers responsible for building (*Bauministerkonferenz*, abbreviated as BMK) decided in autumn 2012 to set up a project group to evaluate a list of fire incidents involving external thermal insulation composite systems (ETICS) based on polystyrene. This list had been provided by the fire brigades. After evaluation of the list, the group should recommend appropriate measures (if necessary) to improve the fire performance of polystyrene-based ETICS. In several meetings, the project group analysed 25 fire incidents in total. Approved ETICS based on polystyrene were involved in 18 out of 25 cases. In a report to the *BMK* in March 2013 [4], the project group characterised the fire incidents as follows:

- In most cases, the ETICS were exposed to fires from outside the building – e.g. burning waste containers or smaller bulky waste piles.

- In some cases, the fire spread from an annex to the facade – sometimes facilitated by damages of the facade or poor workmanship of the connections between the annex and the main building.
- In one case, the fire occurred during the construction phase. The ETICS was not yet completed, and the fire was caused by a large quantity of polystyrene insulation material stored near the building.
- Only in one case, the ETICS had been exposed to flames of a fully developed room fire.
- In one case with unacceptable fire development, the design of the ETICS and the building itself strongly deviated from standard practices. This is why a special permission had been granted for this particular ETICS, which differed significantly from the approval in essential details.

On the basis of this analysis the project group concluded that:

- ➢ with regard to the fire scenario "flames of a fully developed room fire" the small-scale fire tests (as per DIN 4102-1, DIN EN 13501-1) and, where necessary, the large-scale fire test (as per draft E DIN 4102-20), both used in approval procedures for ETICS in Germany, were sufficient for assessing the fire performance of the ETICS. The ETICS can be considered as safe, if the fire safety measures set out in the approvals are observed.
- ➢ However, for certain external fire scenarios, additional fire safety measures may be required.

As a result, the project group recommended conducting research into this last point.

## 3. OLDER RESULTS OF INVESTIGATIONS OF EXTERNAL FIRE SOURCES

Fires outside a building near the facade can start from the adjacent ground or from external building elements, such as balconies or loggias. However, due to the limited dimensions of the latter, the quantity of combustible materials which can be stored there is restricted (mostly seating furniture or packaging). A higher risk is posed by burning cars or waste containers. Concerning such risks, a couple of tests with different fire sources were performed in the late 1990 and the early 2000 years at MFPA Leipzig. For the tests external fire sources were placed in front of an even test wall (no corner configuration). Table 1 gives on overview of the fire sources tests carried out at that time and their results. The flame heights measured show that the first storey is always in danger in case of such external fire. If the façade is made of flammable material there is also a risk that the fire spreads to the second storey.

This summary of results together with some explanations was also available to the project group for its consultations. It provided the basis for the following reference fire scenario to be used in the proposed research program on fire performance polystyrene-based ETICS under fire exposure from outside the building. The project group suggested using a mixed waste container with a maximum heat release of about 2.4 MW as a reference test scenario. It is considered to also be representative of similar fire hazards, like burning small cars, motor bikes or smaller bulky waste piles.

## 4. THE RESEARCH PROGRAM

### 4.1 General

After BMK endorsed the recommendations of the project group (cf. clauses 2 and 3) in spring 2013 *Deutsches Institut für Bautechnik* (DIBt) was mandated to organize a research program for detailed investigations into the fire performance polystyrene-based ETICS classified as class B1 in accordance with DIN 4102-1 [5] or classes B/C in accordance with EN 13501-1 [6], under exposure to external fires starting near the base of the building. These ETICS can be used on facades of buildings of building classes 4 and 5 (buildings up to about 25 metres high) pursuant to the Model Building Code (*Musterbauordnung, MBO* [7]). For technical and scientific steering of the programme, a small group was set up consisting of experts in façade-related fire safety issues.

**Table 1.** Results of investigations in the late 1990 years.

| Fire source | Burnable components | | Dimensions [m]/[m³] | Burning area [m²] | Place of ignition | Maximum heat release [kW] | Time of fully developed fire [min] | Duration of burning * | Flame height [m²] |
|---|---|---|---|---|---|---|---|---|---|
| | mass [kg] | type | | | | | | | |
| **Small car** | 100 | 90 kg mixed fire load (rubber etc.) 10 l fuel | 1.4 × 3.2 × 1.4 **6.7 m³** | 4 | Engine compartment | 1850 | 15 | 35 (25) | 3.3 (2.8) |
| **2 motor bikes** | 14 | Mixed fire load (rubber, plastics etc.) | 0.8 × 1.6 × 1.2 **1.5 m³** | 1 | seats | 370 | 4 | 25 (15) | 2.5 (2.1) |
| **2 waste cans** | 53 | 23 kg plastics 30 kg paper | 0.5 × 1.2 × 1.3 **0.8 m³** | 0.5 | right can, at the top | 530 | 10 | 40 (15) | 2.3 (1.8) |
| **Mixed waste container** | 188 | 64 kg polyamide (container) 60 kg paper 10 kg shoes 50 kg textiles 4 kg plastics | 1.0 × 2.0 × 1.5 **3.0 m³** | 2 | at the top, top cover 20 cm open | 2350 | 20 | 45 (25) | 3.2 (2.7) |
| **Plastic waste container** | 89 | 64 kg polyamide (container) 25 kg plastic packing material | 1.0 × 1.2 × 1.3 **1.5 m³** | 1.2 | at the top, top cover 20 cm open | 820 | 7 | 35 (15) | 3.7 (3.0) |
| **Paper waste container** | 148 | 64 kg polyamide (container) 88 kg paper and cardboard | 1.0 × 1.2 × 1.3 **1.5 m³** | 1.2 | at the top, top cover 20 cm open | 1370 | 7 | 35 (10) | 3.8 (3.1) |
| **Bulky waste (low fire load)** | 50 | Wood and wood based materials (as a crib) | 0.5 × 1.0 × 0.5 **0.25 m³** | 0.5 | deepest layer, with fire accelerant | 720 | 5 | 25 (10) | 2.3 (1.7) |

* Values in brackets describe the duration of the average flame heights.
** Values in brackets are the average flame heights.

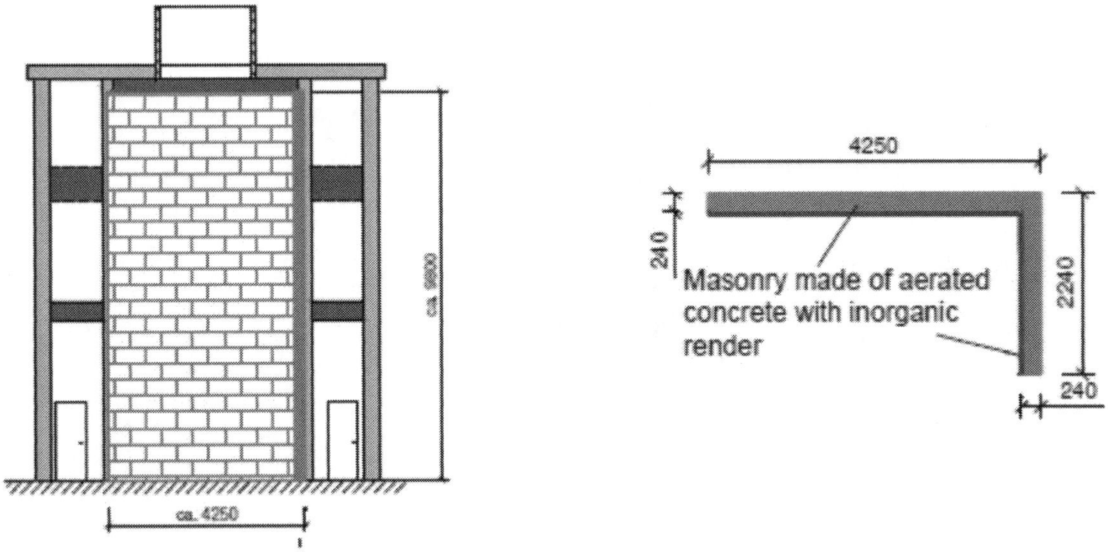


**Figure 1.** Front view and ground view of the test rig.

First of all, the expert group decided to use a special test rig for large-scale ETICS tests. This test rig consists of a main front and a wide wing set up at a 90° angle to each other (similar to the test setup described in draft standard E DIN 4102-20 [2] or standard BS 8414-1 [8]), but with a height of about 10 m and without a fire chamber at the base of the main front. The test rig was made from aerated concrete stones (density $\sim$ 600 kg/m$^3$) and set up in a testing hall of MFPA Leipzig, protected from the weather. Its height results from the expected flame height of the fire source and an additional area needed for observation of the extension of the flame plume and the fire spread on the surface of the ETICS. The height of the test rig roughly equals the height of a 3-storey building section. Figure 1 shows the layout and front view of the test rig.

The investigations comprise three stages, which are described below.

### 4.2 Testswith various waste containers and a 200 kg wood crib

The first stage of the research programme consisted of tests performed on various waste containers. This was considered necessary, because – unlike in the late 1990s – a strict waste separation system has been in place in Germany for some years and mixed waste containers have become very rare. The tests were aimed at confirming that containers filled with homogenous waste do not lead to higher heat release rates than 2.4 MW as assumed in the recommendation of the *BMK* project group (cf. Sect. 3 above).

It was clear from the start that for the large-scale ETICS tests an appropriate, well-reproducible fire source other than a waste container had to be found. The steering group intensively discussed the use of a wood crib with about 200 kg wood as well as a pool fire of isopropanol. The former behaves similarly to a burning waste container, in particular in the starting phase of the fire, because of the gradual fire development. Stable flame plume and high flames for the most part of the test duration are further advantages of the wood crib. By contrast, a pool fire leads to a sudden fire exposure of the facade (untypical of waste containers). But good reproducibility is an advantage of this fire source.

After weighing the advantages and disadvantages of both fire sources, the steering group decided to use a wood crib of 200 kg spruce wood as a primary fire source for the large-scale ETICS tests. For this fire source, the expected averaged heat release was about 2.0–2.3 MW with peaks between 2.5 and 3.0 MW based on available experience, calculations and old internal test results from MFPA Leipzig. The area of the area of the wood crib should be 1.1 m $\times$ 1.1 m. The cross-section of the wood sticks

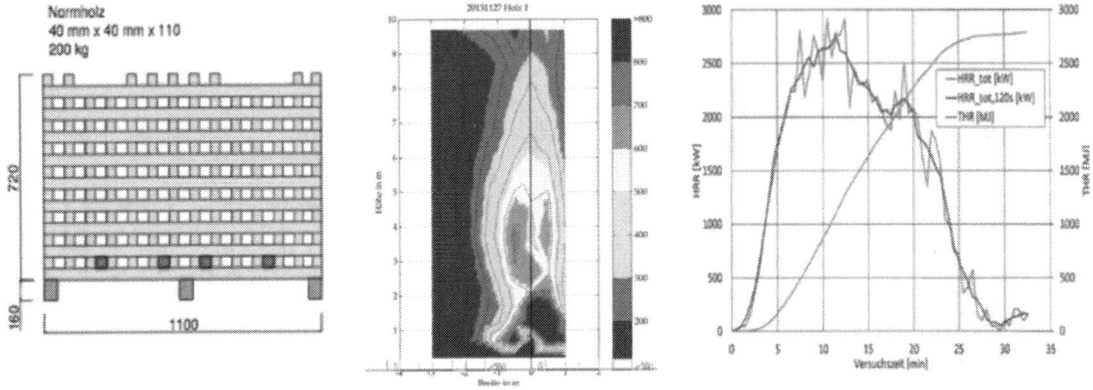


**Figure 2.** Drawing, time-independent maximum isotherms and heat release rate of the wood crib (from left to right).

was 40 mm × 40 mm and the sticks were stacked lengthways and widthways in alternating layers with a wood : air ratio of 1 : 1. The total height of the wood crib was about 70 cm. Figure 2 shows the wood crib.

Six tests had been performed in total within the framework of these investigations:

- 1 paper waste container (made up of about 60 kg polypropylene, filled with about 71 kg paper / cardboard),
- 1 plastic waste container (made up of about 60 kg polypropylene, filled with about 52 kg plastic waste),
- combination of 1 paper waste container and 1 plastic waste container (as described before),
- 3 test with 200 kg wood crib (1 x trial, 1 x repetition, and 1 x demonstration to *BMK* project group).

The following parameters were determined during the tests:

- temperatures in front of the test rig using a grid of thermocouples,
- flame heights and flame plumes with video camera,
- mass loss by using a scale.

Temperatures were used for creating pictures of isotherms. Mass loss data together with mass-weighted averaged gross calorific values served as a basis for calculating heat release rates during the tests. (cf. Fig. 2)

Table 2 gives an overview on the main results of these tests. In December 2013 they had been presented to project group of BMK (including members of fire brigades). The project group confirmed that the 200-kg-wood crib covers more than 90% of external fires affecting facades such as storages within relocations, waste in normal amounts including bulky waste or typical automobiles.

## 4.3 Selection of the ETICS and required laboratory tests

With support of DIBt as the German approval body in the construction sector, the steering group chose an approved ETICS considered to be representative of a wide range of polystyrene-based ETICS available on the German market. On the basis of existing test experience with such ETICS, the following requirements were set out for the test ETICS:

- use of an polyurethane-based adhesive foam for fixing to the substrate,
- white polystyrene with a thickness of 300 mm and an apparent density of about 25 kg/m$^3$,
- base coat with organic binders, 2 mm thick, about 10% organic compounds in dried condition),
- glass fibre mesh as reinforcement with an weight per unit area of about 150 g/m$^2$,

**Table 2.** Overview on results of test with waste containers and the 200-kg-wood crib.

| Fire source | Fire load | Mass-weighted cross calorific value [MJ/kg] | Maximum flame height [m] | Maximum heat release [MW] | Total heat release after 25 minute [MJ] * | Short description of the tests |
|---|---|---|---|---|---|---|
| **Paper waste container** | ~60 kg polypropylene (container) ~71 kg paper and cardboard | 30.4 | 2.5–3.0 (short peaks up to 4) | ~2.0 | ~800 | – fire starts very slowly due to compact filling of the waste<br>– more intensive fire only after 12 minutes<br>– only 60 kg of paper and plastic burnt after 60 minutes |
| **1 Plastic waste container** | ~60 kg polypropylene (container) ~52 kg mixes plastic waste | 41.9 | 2.5–3.0 (short peaks up to 4) | ~5.0 | ~2100 | – fire starts very slowly due to compact filling of the waste<br>– intensive fire starts more than 10 minutes after ignition<br>– development of an intensive pool fire after 21 minutes<br>– only 55 kg in total burnt after 26 minutes |
| **1 Paper waste container + 1 Plastic waste container** | 2× ~ 60 kg polypropylene (containers) ~55 kg plastic waste + 53 kg paper and cardboard | 36.5 | 4.0–4.5 | ~6.7 | ~4000 | – fully developed fire already after 7 minutes<br>– development of an intensive expanded pool fire after 10 minutes |
| **200 kg wood crib (3 test)** | 200 kg spruce wood | 15.8 | 4–5 (short peaks up to 7) | ~2.8/3.2/ 3.4 | ~2750 (in all 3 tests) | – stable flame plume after about 5 minutes<br>– total collapse of the crib after 25 minutes<br>– about 180 kg wood burnt after 30 minutes |

* This period of time reflects the lapse of time between the start of the fire and the latest start of fire-fighting measures (usual in Germany).

- finishing coat with organic binders, 2 mm thick, about 10% organic compounds in dried condition,
- 60 cm high splash water zone made up with polystyrene, 240 mm thick, organic rendering and a profile made of PVC at the lower edge of the ETICS.

An independent local specialised firm was contracted to acquire the components and install the ETICS on the test rig for the large-scale tests to ensure that ETICS manufacturers had no influence on the test results. For the same reason, test specimens of the ETICS, the polystyrene insulation material, the render coats and the PU foam were prepared for small-scale reaction to fire tests as per DIN 4102-1 and further laboratory tests in order to check whether the provisions of the ETICS approval were met. The following tests had been done:

- determination of apparent density and reaction to fire class as per DIN 4102-1 of the polystyrene,
- determination of ash content of the base and finishing coats as per ETAG 004, annex C [9],
- determination of reaction to fire class as per DIN 4102-1 of the ETICS itself and
- determination of reaction to fire class as per DIN 4102-1 of the PU adhesive foam.

All of these laboratory tests were performed twice by two different testing laboratories (MPA Braunschweig and BAM Berlin), independently of each other. The test results confirmed that the approval requirements and the specifications of the steering group for the ETICS itself as well as for its components were met – except for the reaction to fire class of the PU adhesive foam. The latter did not meet the requirements for reaction to fire class B2 as per DIN 4102-1 for unknown reasons. However,

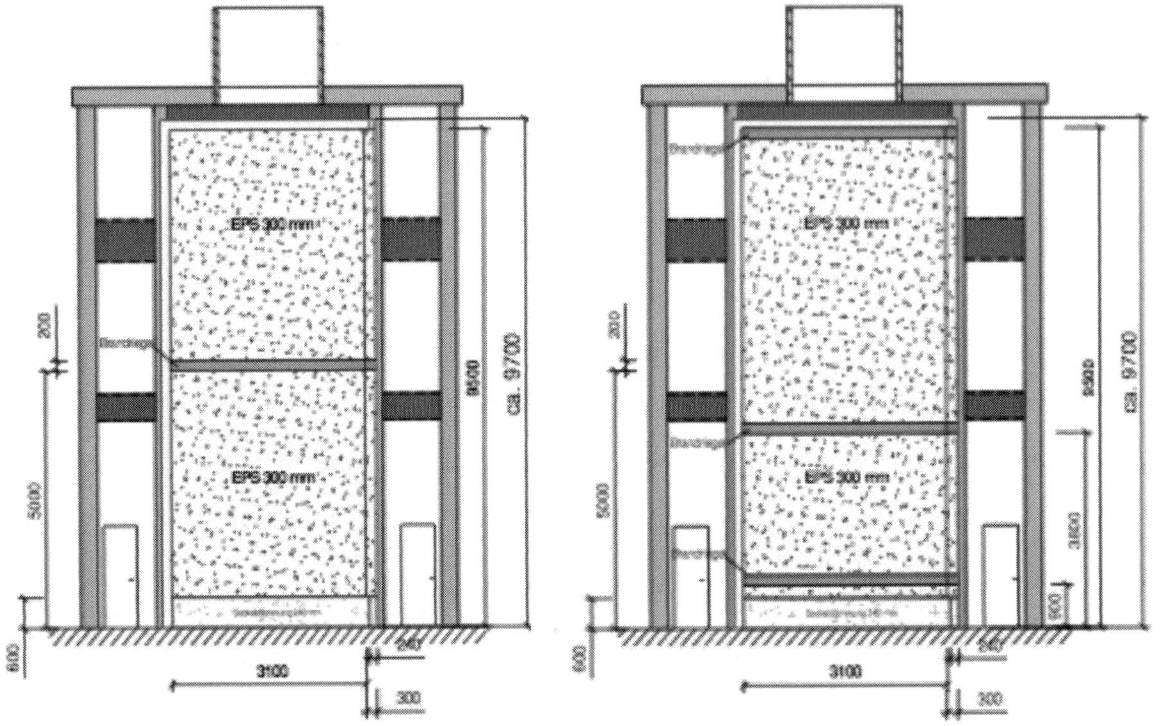


**Figure 3.** Position of the fire barriers (in Green) in test 1 (left) as well as in tests 2 and 3 (right).

with regard to the large-scale tests this means that an especially critical ETICS design (including the fixing) has been tested.

### 4.4 The large-scale tests

Within the framework of the investigation, three large-scale tests had been performed in total in 2014 The minimum test duration had been agreed to be 25 minutes, because – according to the fire brigades – it usually takes 8–10 minutes until an external fire is discovered and another 10–15 minutes until the fire-fighters arrive. The wood crib (as described in Sect. 4.2) was placed directly in the corner of the test rig with a distance of about 10 cm from the surface of the ETICS. It was ignited by a spark pistol and fuelled by 4 small trays each with 200 ml isopropanol placed in the second layer of the wood crib. Assembly of the The ETICS complied with the basic requirements given in Sect. 4.3 for all three tests.

Test 1 was aimed at verifying the fire performance of polystyrene-based ETICS as observed by fire brigades under standardised conditions and determining its causes. The ETICS design used for test 1 included only one fire barrier positioned 5.0 m above the floor of the testing hall. This roughly corresponds to one fire barrier every second storey, as required by approvals in the past. An adhesive (mortar) with organic binders was used for fixing of the fire barrier to the substrate without additional mechanical fixing (anchors). Fire barriers consisted of mineral wool lamella stripes, 200 mm thick, with reaction to fire class A1 as per EN 13501-1, apparent density of about 90 kg/m$^3$ and a melting point of >1000 °C (tested as per DIN 4102-17 [10]). Figure 3 shows the test setup including the position of the fire barriers in test 1 as well as in tests 2 and 3.

Observations during the test and results of test 1:

- ignition and development of an independent fire on the render surface about 7 minutes after the beginning of the test – flame travels upwards leaping over the fire barrier,

- formation of a secondary fire at base of the wood crib, caused by molten polystyrene in the 9$^{th}$ test minute,
- strong deformation of the render, in particular on the short wing of the test rig before the end of the 10$^{th}$ test minute,
- extension of the flames of the primary fire source through energetic contribution of the ETICS up to the top edge of the test rig in the 11$^{th}$ test minute (flame height $\approx$ 10,0 m),
- cracking of the render on the short wing near the inner corner of the L-shaped test rig in the 12$^{th}$ / 13$^{th}$ test minute – subsequently, increasing lateral fire spread and intensive smoke development,
- large parts of the rendering come loose in the 15$^{th}$ test minute – parts of the fire barrier probably come off at this time at the latest,
- fully developed fire after 15 minutes combined with intensive smoke development – temperatures of $\geq$ 1200°C measured in the chimney above the test rig,
- secondary pool fire on the floor of the testing hall is extinguished after the 19$^{th}$ test minute; the wood crib is extinguished after 28 minutes = end of the test.
- After the end of the test, the ETICS including the PU foam layer between the insulation layer and the substrate is found to have been completely combusted.

Results of test 1 were considered as negative and unacceptable with regard to the safety objectives of the building codes. This opinion was shared by both the members of the *BMK* project group as well as the members of the steering group. The findings confirmed observations made by the fire brigades during real fires.

Afterwards for the two further arrangements for improving the fire performance of the ETICS were considered. For that the number of fire barriers was increased and their fixing modified. Additionally, the reinforcement of the base coat was strengthened with a special mesh corner bracket at the inner corner of the ETICS test rig. The first fire barrier was positioned almost immediately above the splash water zone; the second fire barrier in a distance of about 3.0 m above the first one and the third fire barrier directly at the top of the test rig (cf. Fig. 3). Furthermore, an inorganic adhesive (binders: lime/cement) and approved anchors (consisting of sleeves and plates made of plastics and a steel expansion element) were used for their fixing.

Observations during the test and results of the second test:

- intensive exposure of the ETICS to flames shooting up from the wood crib after 4/5 minutes,
- combustion of the insulation material of the splash water zone as well as that of the ETICS, but below the first fire barrier only,
- rendering above the first fire barrier up to the top edge of the test assembly does not crack during the entire test duration,
- only the flames of the primary fire source (wood crib) extend up to a height of about 6–7 m due to combustion of the organic base and finishing coat compounds as well as pyrolysis gases from the polystyrene diffusing trough the render layer,
- development of a secondary fire at the base only after more than 20 minutes,
- fire barriers remain attached to the substrate,
- the wood crib is only extinguished after 35 test minutes – at that time, the fire intensity is clearly decreasing.

Comparison of these results with the results of the pre-tests on waste containers and the wood crib shows only slight differences in flame heights and fire spread on the surface of the ETICS, meaning that the contribution of the ETICS to fire spread was very limited. With regard to the safety objectives set out in the building codes – limiting the number of storeys affected by fire at the very most by the time the fire brigades arrive to 2 storeys – results of test 2 were considered positive.

The third test confirmed the results of the second test. The observations during the test and the damages at the end of the test were similar. Only the flames of the wood crib shot up higher. These differences were essentially caused by different wind and weather conditions at the time of testing

**Figure 4.** Test 1 at the 15th minute (left); test 2 and 3 at the 18th minute (middle and right).

**Figure 5.** Damage of the ETICS assemblies after end of tests 1 to 3 (from left to right).

between test 2 and 3. Figures 4 and 5 give an impression of the three large-scale tests performed within the framework of the research programme.

After the end of the research programme, the steering group and the *BMK* project group jointly discussed whether the scenario "external fire exposure" should also be applied to other external wall cladding systems than polystyrene-based ETICS. The members unanimously came to the conclusion that this was not necessary, since there was no evidence pointing to the involvement of such systems in serious fires. Furthermore, the insulation materials used for these systems (mineral wool, polyurethane foam, phenolic resin foam) do not show the same thermoplastic reaction under fire exposure as polystyrene (including melting and formation of pools near the primary fire source).

## 5. LESSONS DRAWN FROM THE TEST RESULTS

In autumn 2014, the results of the research programme were presented to the Conference of Construction Ministers (*BMK*). At the meeting, the ministers endorsed the proposal to amend all national approvals granted by DIBt for polystyrene-based ETICS rated as class B1 in accordance with DIN 4102-1 or classes B/C in accordance with EN 13501-1 and impose additional fire safety requirements relating to fire exposure from outside the building.

The expert group "Reaction to fire of construction materials" (advising DIBt in fire safety matters related to approvals) agreed unanimously that the "area at risk", i.e. the part of the façade that needed additional protection against external fire exposure comprised the area of the first three storeys of a building, because a critical impact is not to expect above this area caused by such external fire near building's base. Hence, the fire safety measures proved in the large-scale tests should be introduced in the approvals of those ETICS covered by the tests in the following form:

- position of the first fire barrier directly above the splash water zone, no higher than 90 cm above the ground,
- position of the second fire barrier at the height of the ceiling above the first storey, but with no greater distance from the first fire barrier than 3.0 m,
- position of the third fire barrier at the height of the ceiling of the third storey, but with no greater distance from the second fire barrier than 8.0 m,
- integration of a special mesh corner bracket in the reinforced base coat in inner corners of buildings,
- use of further fire barriers in the area of transitions between the external wall and horizontal building elements, e.g. balconies, if these transitions are situated in the "area at risk".

The fire barrier design (type, dimensions, fixing) has to be identical to that tested in large-scale tests 2 and 3. These fire barriers also serve as a protection against possible fire exposure to flames of a fully developed room fire.

However, the consultations of the expert group also showed that the results of the research programme mainly cover the tested ETICS design, but not the entire field of polystyrene-based ETICS. The test results are only valid for ETICS corresponding to the following specifications:

- application on inorganic substrates, such as masonry or concrete,
- fixed with adhesives (mortar or PU foam) only or in combination with anchors or profiles,
- use of polystyrene of class B1 in accordance with DIN 4102-1 or classes B / C in accordance with EN 13501-1, with a thickness of up to 300 mm and a maximum apparent density of 25 kg/m$^3$,
- use of glass fibre meshes with a weight per unit area of at least 150 g/m$^2$ as a reinforcement and
- use of base coats and finishing coats, each with no more than 10% of organic compounds in dried condition, with a total thickness of the rendering of at least 4 mm.

These specifications cover about two-third of polystyrene-based ETICS approved in Germany.

All members of the expert group agreed that the fire safety measures described above are not sufficient for polystyrene-based ETICS:

- in case of application on substrates made of wood or wood-based panels,
- with insulation layers of a thickness of more than 300 mm,
- with ceramic or natural stone claddings (due to the risk of spalling under fire exposure and subsequent damage to the reinforced base coat) or
- without a reinforced base coat.

After intensive discussions and based on technical considerations, it has been decided to require non-combustible external wall cladding systems (usually an ETICS based on mineral wool) in the area of the first two storeys and installation of an additional fire barrier at the height of the ceiling of the third storey in these cases – until further test results are available.

In May 2015, DIBt published an information leaflet [11] on its homepage containing a detailed description and schematic drawings of the additional fire safety measures required for the various types of polystyrene-based ETICS. On the basis of this information, more than 250 approvals were amended in the summer and autumn of 2015. As of 1st January 2016, these amendments have entered into force and have to be considered when installing a polystyrene-based ETICS on site.

## 6. CONCLUSION

A couple of particularly serious fires and extensive media coverage of these events triggered an investigation into fire incidents involving external thermal insulation composite systems (ETICS) based on polystyrene which occurred over the last few years. The investigation showed that the fires and their spread over the facade of the building were often caused by fire exposure from outside the building, e.g. burning waste containers. For this reason, a research programme was carried out in order to determine the intensity of such waste container fires, to verify the reaction of polystyrene-based ETICS observed in real fires under standardised conditions and to test appropriate safety measures to improve the fire performance of these ETICS. The results of the research programme showed that additional fire safety measures against fire exposure from the outside near the base of the building were required – mostly taking the form of fire barriers running around the building and made of non-combustible mineral wool lamella stripes positioned above the splash water zone and at the heights of the ceilings of the first and third storey. These measures were specified in detail and introduced into all national approvals of DIBt for polystyrene-based ETICS granted by DIBt. Since the beginning of 2016, their application has been mandatory in Germany.

At present, consideration of the scenario "fire from the outside near the base of a building" is only necessary for polystyrene-based ETICS. For other external wall cladding systems, there has been no evidence so far pointing to an unacceptable performance of these systems in case of such fires.

### References


[1] ISO 13785-2:2002 Reaction-to-fire test for facades – part 2: Large-scale test.

[2] Draft E DIN 4102-20:2016-03 "Fire performance of construction materials and building elements; part 20: Special determination of the fire performance of external wall cladding systems".

[3] Kottoff, Riemesch-Speer, "Mechanism of fire spread on facades and the new Technical Report of EOTA Large-scale fire performance testing of external wall cladding systems", Proceedings of the 1st seminar "Fire safety of facades", Paris 2013.

[4] Internal progress report of the project group "Fire performance of ETICS with polystyrene insulation" to the meeting of the conference of *Länder* ministers responsible for building (*'Bauminsterkonferenz'* – BMK), held on 22 March 2013 – unpublished.

[5] DIN 4102-1:1998-05 "Fire behaviour of building materials and elements; part 1: requirements and testing of the reaction to fire of building materials".

[6] EN 13501-1:2010 "Classification of building products; part 1: classification of building products with the results of tests to their reaction to fire".

[7] Model Building Code (*Musterbauordnung, MBO*), version 2012, issued by the Conference of *Länder* ministers responsible for building; `www.is-argebau.de`.

[8] Standard BS 8414-1 "Fire performance of external wall cladding systems", part 1 dated 13th December of 2002.

[9] ETAG 004 – "Guideline for European Technical Approvals of External Thermal Insulation Composite Systems (ETICS) with Renderings", published by the European Organization for Technical Approvals (EOTA): edition February 2013.
[10] DIN 4102-17:1990-12 "Fire behaviour of building materials and elements; part 17: Melting point of mineral wool insulation materials; requirements and testing of the reaction to fire of building materials".
[11] Information on ETICS with EPS insulation published by DIBt, version: 27 May 2015, www.dibt.de → Departments → Unit II1.

MATEC Web of Conferences **46**, 03001 (2016)
DOI: 10.1051/matecconf/20164603001


# Fire exposed facades: Numerical modelling of the LEPIR2 testing facility

Virginie Dréan, Renaud Schillinger and Gildas Auguin

Efectis France, Saint Aubin, France

**Abstract.** LEPIR2 testing facility is aimed to evaluate the fire behaviour of construction solutions implemented on facade according with the experimental evaluation required by the French Technical Specification 249 (IT249) of the safety regulation. It aims to limit the risks of fire spreading by facades to upper levels. This facility involves a wood crib fire in the lower compartment of a full scale two levels high structure. Flames are coming outside from the compartment through windows openings and develop in front of the facade. Computational fluids dynamics simulations are carried out with the FDS code (Fire Dynamics Simulator) for two full-scale experiments performed by Efectis France laboratory. The first objective of this study is to evaluate the ability of numerical model to reproduce quantitative results in terms of gas temperatures and heat flux on the tested facade for further evaluation of fire performances of an insulation solution. When experimental results are compared with numerical calculations, good agreement is found out for every quantities and each test. The proposed models for wood cribs and geometry give correct thermal loads and flames shape near the tested facade.

## 1. INTRODUCTION

Development of external insulation facade solutions for construction is in great expansion because of economic and environmental considerations. The requirements provided by the French Technical Specification 249 (IT249) [1] of the safety regulation should help limiting the risks of fire spreading by facades to upper levels in case of fire occurring in a building.

The justification for facades not complying with the IT 249 technical requirements consists in a full scale test called LEPIR2 (for Experimental 2 levels Local For Real Fire) defined by the French ordinance of September 10th, 1970 [2]. The facility consists in two bunk rooms: two complete levels (R0 and R+1) and a starting of a third level (R+2).The lower level (R0) contains the fire sources, consisting in two calibrated wood cribs representing a generalized cellulosic fire. The fire load attempts to simulate the standard ISO fire curve [3] in the first 30 minutes of the test. The floor between the ground level (R0) and the upper level (R+1) can be positioned at different heights.

The implementation of this test is restrictive because it requires the testing of a full scale facade. Moreover its cost may be a limitation for solution technologies development.

The use of numerical simulations is then particularly indicated in order to analyse LEPIR 2 test results for a given facade solution, and in order to assess extending configurations after a validating test. Indeed, to allow for subsequent evaluation of the tested facade, the prediction of the thermal loads imparted to the facade materials is needed.

In this context, a numerical LEPIR2 model was developed with the FDS code (version 5) ("Fire Dynamics Simulator"). This computational fluid dynamics (CFD) code, developed by the American institute NIST [4–6], is widely used in the fire safety community.

This paper presents the preliminary studies performed to develop and validate this virtual testing facility. Any insulation material or cladding is considered on the facade, so that the facility is only made of concrete. So, the presented studies focus only on calibration tests. The following step, consisting in the implementation of an insulation material in front of the facade to evaluate the fire behaviour of facade solution is not exhibited here.

The physical phenomena to model with FDS in the case of LEPIR2 tests are primarily the burning of wood cribs in the fire compartment, then the flames exiting outside the openings and spreading along the facade.

The phenomenon of external flames is described and studied in [6]. For those flames existing outside of a compartment, it is necessary that they be supplied by leaving unburned combustion gases through the opening. The unburnt fuel gases left are then transported outside by convection where they react with the ambient air, itself driven in the outer flame and the thermal plume.

The dimensions of the external flames are mainly dependent on the air flow rates through the compartment openings. The geometry of the openings determines the shape of the stream of hot gases escaping from the compartment. The gas velocity at the opening determines the travelled horizontal distance, before the effect of buoyancy forces deflects the flames in the vertical direction.

The external thermal plume is naturally accompanied by air entrainment which leads the cooling of ascending hot gases. In the case where there is a wall above the opening, air entrainment is limited to one side. It contributes to a plume closer to the wall under the suction pressure effect. Therefore, the reattachment of the outer flames to the facade will be well modelled to properly assess the thermal loads transmits to the tested wall.

## 2. LEPIR 2 TEST FACILITY

### 2.1 Structure

The structure of LEPIR2 facility is regulated by Ordinance [1] modified by an application protocol [2].

It consists in two superposed rooms made of reinforced concrete with a thickness of 200 mm: two complete levels (R0 and R+1), and the beginning of a third level (R+2). An opening is installed at the rear side of the ground level. It is provided with a door, and can be more or less closed during the test in order to control the incoming and outgoing air flow in the compartment. Windows are arranged in front of the two levels and the frames are made of PVC.

The typical construction is presented in Fig. 1.

### 2.2 Fire load

The fire load during the test consists in two wood cribs, standardized by [2] and laid on the floor of the R0 compartment.

Normalized cribs are made of softwood with density 480 +/− 50 kg/m$^3$ and moisture content between 9 and 15%. Each crib consists in 3 different stacks (lower, medium, higher) (Fig. 2) with total dimensions of 1000 × 1000 × 1800 mm (W × L × H) and mass of 300 kg (600 kg for the two cribs). The cribs are located in the lower compartment, at a distance of 100 mm from the front wall.

Steel trays filled with a mixture of heptane and fuel oil are placed under the fire loads for the ignition.

**Figure 1.** View of the Efectis LEPIR 2 facility.

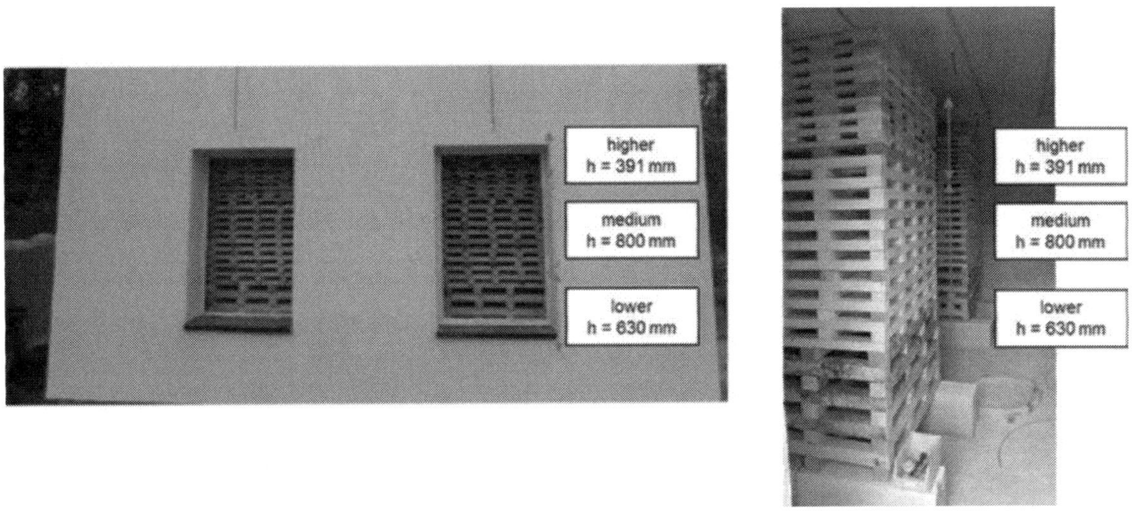


**Figure 2.** Fire load in the LEPIR2 facility.

The temperature of the hot gases near the compartment roof must then approach the ISO curve in the first 30 minutes of the test.

## 2.3 Instrumentation

To measure the evolution of the hot gas temperature and radiation during the test, 8 mm diameter Inconel type thermocouples are located within the fire compartment and on the exterior of the building, 150 mm from the tested facade to observe the outer flames propagation (Fig. 3).

The validation criteria to be controlled are:

- During the 30 minutes of the test, no ignition or flame propagation, lasting more than 20 seconds at the start of the third level facade, so no drilling of the facade is observed,
- No lateral spread of the inflammation of the facade on its entire width,
- No flame passage to the upper floor from the facade / floor junction,

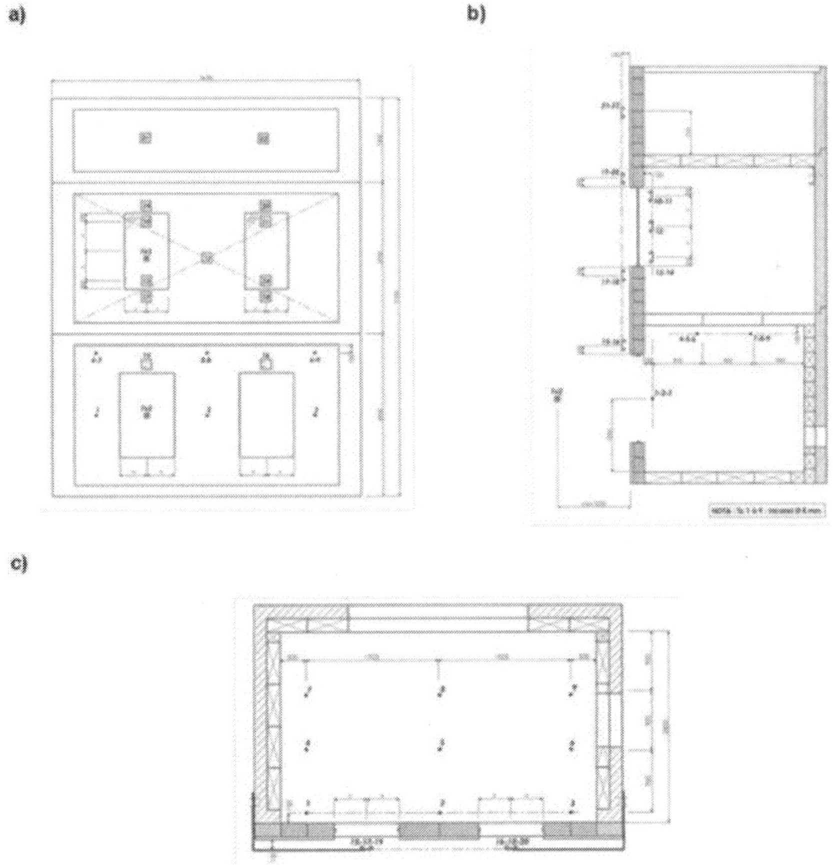

**Figure 3.** Location of instrumentation (thermocouples and plate thermometers) in LEPIR2 facility a) front facade b) lateral view c) top view of the R0 compartment.

- No passing of hot gases having a temperature above 350°C trough the facade / floor junction,
- No temperature rises above 180°C on the unexposed side of the floor 50 mm back from the facade.

### 2.4 Calibration tests

During the two calibration tests performed, called respectively A and B tests, the facade exposed to external flame consists only of inert concrete with a 200 mm thickness.

The simulation of these tests allows validating the proposed model for the geometry (mesh, boundary conditions) and the fire source, without the influence of a facade construction solution.

## 3. NUMERICAL SIMULATIONS

### 3.1 CFD code

FDS is a computational code in fluid dynamics that incorporates a combustion model with a mixture fraction approach and a large eddy scale model (LES) for the description of turbulent flows.

This tool allows 3D modelling of the computational domain. It takes into account heat transfer conditions at walls (1D heat equation solved in solids), and ventilation conditions.

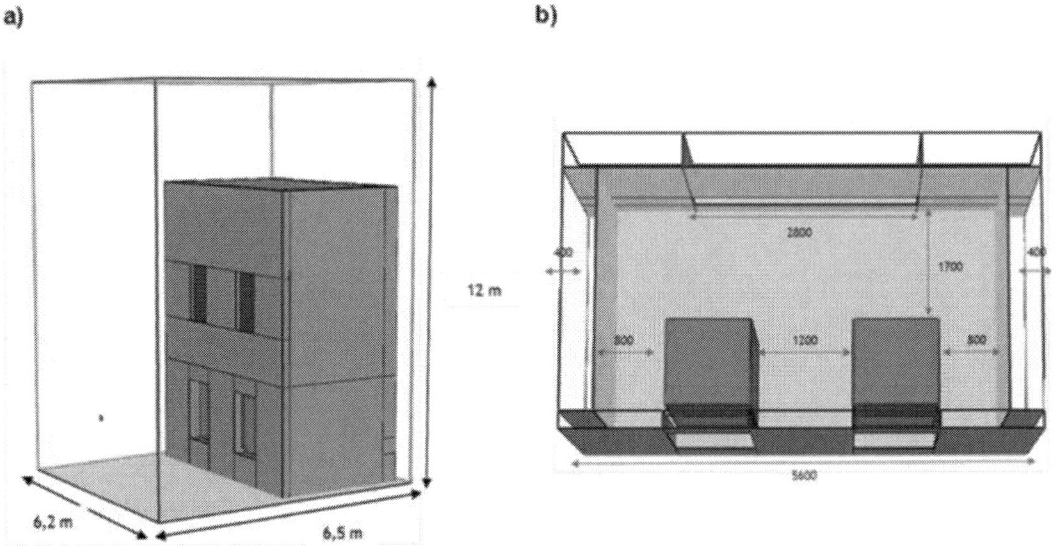


**Figure 4.** Numerical modelling of LEPIR II facility a) construction b) top view of the R0 (fire) compartment.

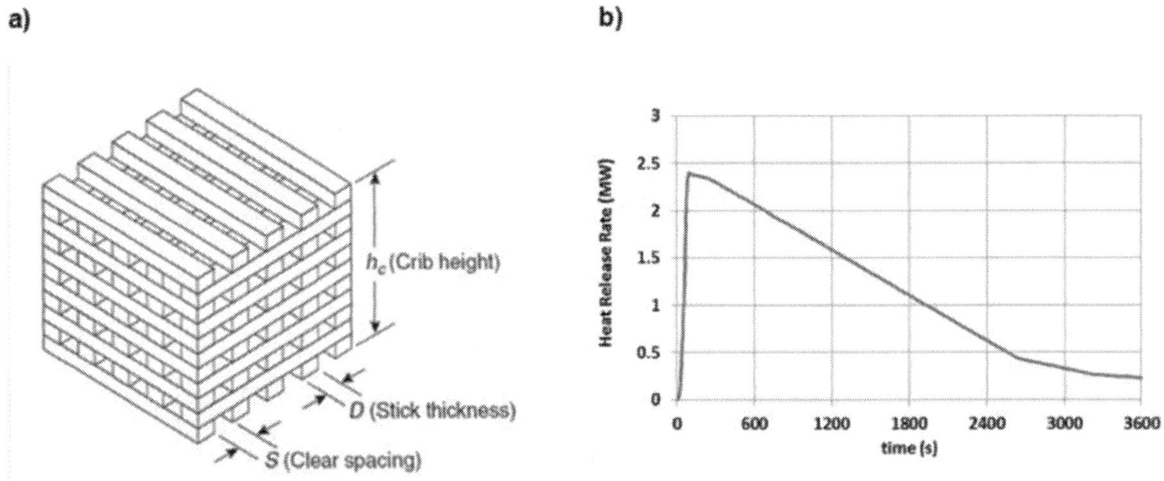


**Figure 5.** a) Wood crib representation for the burning rate evaluation [9] b) evaluated HRR for a single wood crib used in LEPIR 2.

## 3.2 LEPIR2 construction

The size of the computational domain is 6.5 × 6.2 × 12 m approximately. The dimensions of the test construction are adapted regarding the mesh size (Fig. 4). Grid size of 0.1 × 0.1 × 0.1 m is considered for the flame zone (R0 and R+1), and 0.2 × 0.2 × 0.2 m for the flameless zone (R+2). These dimensions permit to capture the main phenomena of combustion and ventilation, with a reasonable calculation time.

A condition of a zero gradient pressure (open) is imposed at the domain boundaries, except on the ground, considered as a 20°C surface. The boundaries of the domain are far enough from the construction to have no influence. The thermal properties for concrete structures are provided in [8].

In Fig. 4, the fire sources are represented by red blocks (R0), the windows by the blue surfaces (R+1). One of the windows is modelled bare inside of the facade, and the other bare outside. Glazing is modelled using thermal properties and radiation absorption coefficient inside the material.

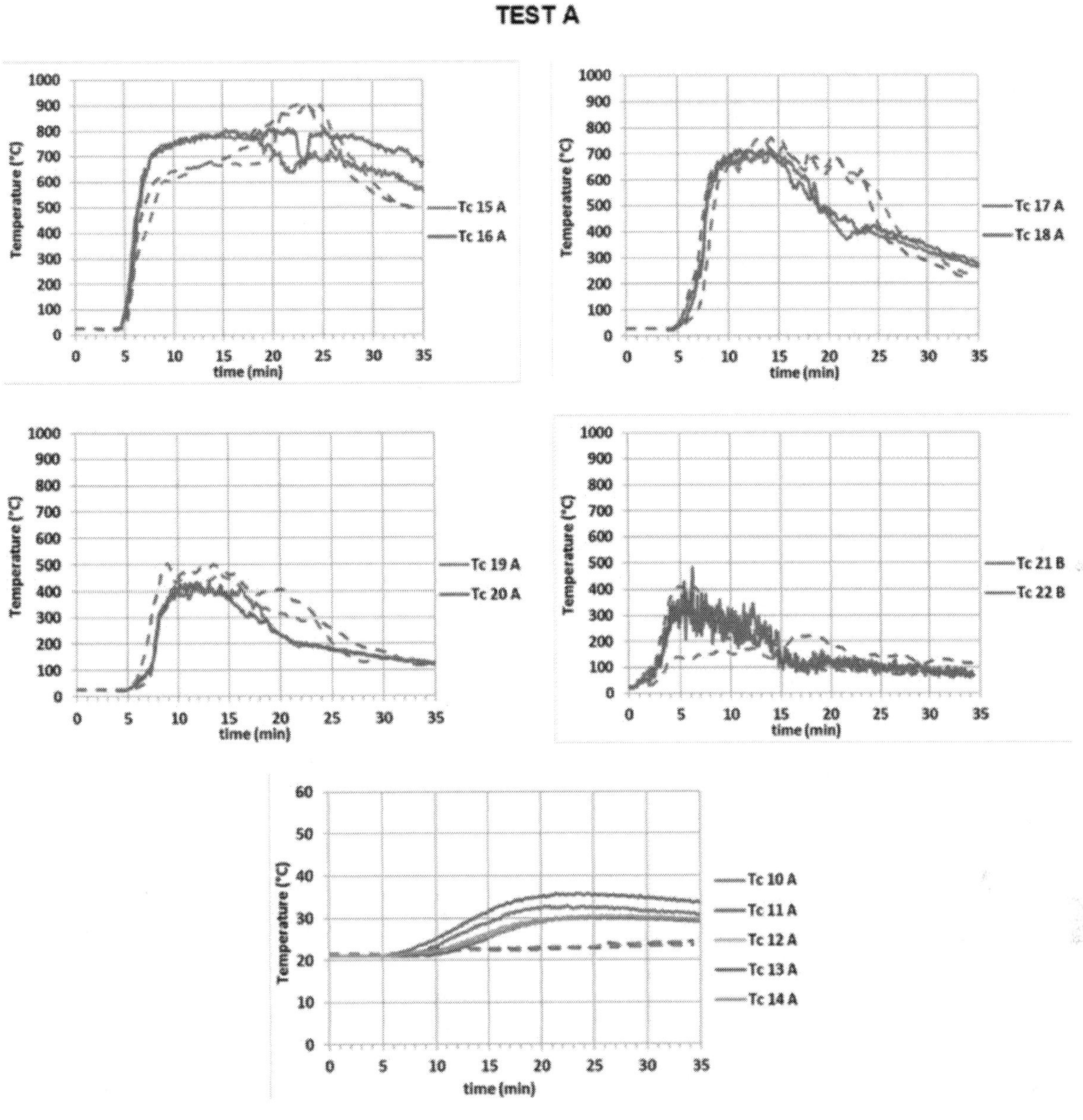


**Figure 6.** Comparison of experimental (dotted lines) and numerical (solid lines) gas temperatures 150 mm in front of the facade and in R+1 compartment for the A calibration test of LEPIR2.

During the tests, the ventilation shutter at the rear side of the fire compartment can be opened or closed to regulate the air inflow and thus the wood cribs burning. The times and rates of opening are indicated in the test reports. However, the simulations are performed with the shutter opened.

### 3.3 Fire load model

The stoichiometric equation for the overall combustion reaction of wood is:

$$C_{3,4}H_{6,2}O_{2,5} + 3,62(O_2 + 3,76N_2) \rightarrow 3,31CO_2 + 3,10H_2O + 0,01CO + 0,07C_{suies} + 13,61N_2. \tag{1}$$

A heat of combustion of 14 MJ/kg is used.

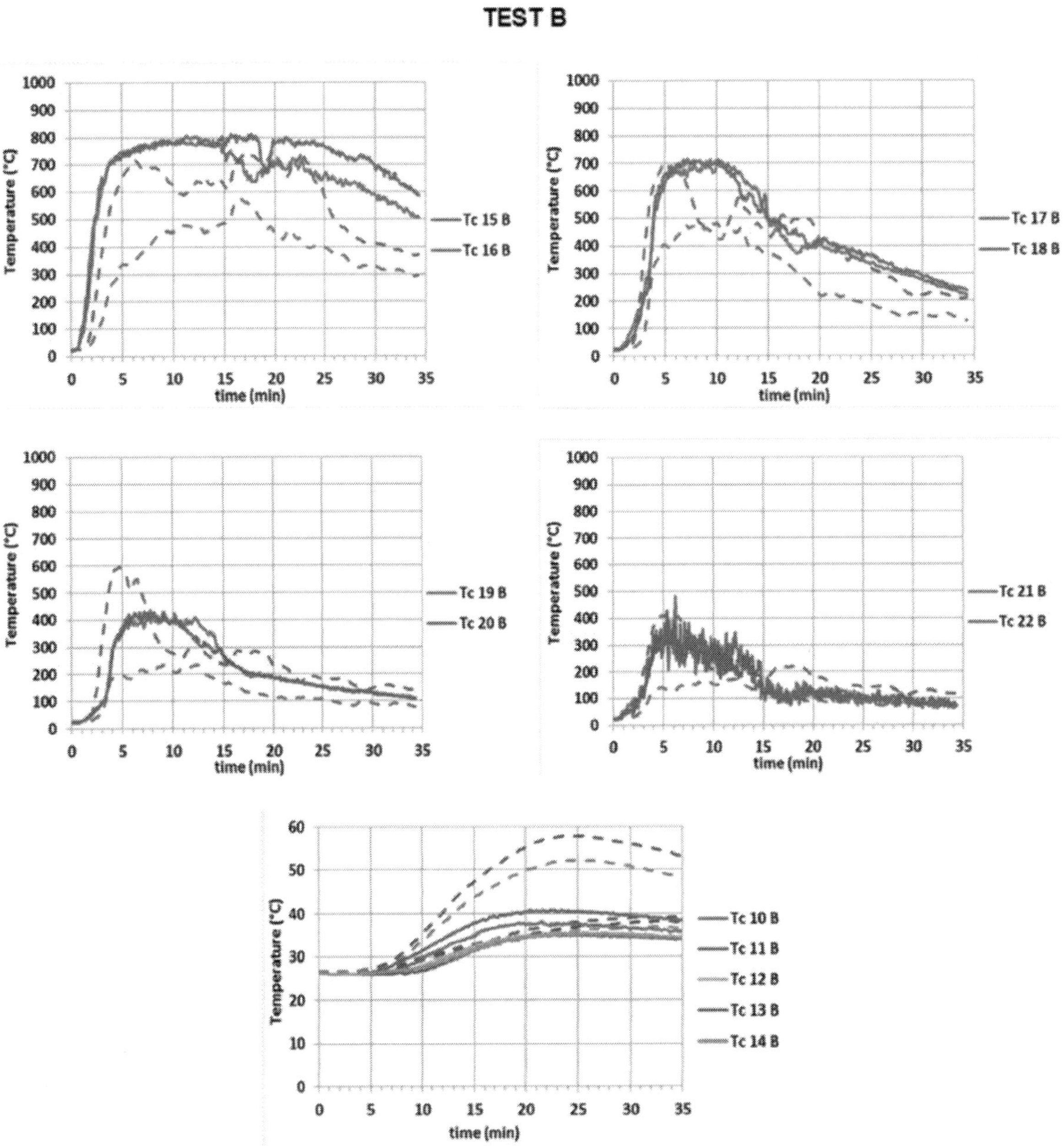


**Figure 7.** Comparison of experimental (dotted lines) and numerical (solid lines) gas temperatures 150 mm in front of the facade and in R+1 compartment for the B calibration test of LEPIR2.

The heat release rate (HRR) is evaluated following the model detailed in [9]. It depends on the characteristics of the stack (Fig. 5), the porosity of the crib and the fuel surface. In all cases, the HRR is limited by the ventilation conditions in the compartment. For each stack composing a crib, an individual HRR is calculated.

The evaluated HRR is then prescribed on both fire source modelled. Almost 5 MW are released by the two wood cribs in the R0 compartment.

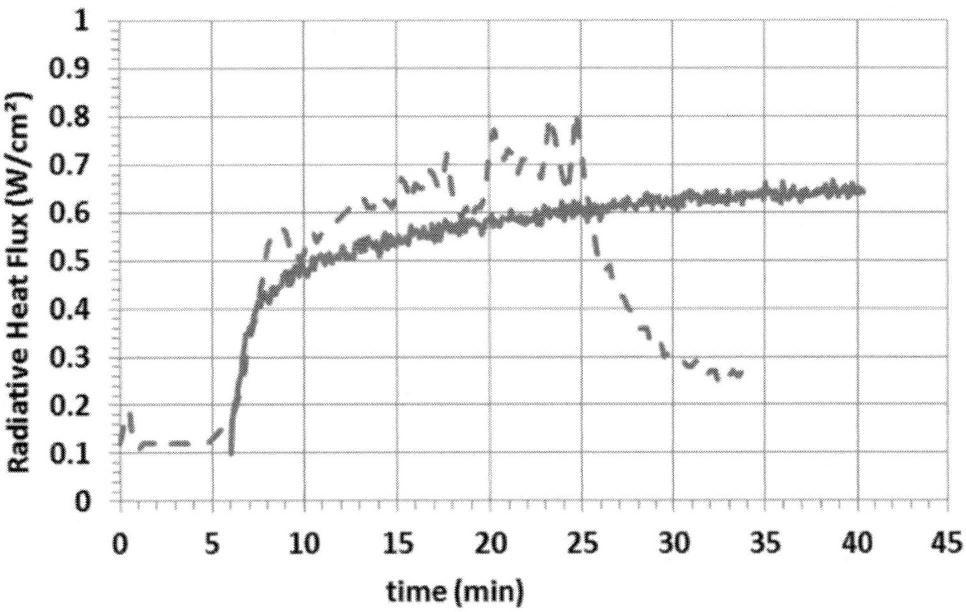


**Figure 8.** Comparison of experimental (dotted lines) and numerical (solid lines) radiative heat flux 4 m in front of the facade for the A calibration test of LEPIR2.

### 3.4 Results

External flames come from unburned fuel inside the compartment escaping through openings. In simulations, the total heat released inside the R0 room is close to 2.6 MW at its maximum level. So, 2.2 MW are released outside the compartment.

Predicted gas temperatures located at the facade (see Fig. 2 and Fig. 3) are compared with experimental measurements for test A and test B in Fig. 6 and Fig. 7 respectively.

During test A, the radiative heat flux is measured 4 m in front of the facade. Numerical results are compared with experimental ones on Fig. 8.

## 4. DISCUSSION

At first, it must be noted that the comparison of experimental results for test A and test B shows an overall good repeatability of the LEPIR2 tests. Gas temperatures are very comparable at each level on the facade. During test B, a slight asymmetry is observed on the front facade above one window. This can be explained by a side wind on the test's day.

When compared with numerical predictions, good agreement is found for every quantity and both tests. The proposed wood crib model gives correct thermal loads and flames shape near the tested facade.

Temperatures with order of magnitude of 700°C are spotted at thermocouples TC 15 to 18. They are located directly in the external flame near the facade, between R0 and R+1 opening. Indeed, the simulation shows that the flames well reattach the facade above the R0 openings, due to the absence of obstacles above the windows.

The most critical zone of the facade is then located between windows of R0 and R+1 levels. The validation criteria to be considered are mainly the temperatures at the junction between floors: R0 to R+1, and R+1 to R+2. Gas temperatures close to the facade are correctly predicted for each zone. For the thermocouples located up to the R+1 opening, i.e. TC 19 to TC 22, very good agreement is found. These thermocouples are not located directly in the flames, but rather in the ascendant hot gases.

Radiative heat flux in front of the facade is correctly reproduced by numerical simulation.

# 5. CONCLUSION

CFD simulations have been carried out with the FDS code for two full-scale experiments performed by Efectis France laboratory with the LEPIR2 facility. The first objective of this study was to validate the numerical model and to evaluate the ability of FDS to correctly predict gas temperatures and heat flux on the tested facade.

The good agreement examined trough the comparison between numerical and experimental results, allows the validation of the proposed wood crib model providing a correct thermal load and a suitable flames shape near the tested facade.

These numerical simulations allow for subsequent evaluation of the tested facade by reproducing the thermal actions imparted to the facade materials. Then, analysis can be done to understand material behaviour or implementation effects in critical zones, thanks to posteriori heat transfer calculations performed with a finite element code.

The next development step for the numerical model will consist in the implementation of the facade construction solution to take into account interaction between the facade product and flames.

## References

[1] Arrêté du 24 mai 2010 portant approbation de diverses dispositions complétant et modifiant le règlement de sécurité contre les risques d'incendie et de panique dans les établissements recevant du public, Journal officiel de la République Française du 6 juillet 2010, texte 31.
[2] Protocole d'application de l'arrêté du 10 septembre 1970 relatif à la classification des façades vitrées par rapport au danger d'incendie.
[3] Fire resistance tests: general requirements; NF EN 1363-1, June 2000.
[4] Fire Dynamics Simulator FDS, version 5.5.1, www.nist.gov
[5] Fire Dynamics Simulator (Version 5) User's Guide, NIST Spec. Publ. 1019-5, 2007.
[6] Fire Dynamics Simulator (Version 5) Technical Reference Guide, NIST Spec. Publ. 1018-5, 2007.
[7] Deshangère, S., "Determination of heating conditions for external structures submitted to fire", PhD Thesis, INSA (Institut National des Sciences Appliquées) Rouen, 2006.
[8] Eurocode 2: concrete structures, Part 1–2, NF EN 1992-1-2, December 2004.
[9] SFPE Handbook of Fire Protection Engineering, Third Edition, NFPA, 2002.

MATEC Web of Conferences **46**, 03002 (2016)
DOI: 10.1051/matecconf/20164603002


# Uncertainties in façade fire tests – measurements and modeling

Johan Anderson, Lars Boström, Robert Jansson McNamee and Bojan Milovanović*

SP Technical Research Institute of Sweden, Fire Research, Borås, Sweden
* Faculty of Civil Engineering, Department of Materials, University of Zagreb, Croatia

**Abstract.** In this paper a comparison between test and modelling results are performed for two large-scale façade fire testing methods, namely SP Fire 105 and BS 8414-1. In order to be able to compare tests and modelling the uncertainties have to be quantified both in the test and the modelling. Here we present a methodology based on deterministic sampling to quantify uncertainties in the modelling input. We find, in general good agreement between the models and the test results. Moreover, temperatures estimated by plate thermometers is indicated to be less sensitive to small variations in model input and is thus suitable for these kind of comparisons.

## 1. INTRODUCTION

The SP Fire 105 [1] and the British Standards BS 8414-1:2002 [2] are two among a variety of different test methods that are used to evaluate fire spread from floor to floor via external walls for different wall claddings, insulations and geometrical considerations [3–6]. Internal fire spread in enclosed burnable components in the system as well as recording falling down of parts and the occurrence of burning droplets are also partly evaluated. The SP Fire 105 test method for façade systems was defined in 1985 [1] and simulates a three-story apartment building (height 6.7 m, width 4 m and depth 1.6 m) i.e. a large scale façade test method, see Fig. 1. The BS 8414-1 is similar however it includes a return wall or wing on one side. The fire exposure conditions represent a fully-developed fire in a room, venting through an opening such as a window aperture that exposes the cladding to external flames. The main differences between the methods are, firstly the fuel used and secondly the height of the combustion chamber and the fact that the British test can be performed outside. In BS 8414-1 the square opening of the combustion chamber has side length of 2 m and the fire source is a wood crib with a nominal total effective energy release of 4500 MJ over 30 minutes at a peak rate of 3±0.5 MW. In SP Fire 105 the fuel consists of 60 l heptane burning in a special purpose tray to stabilize the flames and the height of the combustion chamber is 710 mm. The fuel consumed in the SP method corresponds to a total energy release of 1830 MJ and the peak rate is approximately 2.0 MW. The SP Fire 105 method has then been presented internationally for different standardization committees and has been classified as a valid test method in Sweden, Denmark and Norway although it is not presently a European standard.

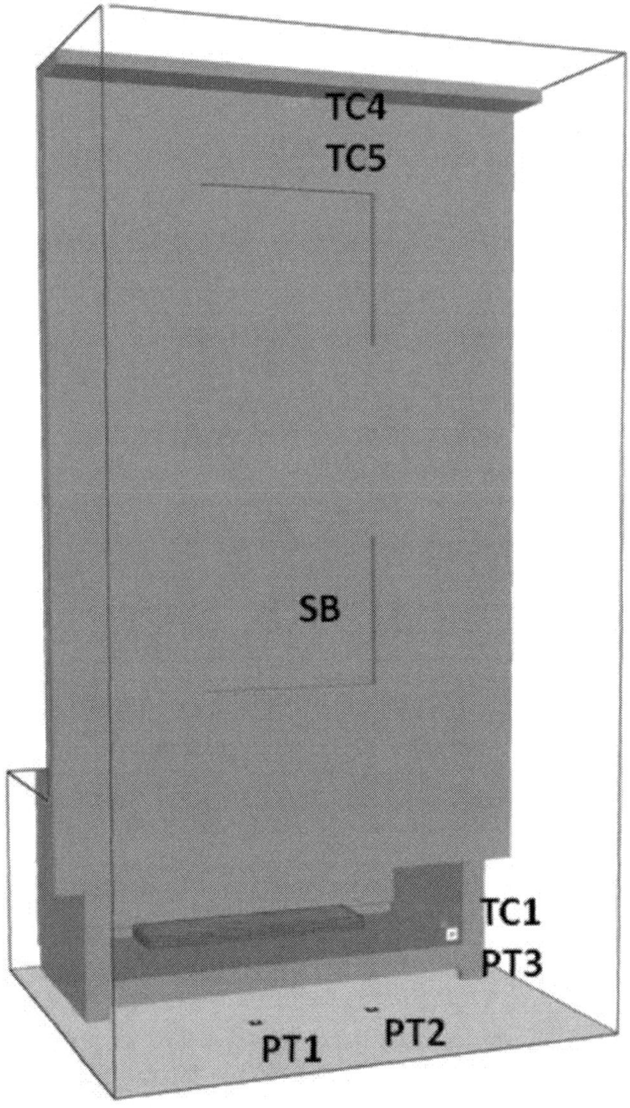


**Figure 1.** The model of the SP Fire 105 façade fire test.

In a previous paper by Anderson et al. [7] results regarding two series of tests according to BS 8414-1 that have been carried out outdoors in Zagreb, Croatia are presented. In each test three façade rigs were used with different test specimens, i.e. three different types of external thermal insulation composite systems (ETICS) was tested at the same time. In addition to the measurements described in the standard, temperatures were measured at different heights and in front of the combustion chamber with different types of thermocouples. In comparison, nine tests according to SP Fire 105 were carried out indoors in a fire testing hall, and thus the ambient conditions are more controlled. The nine tests presented were, four tests of combustible insulation with a protecting render, three different wood façade systems with wood impregnated with fire retardants, one test with combustible insulation (phenolic resin) protected with bricks and one where the insulation of phenolic resin was directly exposed. As in the Zagreb tests described above the thermal exposure was measured 0.5 m from the combustion chamber using three additional plate thermometers not specified in the test method. Both methods define an amount of fuel to be used, in BS 8141-1 a volume of wood and in SP Fire 105 a volume of heptane, as well as the geometry of the combustion chamber. It is thus not possible to regulate the heat exposure to the façade surface during the test, and it may be different from test to test due to factors such as air movements around the combustion chamber and geometry of the test specimens. For example the thickness of the test specimen will have an effect since energy will be absorbed by the boundaries and the flame will be mixed with more cool air before the fire reaches the façade surface.

The current paper will be divided into two parts where the first aim at quantifying the uncertainties induced by modelling and unknown conditions such as material properties and numerical input parameters. The second part will be devoted to comparisons to experiments. In this paper the previous study [7] is extended to accommodate for more analysis as well as numerical investigations. The fire dynamics is investigated by numerical means similar to what was done in several recent papers [8–10], where tests according to the test method SP Fire 105 were modelled with good correspondence between measurements and experimental findings. The numerical model was constructed in the CFD program Fire Dynamics Simulator (FDS) version 6, McGrattan, et al. (2013) [11], with analogous geometry and instrumentation as in the described fire tests. Here we present the results from a series of simulations with varying thicknesses of the façade specimens and varying thermal properties as well as a number of different fuels.

### 1.1 Theoretical framework for evaluating uncertainties by deterministic sampling

Due to the lack of knowledge of the precise details of the system, seemingly random behavior can be governed by deterministic, though non-linear models. Thus, in such systems small variations may have a significant change in the outcome. Most of the calculations both numerical and analytical are often performed with very little thought given to these uncertainties and thus designs are many times based on potentially inaccurate information. Moreover, modeling is often performed in stages by different numerical or analytical tools where the uncertainties may propagate through the system without much control [12, 13]. It is usually a very difficult task to objectively establish the confidence levels in numerical predictions. Uncertainty Quantification (UQ) is the science of quantitative characterization and reduction of uncertainties in numerical studies and real world experimentation. UQ aims at determining how likely certain outcomes are if some aspects of the system are unknown. Deterministic Sampling (DS) [14] is a relatively new method used for Uncertainty Quantification (UQ), is employed in this paper to offer an efficient alternative. The method is based on the idea that a continuous probability density function can be replaced by an ensemble of discrete deterministic samples, provided that the two representations have the same statistical moments.

Quality assurance in CFD simulations [15] is of great importance in order to provide with reliable simulations and with the possibility of validation against experimental results. In this section we shortly summarize the theoretical background that is the basis for the explicit choices of ensembles used in this work. We will construct an example where seven parameters are modeled uncertain since this is the needed input to the numerical model however to reduce to computational time we will reduce the study to include only a few particular parameters of interest. In the absence of correlations, the ensemble of $m$ samples is given by,

$$\Sigma = \langle\theta\rangle \otimes 1^{1\times m} + \operatorname{diag}\left(\langle\theta\rangle \circ \operatorname{std}\left(\frac{\theta}{\langle\theta\rangle}\right)\right) \cdot \hat{V},$$

where $1^{1\times m}$ denotes a row vector of $m$ "ones", $\otimes$ outer product, $\circ$ element-wise multiplication, and $\operatorname{diag}(X)$ is the diagonal matrix with the vector X on its diagonal. The excitation matrix $\hat{V}$ contains all variations; each column describes one "normalized" model sample variation from its mean.

The suggested ensembles formed by the given the information above where it appears that there are $n = 7$ parameters that need to be modeled uncertain. For instance the standard (STD) ensemble can be used,

$$\hat{V}_{STD} = \sqrt{n} \cdot (I_{n\times n} - I_{n\times n})$$

$$= \begin{pmatrix} \sqrt{7} & 0 & 0 & 0 & 0 & 0 & 0 & -\sqrt{7} & 0 & 0 & 0 & 0 & 0 & 0 \\ 0 & \sqrt{7} & 0 & 0 & 0 & 0 & 0 & 0 & -\sqrt{7} & 0 & 0 & 0 & 0 & 0 \\ 0 & 0 & \sqrt{7} & 0 & 0 & 0 & 0 & 0 & 0 & -\sqrt{7} & 0 & 0 & 0 & 0 \\ 0 & 0 & 0 & \sqrt{7} & 0 & 0 & 0 & 0 & 0 & 0 & -\sqrt{7} & 0 & 0 & 0 \\ 0 & 0 & 0 & 0 & \sqrt{7} & 0 & 0 & 0 & 0 & 0 & 0 & -\sqrt{7} & 0 & 0 \\ 0 & 0 & 0 & 0 & 0 & \sqrt{7} & 0 & 0 & 0 & 0 & 0 & 0 & -\sqrt{7} & 0 \\ 0 & 0 & 0 & 0 & 0 & 0 & \sqrt{7} & 0 & 0 & 0 & 0 & 0 & 0 & -\sqrt{7} \end{pmatrix}$$

Another example is the binary (BIN) ensemble,

$$\hat{V}_{BIN} = \begin{pmatrix} +1 & -1 & +1 & -1 & +1 & -1 & +1 & -1 \\ +1 & +1 & -1 & -1 & +1 & +1 & -1 & -1 \\ -1 & +1 & +1 & -1 & -1 & +1 & +1 & -1 \\ +1 & +1 & +1 & +1 & -1 & -1 & -1 & -1 \\ -1 & -1 & +1 & +1 & +1 & +1 & -1 & -1 \\ -1 & +1 & +1 & -1 & +1 & -1 & -1 & +1 \\ +1 & -1 & +1 & -1 & -1 & +1 & -1 & +1 \end{pmatrix}$$

Note: The maximum variation of the standard ensemble is $\sqrt{7}$, while it is only 1 for the binary. By varying all parameters in all samples of the binary ensemble, its maximum variation is minimized. The risk of "saturating" the model is thereby minimal with the BIN ensemble. The standard ensemble is easily generalized, but the binary ensemble has a more complex construction. Calculate the *model ensemble* by evaluating the model $H$ for every sample (column) of $\Sigma$, producing the row vector $H(\Sigma)$ of m results. The expected result is given by,

$$\langle H \rangle = \langle H(\Sigma) \rangle = H(\Sigma) \cdot (1/m)^{m\times 1}.$$

Compare with $H(\langle \theta \rangle)$ (ordinary deterministic result) to have an idea about non-linear effects. If the difference is large, non-linear effects are significant. If the difference is small, it is not sure that the model can be approximated to be linear, even though it is likely.

The variance of the model result is given by,

$$\text{var}(H) = \left\langle (H(\Sigma) - \langle H(\Sigma) \rangle)^2 \right\rangle = \left( H(\Sigma) - \langle H(\Sigma) \rangle \cdot 1^{1\times m} \right)^2 \cdot (1/m)^{m\times 1}.$$

Assuming a coverage factor $k = 2$ for also the result, the modeling uncertainty will be,

$$\text{unc(H)} = 2\sqrt{var(H)}.$$

In summary, the confidence interval of the modeling result is given by,

$$\left[ \langle \text{H} \rangle - 2\sqrt{var(H)}, \langle \text{H} \rangle + 2\sqrt{var(H)} \right].$$

The methodology will now be utilized to assess uncertainties in a CFD model of the SP Fire 105 façade fire test method.

**Table 1.** The seven parameters to be modelled uncertain.

| Parameter | Mean | Deviation |
|---|---|---|
| Radiation angles | 100 | ±50 |
| Placement of PT3 | | ±5 cm (one cell) |
| Placement of TC1 | | ±5 cm (one cell) |
| HRR (Heat Release Rate) | Babrauskas see Fig. | ±10% |
| Thermal conductivity of specimen | 0.242 | ±30% |
| Specific heat of specimen | 1000 | ±30% |
| Density of specimen | 975 | ±30% |

## 1.2 Estimating uncertainties in a CFD model

In general we find good agreement and qualitatively the same behavior in comparing simulation results to the experimental data, see Ref. [7–10] however there are lot uncertainties accompanied to the input to simulations. Thus we have investigated the effects of uncertainties in seven parameters in the model [7–10] of the SP Fire 105 façade fire test of previous simulations. In Ref. [10] a grid resolution study was performed where it was concluded that a cubic grid of 5 cm × 5 cm × 5 cm is suitable for this model. Note that we will first discuss uncertainties stemming from the simulations and then compare to experimental data. The parameters with assumed uncertainties are summarized in Table 1 below, note that all uncertainties are assumed and are only used in this model to investigate how they propagate through a non-linear model. The general problem of uncertainty quantification in CFD models is usually limited by long simulation times and thus ensembles with relatively few samples/simulations however employing deterministic sampling methods we are able to perform adequate uncertainty quantification with fewer simulations. Here we can estimate the variance in the output with only eight simulations where seven parameters to be modelled uncertain. Although, the assumed variations in the material parameters are larger than can be assumed from measurements however this variation is used to assess the dependence of different material properties of the façade on the computed results. The parameters in Table 1 are given with max deviation around the mean value.

Using the parameters in combination with the BIN ensemble we will compute the variation in the output. The Heat Release Rate (HRR) as a function of time is presented in Fig. 2 as adopted from Ref. [16].

The results of the uncertainty quantification will be assessed for the thermocouples TC1 (thermocouple 1) and TC4 (Thermocouple 4), the plate thermometer PT3 and the heat flux gauge SB using the HRR shown in Fig. 2.

The results in Figs. 3 to 6 show that uncertainties can vary quite a lot depending on what output parameter is considered. The mean value of the eight simulations is used in computing the standard deviation. Here it is interesting to note that this mean values almost exactly coincides with the mean found by performing one simulation taking the mean values for all seven parameters however somewhat less fluctuations in the output is found. In Figs. 3 to 6 we present the deviations using one standard deviation. In Fig. 3, the result for TC1 (Thermocouple 1 in the lower right corner of the fire room) is displayed and variations up to approximately 10% which is of the order of the variation in HRR can be found. In the figure TC1+ and TC1− denote the upper and lower bounds. However, it is instructive to investigate which parameter actually influence the output the most, it is found that placement mostly affects the uncertainty and to somewhat lesser degree the HRR. The uncertainty in the simulation data of the PT3 (in the lower right corner of the fire room) result is shown in Fig. 4. Here a very small variation is found, around 3% significantly smaller than the variation in HRR. Note that in this position the simulation significantly over predicts the temperature whereas the comparative TC values are rather well represented, see Ref. [10]. We find that these additional uncertainties cannot explain the excess simulated temperature data. One possible reason for the discrepancies is the flame suppressing grid on the fire source and that may change the local dynamics compared to that of burning from a liquid surface.

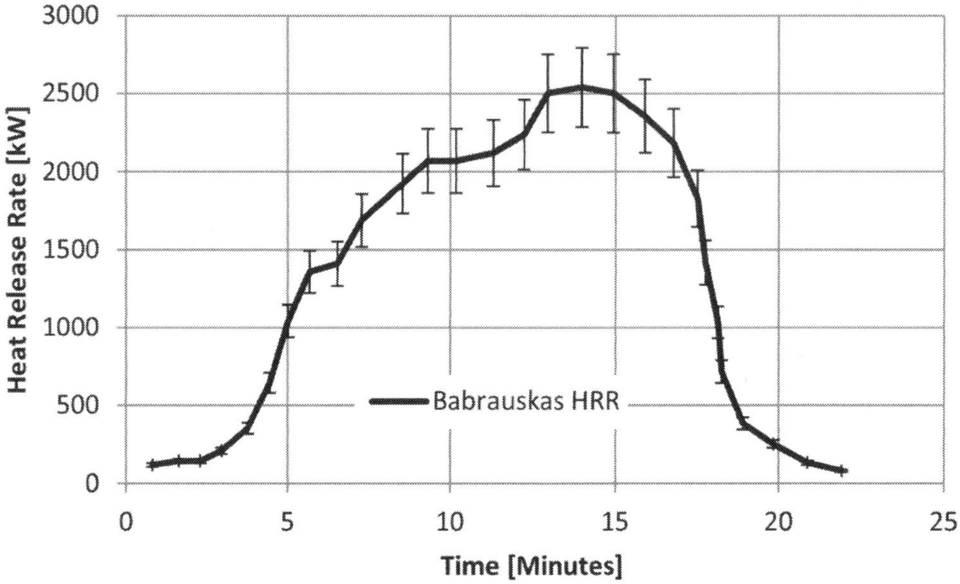


**Figure 2.** The heat release rate as a function of time with uncertainty bars.

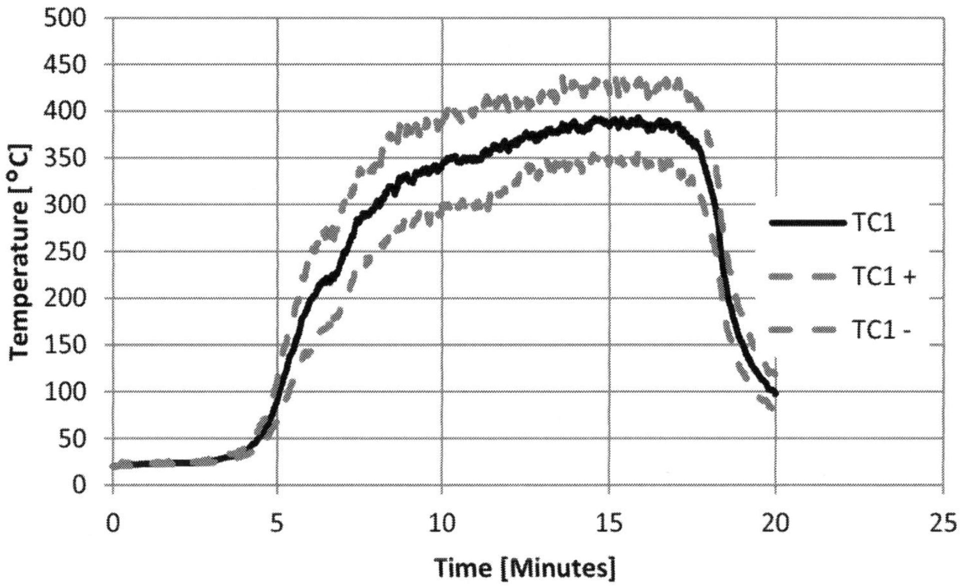


**Figure 3.** The temperature of TC1 (in the lower right corner of the façade) with minimum/maximum boundaries on uncertainties in the numerical simulation.

In Fig. 5, the uncertainties in the temperature at the eave will be discussed. We find a medium deviation of around 10%. The heat flux computed in the middle of the first fictitious window is shown in Fig. 6 where rather large uncertainties of maximum 30% can be observed. The determining factors for this uncertainty are the change in HRR however the number of angles in the radiation model is also very important. Note that that heat flux is the most sensitive quantity to small changes in input data and thus also from a computational point of view less practical to make comparisons with. Next we present a detailed comparison with experimental data where similarities and deviation between experimental and theoretical work will be discussed.

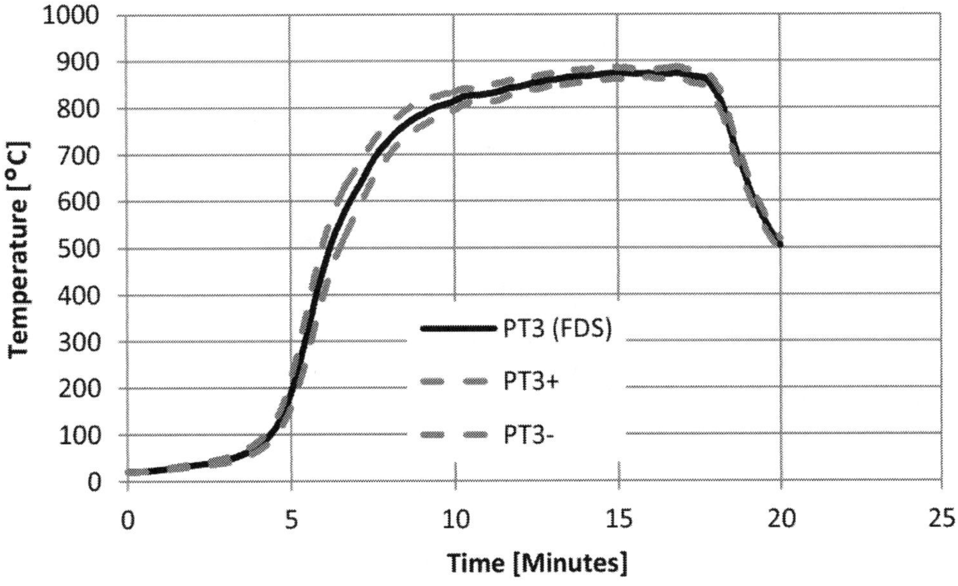


**Figure 4.** The temperature of PT3 (in the lower right corner of the façade) as a function of time with minimum/maximum boundaries on uncertainties in the numerical simulation.

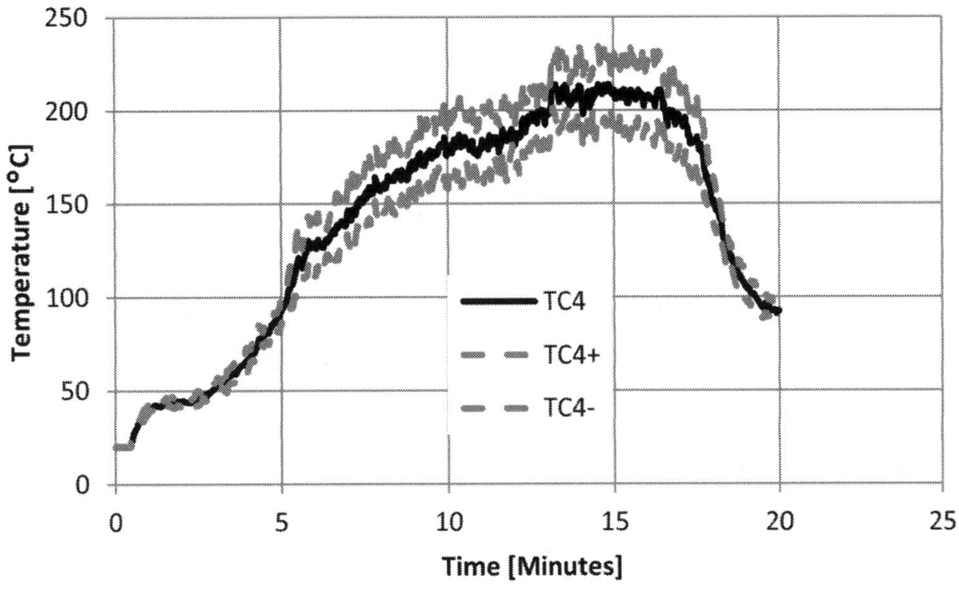


**Figure 5.** The temperature at TC4 (at the eave) as a function of time with minimum/maximum boundaries on uncertainties in the numerical calculation.

## 2. COMPARISON WITH EXPERIMENTS

We will here compare the measured and simulated temperatures measured by PTs in front of the façade and one just beside the heat flux gauge in the first fictitious window. It was found in Ref. [7] that varying the thickness of the façade specimen changed the heat exposure on the façade and in front of the façade; in this work we have explored this difference in the numerical model. The SP Fire 105 tests are carried out indoors in a fire testing hall, and thus the air movements are kept on a very low level.

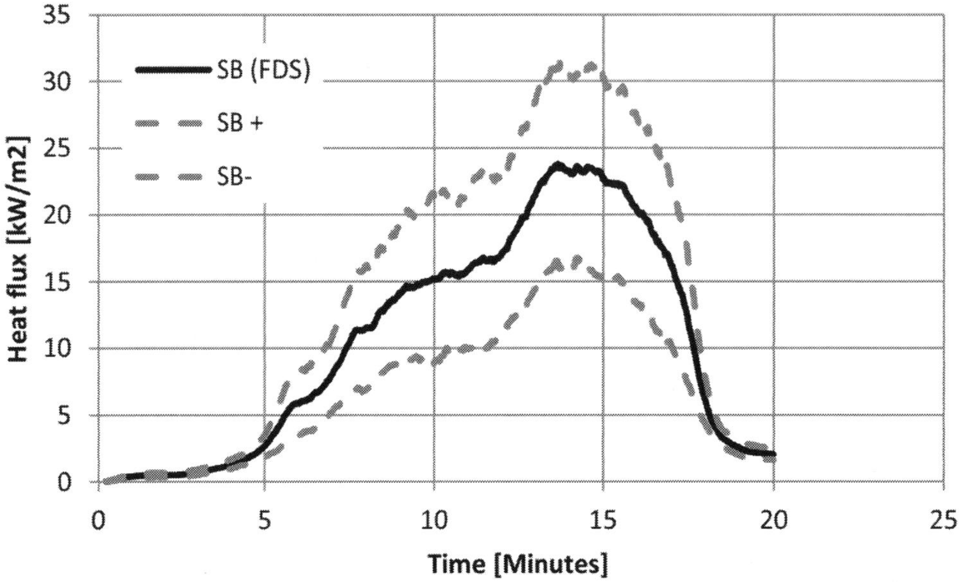


**Figure 6.** Heat flux (at the center of the first fictitious window) as a function of time with minimum/maximum boundaries on uncertainties in the numerical calculation (averaged over 30 s).

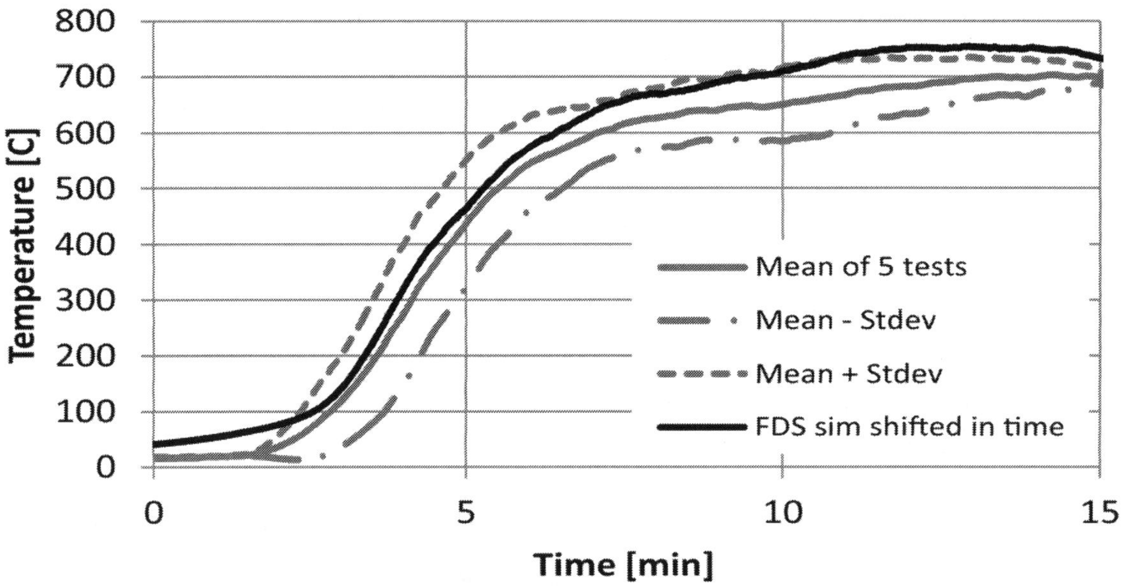


**Figure 7.** The resulting temperatures from the PT in front of the façade, 0.5 m from the fire room.

In Figs. 7 and 8 we have gauges the fire source by PTs 0.5 m from the fire room and one PT pointing outwards 2.1 m above the upper edge of the fire room (beside the heat flux meter in the fictitious window). These measurements are in addition to the standardized measurements specified in the method. The experimental data are taken from tests with three different wood with façade systems containing fire retardants, one brick wall with combustible insulation and one directly exposed insulation of phenolic resin. The data from the five tests are used to compute mean value and standard deviation and then compared to the results from the simulation. We have adjusted the time (30 s) for the simulation to roughly match the exposure in Fig. 7 finding a quite good qualitative and quantitative agreement

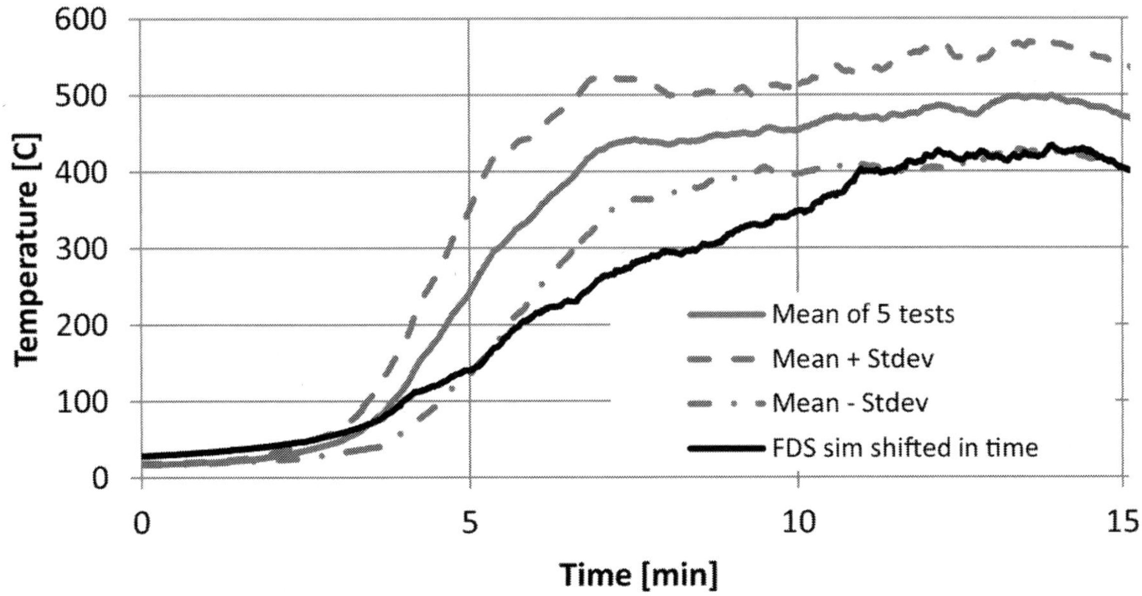


**Figure 8.** The resulting temperatures from the PT in the first fictitious window, 2.1 m above the fire room.

mostly within one standard deviation. We have approximately the same amount of uncertainty around 3% as shown Fig. 4. In Fig. 8, the exposure on the façade as estimated by a PT where we find that the temperatures in the simulation is significantly lower than those in the test however there is a significant contribution to the HRR from the wood and phenolic resin façade systems yielding increased measured temperatures.

One test deviates from the other and that is the very thick façade with bricks, see Ref. [7]. The measurements show that the plate thermometer temperature from the combustion chamber is higher than the other tests, and that the temperature 2.1 m above the combustion chamber is lower. This can to some extent be explained by the thickness of the façade system. The radiative heat depends on the thickness of flames. When the thickness of the façade increases, also the distance from the initial fire room to the façade surface is increased, and thus the there is a change in where the actual combustion takes place influencing the measurements by the plate thermometers. This underlines the fact that plate thermometers measure a type of effective surface temperature on a small reference specimen. This effective temperature is a much more representative quantitative measure of the thermal exposure on flat objects than the gas temperature. Furthermore the uncertainties are significantly smaller compared to the heat flux measurements. The thickness of the façade may change distance travelled by the plume, the radiation exchange with the exterior while some of the heat will be absorbed by the bottom edge, i.e. the edge facing downwards and can be seen as an extension of the fuel chamber. The thicker the specimen is, the more these factors will influence the dynamics and more heat will be absorbed by the bottom edge.

In Fig. 9 the simulated temperatures are shown for a regular façade (100 mm) and a thicker façade (330 mm) system. The same qualitative behavior as that in the experimental comparison performed in Ref. [7] where an increase temperature in front of the façade whereas a decreased temperature on the façade 2.1 m above the fire room is observed. Note that by varying the basic chemistry and soot yield in the simulation we can simulate the effects of varying fuel used in the fire test. We have simulated use of heptane (standard fuel), wood and propane. The results show very small differences between the different fuels on the temperatures measured by the PTs in front of the façade (0.5 m from the fire room) and the PT (2.1 m above the fire room) on the façade.

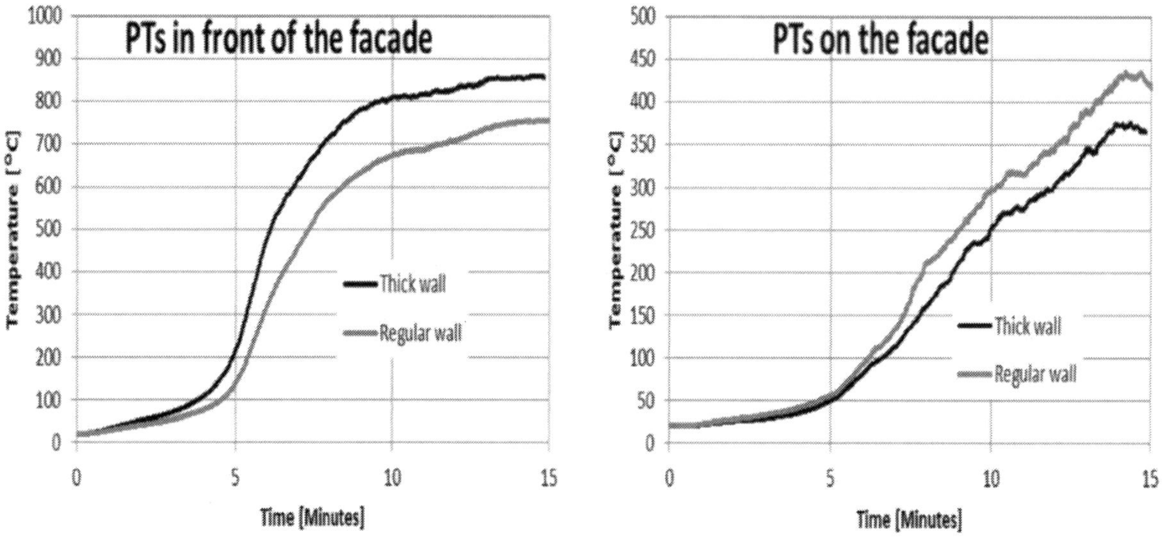


**Figure 9.** The results from the PT in front of the façade (0.5 m from the fire room) and the PT on the facade.

Next we will investigate how these experimental and numerical results relate to similar results obtained by testing in the British standard, see the model adaptation of the façade testing method in Fig. 10. Two tests have been carried out outside Zagreb, Croatia, one in March 2014 and one in May 2014. In each test three façade rigs were used with different test specimens, i.e. three different façade systems. The three façades are prepared with different types of external thermal insulation composite systems (ETICS). Wall 1 consisted of noncombustible mineral wool insulation, Wall 2 of EPS insulation and fire stops at different heights of rock-wool insulation, and Wall 3 of EPS insulation. All three walls had an organic rendering on the surface and classified for the reaction to fire as B-s2,d0 or better. The tests were made in accordance with BS 8414-1:2002 [2], i.e. a large scale façade test method. The two tests were exactly the same, except the climate conditions since the tests were carried out outdoors. In the first test the temperature was around 15°C and the wind speed varied between 2 and 5 m/s. In the second test the temperature was around 25°C and the wind speed varied between 0 and 2 m/s. For more details on these tests see Ref. [7] where also the HRR used in the simulations are presented.

In Figs. 11 and 12 a comparison between numerical results and measured data measured on the Wall 1 with non-combustible insulation material is shown. The effect of the outdoor climate is clearly visible where the temperatures found in the measurement have large fluctuations and show consistently lower temperatures compared to the simulated values. We note that there are significant differences between the two experimental data sets sampled at different days for the same set-ups. Qualitatively similar trends are found in the simulations and experiments.

## 3. DISCUSSION

In this paper we present results from simulations and measurements concerning two large scale fire façade test methods. The numerical model representing the SP Fire 105 façade and BS 8414-1 test rigs were made in FDS however no self-consistent fire source is used but rather measured HRR are used as input. We find good repeatability between similar test set-ups however one particular test in SP Fire 105 with a thick wall stood out. In general good agreement between simulations and test results within one standard deviation are found; however there seems to be a time lag in the simulations. A similar trend with higher temperature in front of the fire room while lower temperature on the façade for the thick walled façade compared to a regular specimen was seen in both simulations and tests. In addition small

**Figure 10.** The model of the experimental set-up for the British standard 8414 with a velocity slice file is displayed. The positions of the PTs in front of the fire room are visible in white.

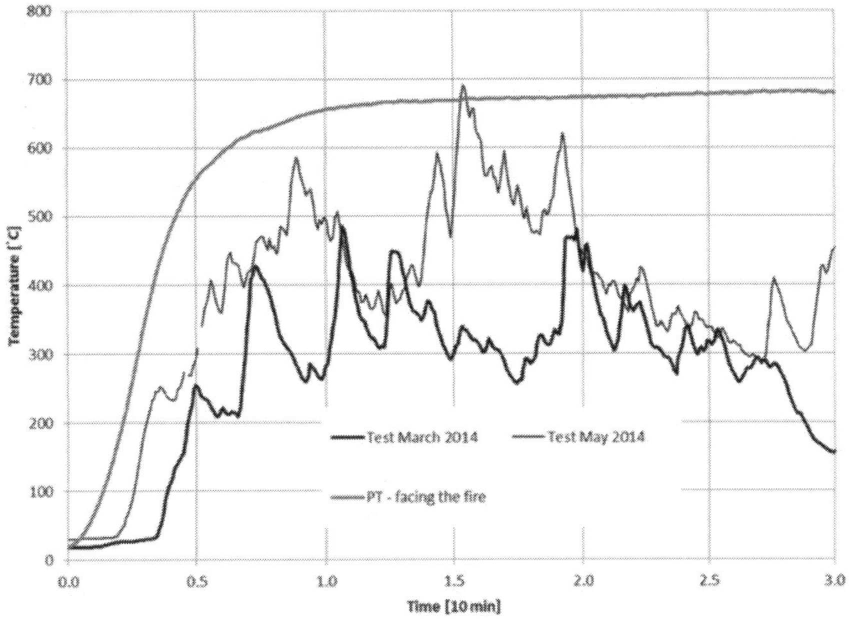


**Figure 11.** A comparison of temperatures measured by PTs located 0.5 m from the fire room in the BS8414-1 test method and simulation.

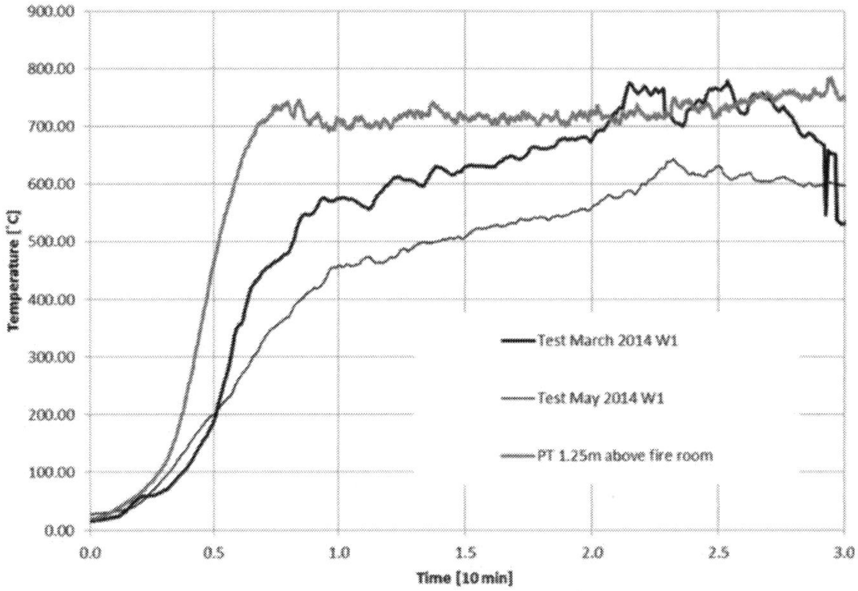


**Figure 12.** A comparison of temperatures measured by PTs located 1.25 m above the fire room in the BS8414-1 test method and simulation.

effects of varying the fuel chemistry and soot yield were seen on the computed temperatures. In order to ensure a robust, repeatable and reproducible test method for façades there are several issues that must be resolved. The heat exposure on the façade can for example be measured with plate thermometers, which is more linked with the thermal exposure of the façade surface compared to conventional thermocouples. A possible solution is to define a heat exposure curve, as in fire resistance testing, and use gas burners that can be regulated instead of using a defined amount of free burning fuel. The appropriate level and nature of the fire source has been debated for a long time and is in need of reassessment. Also a numerical tool for estimating the effects of design changes to the test methods is presented. The use of simulations may reduce the amount of large scale development testing which is inhibited by large costs.

The Authors is indebted to Dr. Hessling for constructing ensembles that made uncertainty quantification in the FDS model viable. The work presented here is sponsored by Brandforsk the Swedish Fire Research board.

## References

[1] SP FIRE 105, – Method for fire testing of façade materials, Dnr 171-79-360 Department of Fire Technology, Swedish National Testing and Research Institute.

[2] British Standard BS 8414-1:2002, Fire performance of external cladding systems – Part 1: Test method for non-loadbearing external cladding systems applied to the face of the building (2002).

[3] White, N. and Delichatsios, M., Fire Hazards of Exterior Wall Assemblies Containing Combustable Components, National Fire Protection Association (NFPA) Fire Protection Research Foundation Report (2014).

[4] Yokoi, S. Study on the prevention of fire-spread caused by hot upward current. Japanese Ministry of Construction, Building Research Institute Report **34**, 1960.

[5] Ödeen K. & Nordström Å., Fire and smoke spread along facades and in ventilation ducts (in Swedish), Arbetshandling 12:1967 från Byggforskningen, Statens Provningsanstalt (SP Technical Research Institute of Sweden), 1967.

[6] Ondrus, J. and Pettersson, O., Fire hazards of façades with externally applied additional thermal insulation, LUTVDG/(TVBB-3025) (1986).
[7] Anderson, J., Boström, L., Jansson, R., Milovanović, B., Fire dynamics in façade fire tests: measurement, modelling and repeatability, Applications in Structural Fire Engineering, 15–16 October 2015.
[8] Anderson, J. and Jansson, R., Façade fire tests – measurements and modeling, MATEC Web of Conferences 9, 02003 (2013).
[9] Anderson, J., Jansson, R., Fire Dynamics in Façade Fire Tests: Measurement and Modeling, Proceedings of Interflam 2013, page 93, Royal Holloway College, University of London UK, (2013).
[10] Jansson, R., Anderson, J., Experimental and numerical investigation of fire dynamics in a façade test rig, Proceedings of Fire Computer Modeling, Santander, Spain, 18–19th October 2012, p. 247.
[11] McGrattan, K., McDermott, R. Hostikka, S. Floyd, J., Fire Dynamics Simulator (Version 6) – user's guide. NIST Special Publication 1019, 2015.
[12] Guide for the Verification and Validation of Computational Fluid Dynamics Simulations (AIAA G-077-1998 (2002)).
[13] Quality Management in Engineering Simulation, A Primer for NAFEMS QSS, NAFEMS 2014.
[14] Hessling, J.P., Deterministic sampling of propagating model covariance, SIAM/ASA J. Uncertainty Quantification **1**, pp 297–318 (2013).
[15] Hermansson, K., Quality assurance and the simulation of fires – A practical application for automated validation of user-generated input data for FDS, Thesis Report 5501 Lund University (2015).
[16] Babrauskas, V., Facade Fire Tests: Towards an International Test Standard, Fire Technology **32**, 219–230 (1996).

MATEC Web of Conferences **46**, 03003 (2016)
DOI: 10.1051/matecconf/20164603003


# Numerical simulations of the ISO 13785-2 façade fire tests

Simo Hostikka and Gleb Bytskov

Aalto University, Espoo, Finland

**Abstract.** In this work we created a numerical model of the ISO 13785-2 test setup for testing the fire behaviour of building façade systems, and used the model to simulate the thermal environment on the façade. The model, created using Fire Dynamics Simulator – software, was first validated using the experimental data by Yoshioka *et al.* (2012). Next, the sensitivity of the façade heat fluxes on the geometrical and model parameters was studied, revealing for instance that the size of the combustion chamber window will influence the thermal exposure high above the window. Finally, the model was used to estimate the thickness of non-combustible insulation layer that is needed to protect combustible materials from melting or decomposition.

## 1. INTRODUCTION

The development of test standards and fire safety classification basis for building façade products is currently an active topic of research and standardization activities in Europe and globally. The development of such standards requires investigation of the test characteristics, sensitivities of the produced fire exposure to the geometrical and environmental parameters, and the suitability of the test for the range of intended products. While the full-scale tests with alternative details cannot be avoided, many of the phenomena can be studied using Computational Fluid Dynamics simulations.

The goal of this work was to prepare a validated model for the ISO 13785-2 test setup and then use the model to investigate the sensitivity of the fire exposure at the façade to the physical and numerical parameters. An additional objective was to determine is the model can be used to determine geometrical and material requirements for the protection of combustible insulation materials. This research did not explore the hazards related to real fires, such as the fire spreading to the apartments above the fire or to adjacent buildings.

## 2. METHODS

### 2.1 ISO 13785-2 standard test

ISO 13785-2 is a relatively recent test standard for large-scale testing of building facade materials and products [1]. The test facility consists of a combustion chamber, facade specimen with side wall, opening and measurement devices. The combustion chamber has volume in range 20–100 m$^3$, and it has a 2 m wide by 1.2 m high opening. The façade wall extends up to 4 m above the top edge of window, to the

**Figure 1.** FDS model of the ISO 13785-2 test.

total height of 5.7 m. The vertical side wall (wing) is 1.2 m wide with same height as the main façade specimen. Figure 1 shows the FDS model of the test setup with flames emerging from the combustion chamber to the façade.

According to ISO 13785-2 standard, any fuel can be used to produce a window flame which exposes the test specimen to heat flux of $55 \pm 5\,kW/m^2$ at a height 0.5 m above the opening, and $35 \pm 5\,kW/m^2$ at a height of 1.5 m above the opening. It is not necessary to perform ISO 13785-2 test with maximum heat release rate, if the exposure level on façade wall can be achieved with lower HRR. Annex A of the standard suggest that propane gas with 95% purity can be used for testing. Propane burner consists of perforated steel pipes, positioned uniformly in the room. Fuel flow rate reaches its maximum 5 minutes from the test start and decays to zero between 20 and 25 min.

Heat fluxes are measured at 3.5 m above the top of the window, and thermocouples are installed at the top of the test specimen and at the top of the window opening. Evaluation or performance criteria are not included in the standard.

## 2.2 Numerical

The façade fire tests were modelled and simulated using Fire Dynamics Simulator (FDS), version 6.1 [2]. The current simulations were made with the modelling options that are commonly used in Fire Safety Engineering (FSE) applications, assuming, e.g., simple chemistry for combustion reactions and grey radiation. Wall-functions and empirical heat transfer coefficients were used for the wall stress and heat flux, respectively. Post-processing of the simulation results was performed with the Smokeview program and Matlab-routines.

In the context of compartment fire simulation, the quality of the computational mesh is commonly assessed using the non-dimensional $D^*/\delta x$ ratio, where $\delta x$ is the size of the grid cells and $D^*$ is a characteristic fire diameter

$$D^* = \left(\frac{\dot{Q}}{\rho_\infty c_p T_\infty \sqrt{g}}\right)^{2/5} \qquad (1)$$

where $Q$ is the HRR, and $\rho_\infty$, $c_p$, $T_\infty$ and $g$ are the ambient density, specific heat, temperature and gravitation acceleration, respectively. Values of 10 or more for the $D^*/\delta x$ are required to obtain reliable predictions of the radiative heat flux [3]. Substituting the lowest considered value of HRR (25% of the nominal value 5600 kW) to Eq. (1) indicates that the cell size should be order of 10 cm.

The model included the chamber, façade and 3.0 m of empty space in front of the façade. The vertical and top boundaries of the external space were defined as open pressure boundaries. The burners were modelled as rectangular surfaces of propane inflow. The 10 cm cell size did not enable us to simulate the flows and heat transfer in the gap between the façade and chamber, but the role of this simplification is expected to be minor.

### 2.3 Validation experiments

For the validation of the FDS model, we used the experimental data measured at the Tokyo University of Science (TUS) in a research project with partners from the Japanese National Institute of Land and Infrastructure Management, and Building Research Institute. The experiments have been summarized by Yoshioka et al. [4]. They measured nearby gas temperatures and total heat fluxes to a non-combustible façade specimen at different heat release rates and in different ventilation configurations.

The test apparatus consisted of the combustion chamber and the façade test specimen, dimensioned according to the standard. Combustion chamber walls and ceiling were made from (starting from outside) 6 mm steel plates and 50 mm ceramic fibre boards. Floor of combustion chamber was constructed of (starting from outside) 6 mm steel plate, 50 mm ceramic fibre board and 25 mm ceramic fibre blankets. The dimensions of the combustion chamber were 4.0 × 4.0 × 1.7 meters. (Note that the chamber size, as well as some other features were reported incorrectly in [4].) The façade specimen consisted of a main wall with dimensions 3 × 5.7 m (width and height) and a 1.2 m wide side wall with the same height. The façade wall specimens were constructed of 25 mm ceramic fibre blankets on the surface exposed to fire and 25 mm calcium silicate boards connected to the steel frame.

There were three series of tests with propane gas used as fire source. Each series of tests was done with some modifications to the standard ISO test. According to the ISO 13785-2, it is possible to make vents on the opposite side of the combustion chamber in order to enhance the buoyancy-driven ventilation. In the TUS tests, the combustion chamber was equipped with four of these openings. First group of tests, marked with I-letter, was done with regular façade window and with one rear side opening opened. The second group of tests, marked with ①-sign, were performed with all four vents opened. In the last group of tests, marked with ②-sign, all of supplemental openings were closed, and the façade window dimensions were modified to investigate the effect of the flame characteristics. In each series, two or three consecutive tests were performed with different propane flow rates and, consequently, heat release rates. The tests conditions are summarized in Table 1. The tests were carried out one after another, so that there was a 10 min cooling period between the 5 min tests.

The simulated HRR and mass burning rate histories are shown in Fig. 2. Spatial resolution of 10 cm was used. An additional empty space was added to the back of the combustion chamber for accurate computation of the flows through the rear vents. Initial temperature was set to 40°C based on the measurement data, although the vertical stratification of the test hall was not taken into account.

The instrumentation of the experiments was more comprehensive than what is required by the ISO standard. Thermocouple and heat flux gauge positions on the façade are shown in Fig. 3. Thermocouples were simulated using the FDS's built-in thermocouple models, with bead diameters 3.2 mm (on the façade) and 1.6 mm inside the chamber. For heat flux, a numerical model for the water-cooled heat flux gauges was used.

## 3. RESULTS AND DISCUSSION

### 3.1 Validation results

Performing a rigorous model validation using experimental data is a long and laborious exercise. Despite the relative quick throughput of the simulation process, it takes several iterations to confirm experimental

**Table 1.** Validation tests from [4].

| Case | Heating Load | Test duration | Gas flow (m3/h) | HRR (MW) | Opening | | | Vents at rear of chamber |
|---|---|---|---|---|---|---|---|---|
| | | | | | B(m) | H(m) | B/(H/2) | |
| I-1 | 1/2 of ISO | 5 min | 225 | 2.8 | 2 | 1.2 | 3.3 | 3: Closed<br>1: Open |
| I-2 | 3/4 of ISO | 5 min | 337.5 | 4.2 | 2 | 1.2 | 3.3 | 3: Closed<br>1: Open |
| ①-1 | 1/4 of ISO | 5 min | 112.5 | 1.4 | 2 | 1.2 | 3.3 | All (4) open |
| ①-2 | 1/2 of ISO | 5 min | 225 | 2.8 | 2 | 1.2 | 3.3 | All (4) open |
| ①-3 | 3/4 of ISO | 5 min | 337.5 | 4.2 | 2 | 1.2 | 3.3 | Former: All open<br>Latter: 2 closed |
| ②-1 | 1/8 of ISO | 5 min | 56.25 | 0.7 | 2 | 0.2 | 20 | All (4) closed |
| ②-2 | 1/4 of ISO | 5 min | 112.5 | 1.4 | 2 | 0.5 | 8 | All (4) closed |
| ②-3 | 1/2 of ISO | 5 min | 225 | 2.8 | 2 | 1 | 4 | All (4) closed |

details, to observe and fix the mistakes and errors in the publications presenting the experimental results, and to estimate the unknown input parameters. The results presented below represent a *work-in-progress* in this respect.

The simulation results were used to quantify the uncertainty of three FDS output quantities: gas temperature within the combustion chamber, gas temperature at the façade, and surface heat flux at the façade. Figure 4 shows simulated chamber temperatures in tests I-1 and I-2 for "Pole A", a vertical tree of thermocouples in the corner of the chamber that was near the front wall on the left. As we can see, the model uncertainty is of the same order or even less than the current FDS validation database estimate for the hot gas layer temperature in naturally ventilated fires (Bias = 1.07 and relative standard deviation of 7%). Unfortunately, much worse results were obtained for some other tests of the TUS series. This can be seen in Fig. 5 that shows the scatterplot of the predicted vs. measured peak temperatures for all eight tests and $2 \times 5$ thermocouples. In average, the simulations overestimate the temperatures by 16% and the relative standard deviation is 26%. The current uncertainty metric is not fully comparable to that of FDS Validation Guide because in there, the comparison is made using an average of several thermocouples for each test, thus reducing the influence of local fluctuations and details.

The comparisons of the façade gas temperatures were made at different heights above the window, ranging from 0.05 m to 4.0 m. For the heat flux, the comparisons were made between 0.9 and 2.0 m above the window. Figure 6 shows a comparison of the gas temperatures and heat fluxes at three different vertical lines (a on the left, b in the middle, c on the right) for Test I-2. These results are among the best predictions and do not represent a "typical" behaviour. Different colours of the curves correspond to different heights. Solid lines are experimental data and dashed lines are FDS results. Highest temperatures and heat fluxes are, in this test, obtained in the line middle of the window (line b). Combining the results for all tests and all thermocouples/heat flux gauges provides us with the following uncertainty metric: The temporal peak values of the plume temperatures in different locations are, in

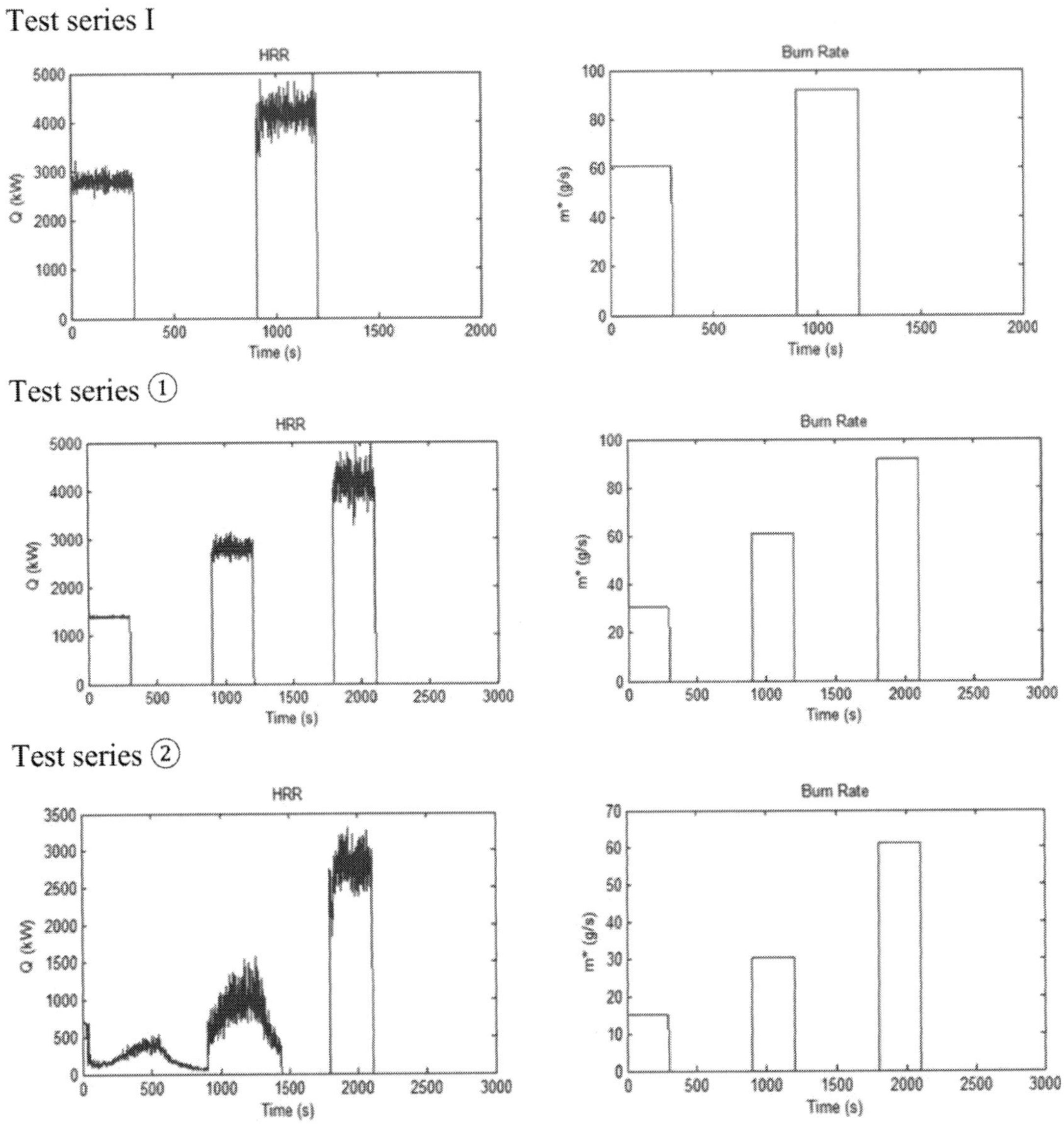


**Figure 2.** HRR and burning rate histories of the validation simulations.

average, underestimated by 18% and the heat fluxes overestimated by 14%. These bias factors are not significantly greater than those determined from the current validation database. The relative standard deviations are, however, much greater: 56% for temperature rise and 66% for the heat flux. More detailed analysis is needed to understand what is behind these uncertainties. It is clear that some of the simulation results are even qualitatively different from the corresponding measurement. For instance, in some tests, the heat fluxes on one of the three lines (a, b and c) were overestimated while others were underestimated. The asymmetric geometry of the tests appears to cause difficulties as the horizontal location of the window plume has a strong influence on the observed temperatures and heat fluxes. In symmetrical geometries, both experiment and simulation would be predictable as the peak temperatures would be observed on the symmetry axis of the setup. Here, a slight difference in the actual plume position can cause a significant difference between the experiment and simulation.

In addition to the plume location issues, the exact characteristics and operational procedures of the gas burners was not known for the simulation boundary condition. Also, the hydrodynamic conditions

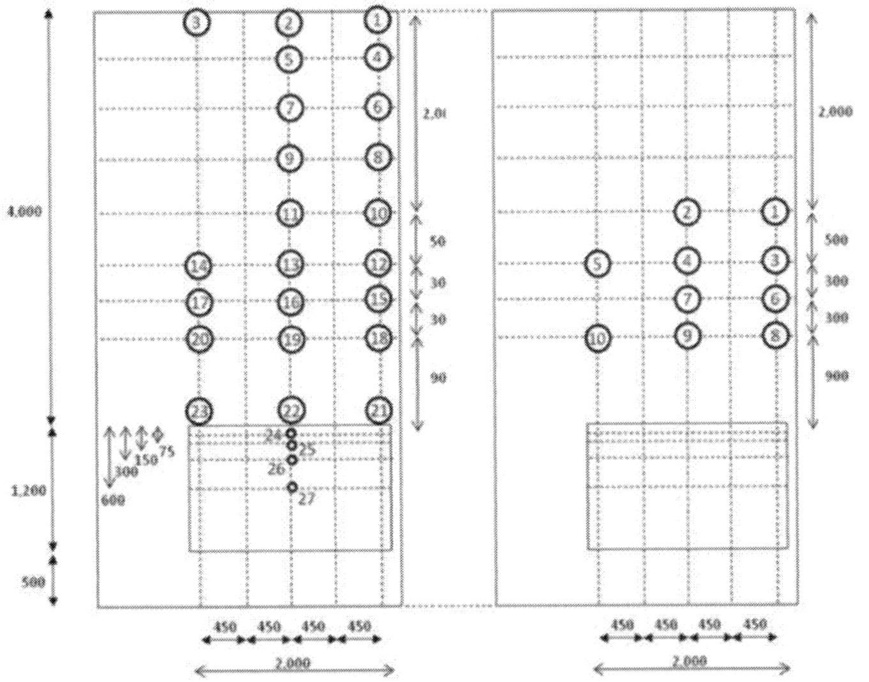

**Figure 3.** Facade instrumentation in the TUS tests [4]: thermocouples (left) and heat flux gauges (right).

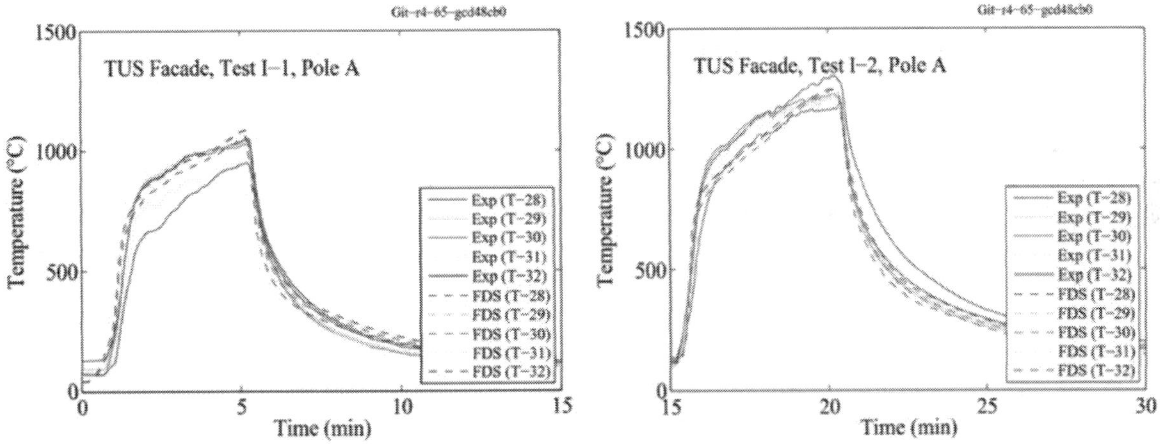


**Figure 4.** Combustion chamber temperatures in tests I-1 and I-2.

in the tests with two-sided ventilation included some uncertainties which may be reflected in the temperatures of combustion chamber internal temperatures.

### 3.2 Sensitivity study

The sensitivity of the ISO 13785-2 simulation predictions on the physical and numerical parameters was next investigated. Simulations were made with the following characteristics:

- Maximum heat flux corresponds to fuel supply flowrate of 120 g/s.
- Fuel flowrate increased linearly from zero to maximum in first 5 minutes of simulation.
- Simulation was extended for two minutes with maximum flowrate.

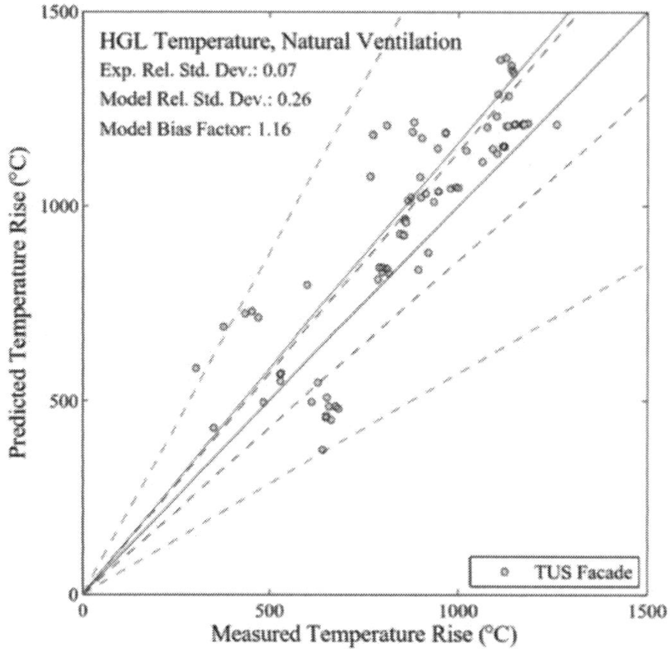


**Figure 5.** Uncertainty statistics for all the combustion chamber gas temperatures of the TUS series.

The results are summarized by presenting the maximum surface heat fluxes at two heights above the window for different alternations of test setup and the radiative fraction (Fig. 7). The geometrical features of the test setup were varied within the range allowed by the standards. The results show that both widening the combustion chamber and introducing additional ventilation to the back of the chamber will reduce the heat fluxes. These findings can help in the design of test sites, but they do not significantly affect the interpretation of the test results because the HRR of the fire would in any case be calibrated to produce the requested heat flux ($55 \pm 5\,kW/m^2$) on the façade.

Low left hand – side plot of Fig. 7 shows that the variations in the window size did not affect the heat flux at 0.6 m above the window, but at 3.6 m above the window, a larger window would cause lower heat fluxes. This is an important finding because the overall exposure to the façade seems to be dependent on the window sizing while the flux at the calibration height is not.

Finally, the effect of the combustion reaction radiative fraction was investigated. The results shows that increasing radiative fraction will also increase the heat flux. This is due to the fact that the test typically shows a large flame emerging out of the window, and thus the heat flux is mainly by radiation. Similar result was obtained by increasing the soot yield of the combustion reaction. The quality of the fuel can therefore influence the heat fluxes, and must be kept in mind in the calibration. In reality, however, the radiative behavior can be different inside and outside the chamber if the fire is in the ventilation controlled regime.

The sensitivity of the heat fluxes on the numerical resolutions of the simulation was also investigated. Reducing the spatial grid cell size from 10 cm to 5 cm increased the heat fluxes few percent. The angular resolution of the radiation solution, in turn, did not have a significant impact on the heat fluxes. Nor had the resolution of the solid-phase heat conduction solver.

## 3.3 Application example

The validated simulation model was used to investigate the thermal protection requirement for the insulation products that are based on synthetic polymers. The heat flux distribution on the whole façade

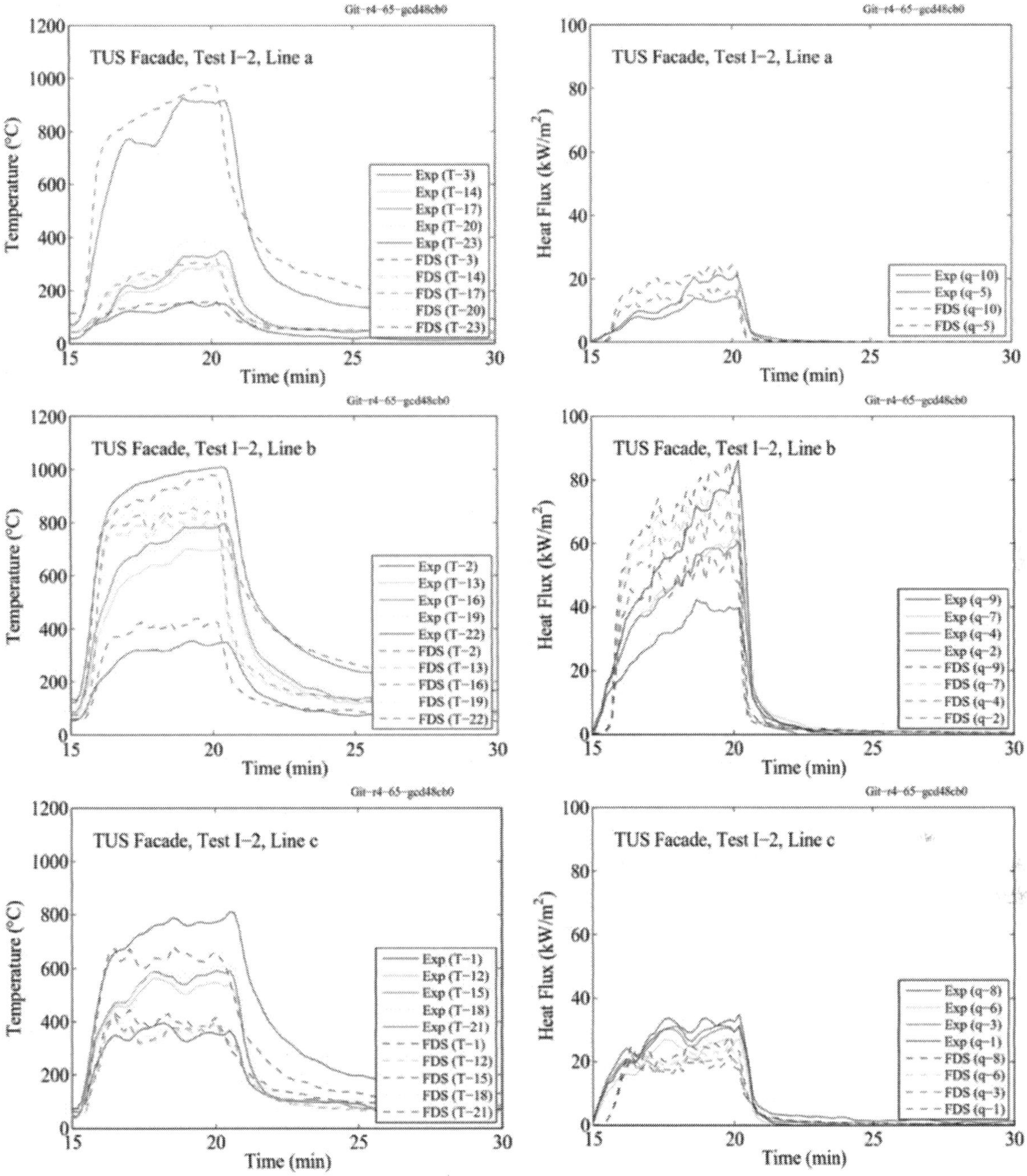


**Figure 6.** Façade gas temperatures (left column) and heat fluxes (right column) for Test I-2.

surface was first simulated over the time period required by the standard. An instantaneous distribution is shown in Fig. 8. The highest heat fluxes are observed just above the window's left part, and on the right-wing wall close to the window. Higher above the window, the plume bends towards the corner focusing the high-heat flux region.

Heat flux histories were saved from the region above the window, and used as a boundary condition for one-dimensional heat conduction solver. This procedure can be repeatedly used to investigate different materials or their combinations without the actual CFD simulation. By setting the thermal properties of the façade material to those of stone wool, we investigated the temperatures at different

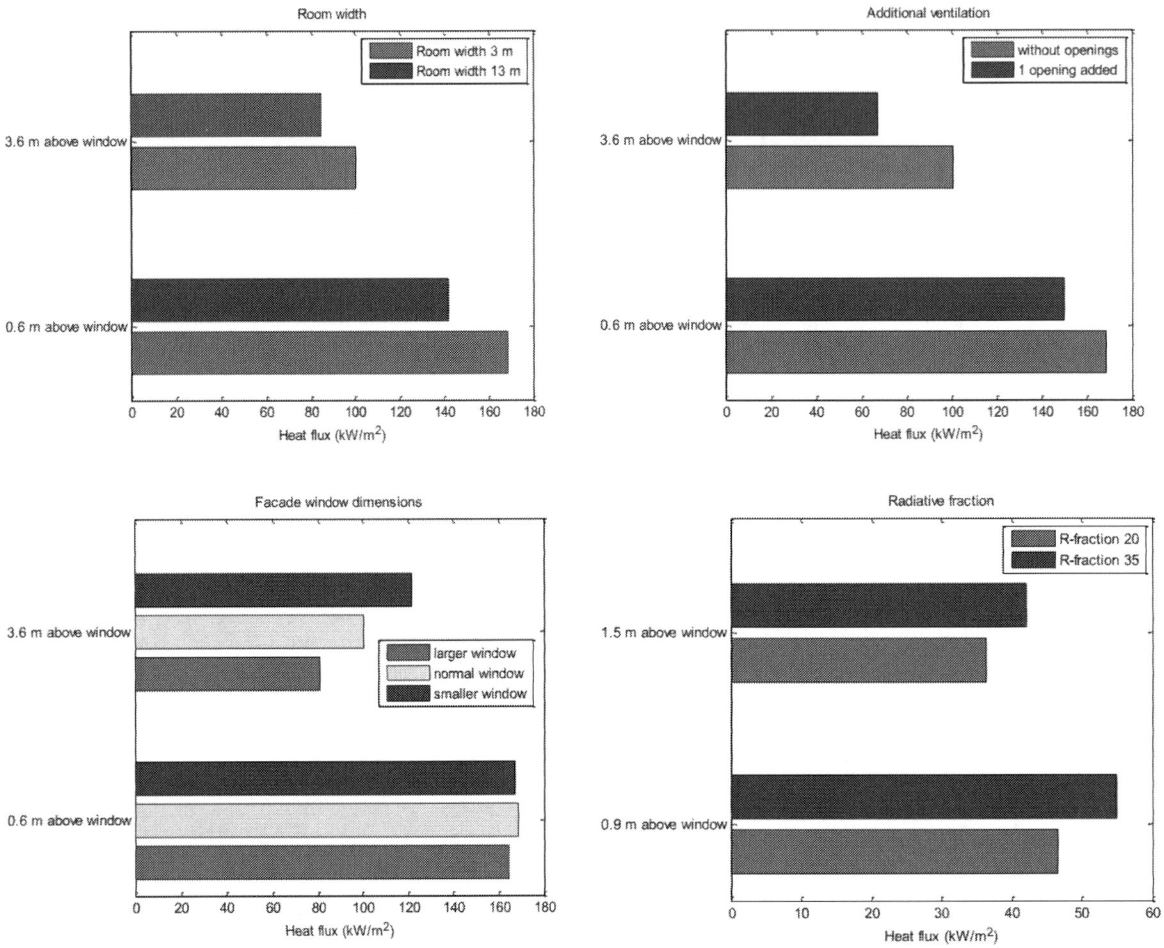


**Figure 7.** Sensitivity of the predicted heat fluxes on the geometrical alterations and radiative fraction.

**Table 2.** Required stone wool protection thickness to protect combustible insulation materials.

| | Required Stone wool thickness against | |
|---|---|---|
| EPS | Melting (200 °C) | 70 mm |
| | Burning (300 °C) | 60 mm |
| PIR | Slow degradation/burning (100 °C) | 90 mm |
| | Fast degradation/burning (300 °C) | 60 mm |

depths. Maximum (over time) temperatures behind 60 mm and 90 mm layers of stone wool are shown in Fig. 9. The peak temperatures at these depths are 277°C and 97°C, respectively.

Using the simulations results, and assuming certain critical temperatures for the plastic insulation materials, we can estimate the thickness of stone wool that is needed to prevent the damage. The results are summarized in Table 2. The two materials are Expanded polystyrene (EPS) and Polyisocyanurate (PIR). The threshold temperatures were taken from analytical experiments of individual products, and therefore, the results cannot be generalized for other products using similar materials, as the critical temperatures can vary. We can see that 70 mm of stone wool is needed to prevent the melting of EPS. For PIR, about 60 mm of protective material is needed to prevent the main thermal degradation reaction at 300°C, but as much as 90 mm is required to prevent the slower degradation reaction around 100°C.

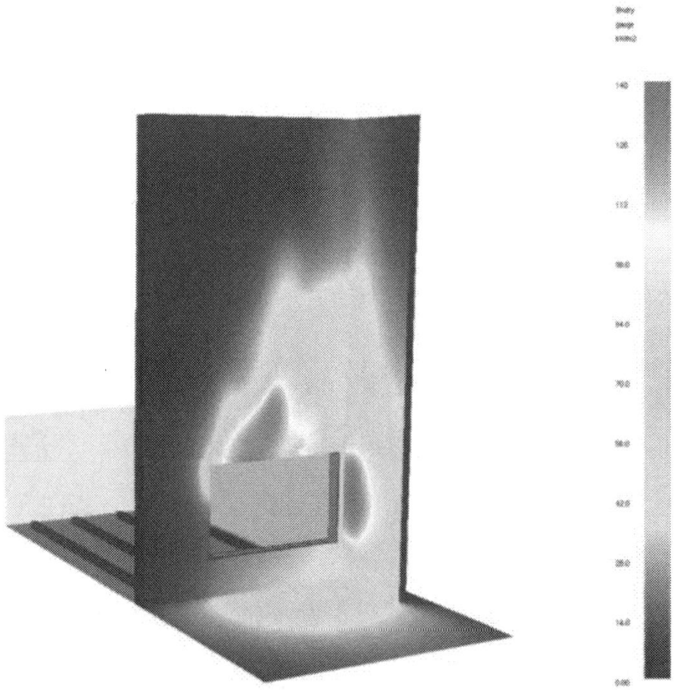

**Figure 8.** Instantaneous heat flux distribution on the façade.

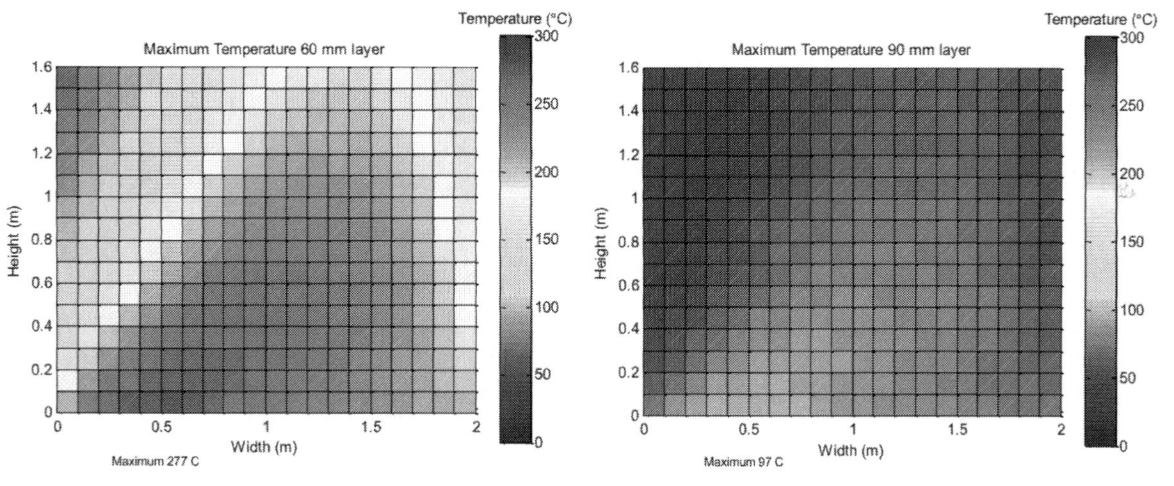


**Figure 9.** Maximim temperatures behind 60 mm (left) and 90 mm (right) layers of stone wool.

## 4. CONCLUSIONS

Numerical simulations of the ISO 13785-2 façade fire test were used to investigate the sensitivity of the thermal exposure on the façade to the geometrical and model-related parameters. The simulation model was first validated against experimental data. In some tests, the model uncertainty was very low, but the overall standard deviation in the simulation data was higher than the current state of the model validation database. In average, the façade temperatures were underestimated by 18% and the heat fluxes overestimated by 14%. Further investigation of the simulation setup and determination of the partially unknown experimental details continues. The current findings show, already, that the experiments used for CFD validation should be designed symmetric to avoid large local errors in thermal predictions due to the otherwise insignificant uncertainties in fire plume position.

The sensitivity study indicated that the heat fluxes in the upper part of the façade are affected by the windows size although the values in the calibration region are not. In addition, the sootiness and high radiative fraction of the fuel were found to increase the façade heat fluxes.

The model was then used to estimate the thickness of non-combustible insulation layer that is needed to protect combustible materials from melting or decomposition. In addition, a procedure was developed for the calculations of the material behaviour without repeated large-scale CFD simulations. Such a procedure can be efficiently applied in product and material R&D.

## References

[1] ISO 13785-2:2002. Reaction-to-fire tests for facades – Part 2: Large-scale test. International Organization for Standardization, 2002.

[2] McGrattan, K., McDermott, R., Floyd, J., Hostikka, S., Forney, G., & Baum, H. "Computational fluid dynamics modelling of fire". *International Journal of Computational Fluid Dynamics*, 1–12, 2012.

[3] Lin, C.H., Ferng, Y.M., & Hsu, W.S., "Investigating the effect of computational grid sizes on the predicted characteristics of thermal radiation for a fire", *Applied Thermal Engineering*, **29**(11–12), 2243–2250, 2009.

[4] Yoshioka, H., Ohmiya, Y., Noaki, M., & Yoshida, M. "Large-scale Facade Fire Tests Conducted Based on ISO 13785-2 with Noncombustible Facade Specimens", *Fire Science and Technology*, **31**, 1–22, 2012.

MATEC Web of Conferences **46**, 04001 (2016)
DOI: 10.1051/matecconf/20164604001


# Investigation into the use of a water curtain over openings to prevent fire spread

Matt Turco[1], Paul Lhotsky[2] and George Hadjisophocleous[1]

[1] Department of Civil and Environmental Engineering, Carleton University, Ottawa, ON, Canada
[2] CIVELEC Consultants Inc. Montreal, Quebec

**Abstract.** This paper presents the results of several full-scale experiments to study heat radiation attenuation and shielding using sprinklers to create a water curtain between a compartment fire and a target wall. This work builds on the concept that water absorbs and scatters heat radiation and applies it as a shielding mechanism, which can be used to protect personnel and property exposed to an opening during a compartment fire. The purpose of the study is to determine the effectiveness of a water curtain created using sprinkler heads in attenuating the heat flux from flames exiting a compartment fire, shielding the target wall and preventing fire spread. The results show that the water curtain reduces heat flux to the target wall which may allow for decreasing the building distance from the property line. Heat fluxes along the target wall are compared to heat fluxes required for the piloted ignition of wood products. The effects of fire size, sprinkler system flow rate, and separation distance to the target wall are discussed.

## 1. INTRODUCTION

During a compartment fire radiation heat exiting the compartment opening poses a risk to adjacent properties. If the heat flux leaving the compartment exceeds the heat flux required for piloted ignition of the building materials the fire could spread from the compartment to the adjacent properties. The radiation heat flux is affected by many factors such as the flames exiting the compartment, compartment size, and opening size. Several studies have been completed on this [1–6]. In order to reduce the radiation heat flux leaving a compartment this study utilized a water curtain created using sprinklers placed over the opening. Cheng [6] studied heat radiation exiting a compartment and impinging on an adjacent target wall. This was done using a built compartment with two different fire sources for comparison. The effect of water droplets on the suppression of fire and as a shielding mechanism has been the subject of many studies [7–17]. Kermida [7] found the ability of water droplets to suppress fires is due to the following four methods; gas phase cooling, oxygen dilution, fuel bed cooling, and radiation attenuation. This study utilized the ability of water to attenuate radiation to reduce the heat flux leaving the compartment. The main mechanisms involved are radiation absorption and scattering by water droplets [7–13]. Chow [14] and Choi [15] performed similar experiments in order to determine if a room could be protected from a fire in an adjacent room with the use of a water curtain. Studies have investigated the ability of water curtains as a shield to protect property [11, 12, 17, 18]. This research was undertaken to determine if the

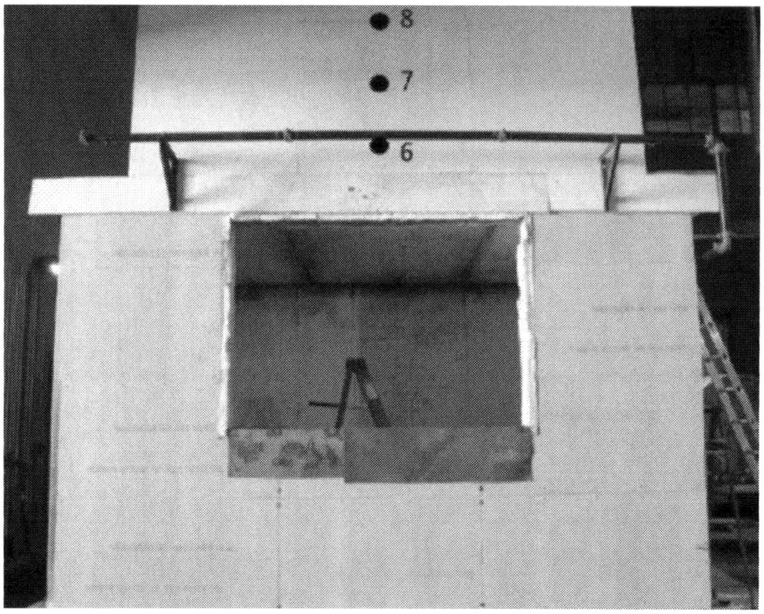


**Figure 1.** Compartment with sensor locations.

use of a water curtain over an opening can be utilized to add another factor of safety to limit the distance of adjacent properties. The main objective of this research is to determine the effectiveness of a water curtain at attenuating heat radiation leaving a compartment to prevent fire spread to adjacent properties and to model the effect that the water curtain has on the radiation exiting the compartment in order to develop a simple mathematical model for radiation that accounts for the reduction caused by the water curtain.

## 2. EXPERIMENTAL SETUP

To address this issue, a series of 24 full scale experiments to investigate the use of a water curtain over a compartment opening to prevent fire spread to adjacent properties was performed at the Carleton University Fire Research Lab. The tests utilized a small compartment and a target wall which can be moved to a desired distance away from the compartment. For the fire, propane burners placed inside the compartment were used.

### 2.1 Compartment design

Figure 1 shows the compartment, built using steel studs protected with half inch cement board and two inches of fire resistant Fibrefrax insulation on all interior surfaces. The floor was layered with fire resistant bricks which supported the propane burners. The compartment was 3 m by 4 m by 3 m high with an opening of 2 m by 1.5 m high. The front wall extended 3 m above the compartment. The front face of the compartment was protected with a layer of cement board.

### 2.2 Water curtain

To create the water curtain 3 standard pendent ($k = 5.6$) sprinklers were used outside of the compartment at spacing's of 2 m. The sprinklers were placed 0.46 m above the opening and 0.3 m away from the front of the compartment. The sprinklers were centred in the middle of the compartment opening. A pump

**Figure 2.** Target wall with sensor locations.

placed within a reservoir inside the lab provided water to the sprinkler system. The flow rate of the sprinklers was monitored using a flow meter and a pressure gauge located along the pipe next to the pump. The flow was controlled using a valve next to the flow meter.

## 2.3 Propane burners

To create the fire, three propane burners were placed in the centre of the compartment 0.06 m above the floor. The pipes for the burners enter the compartment through openings at the rear of the compartment which were sealed after the burners were installed. The propane burners were controlled using valves placed along the pipes outside the rear of the compartment and monitored using pressure gauges and a propane flow meter. This ensured that a constant fire heat release rate was produced and maintained during the tests.

## 2.4 Target wall

Figure 2 shows the target wall constructed of a wood frame built upon rollers with a layer of cement board covering the front face. The wall was 4.9 m by 4.9 m and could be moved to the desired distance away from the compartment.

## 2.5 Instrumentation

Five heat flux gauges and five thermocouples were placed along the target wall beginning 2.2 m above the floor in line with the centre of the compartment opening. The next sensor was 0.2 m above the first and subsequent sensors were spaced at 0.5 m. Three heat flux gauges and thermocouples were placed on the front wall extending above the opening to the compartment. The first sensor on the extension wall was 0.55 m above the opening and subsequent sensors were spaced at 0.5 m. The heat

**Table 1.** Heat flux (kW/m$^2$) results for the 2-MW fire tests at a separation distance of 3 m.

| Sensor Location | 0 L/min | 57 L/min | % Reduction | 76 L/min | % Reduction | 91 L/min | % Reduction |
|---|---|---|---|---|---|---|---|
| 1 | 8.20 | 3.20 | 0.61 | 3.40 | 0.59 | 4.10 | 0.50 |
| 2 | 8.20 | 3.90 | 0.52 | 3.30 | 0.60 | 3.80 | 0.54 |
| 3 | 8.20 | 4.60 | 0.44 | 4.40 | 0.46 | 4.40 | 0.46 |
| 4 | 7.50 | 4.60 | 0.39 | 4.50 | 0.40 | 4.50 | 0.40 |
| 5 | 5.10 | 3.10 | 0.39 | 3.00 | 0.41 | 3.80 | 0.25 |
| **Target Wall Average** | 7.44 | 3.88 | 0.47 | 3.72 | 0.49 | 4.12 | 0.43 |
| 6 | 13.30 | 3.40 | 0.74 | 3.10 | 0.77 | 2.80 | 0.79 |
| 7 | 9.30 | 2.90 | 0.69 | 2.70 | 0.71 | 2.10 | 0.77 |
| 8 | 6.30 | 3.30 | 0.48 | 2.90 | 0.54 | 2.20 | 0.65 |
| **Façade Wall Average** | 9.63 | 3.20 | 0.64 | 2.90 | 0.67 | 2.37 | 0.74 |

flux gauges used were Gardon Gauges accurate to within +/−3%. The heat flux gauges were calibrated by the Medtherm Corporation in compliance with ISO/IEC 17025, ANSI/NCSL Z540-1 and MIL-STD-45662A to Medtherm PI-20 with traceability to the National Institute of Standards and Technology. The thermocouples used were standard K-type thermocouples which are accurate to within +/− 0.75%.

### 2.6 Test series

Two test series were run, one with a separation distance between the compartment and the target wall at 3 m and one at 4 m. The test series considered fires with sizes of 2-MW, 3-MW, and 4-MW. Tests ran until the heat flux readings along the target wall reached steady state. Tests initially ran without the use of sprinklers in order to determine a baseline heat flux along the target wall and then the sprinklers were activated and run at flow rates of 57 L/min (15 gpm), 76 L/min (20 gpm), and 91 L/min (24 gpm) per sprinkler (Note that flow is in US gallon per minute).

## 3. RESULTS

### 3.1 First test series: 3 m separation distance

The heat flux results for the 2-MW fire at 3 m are presented in Table 1. The initial average heat flux experienced by the target wall when no sprinklers were active was 7.44 kW/m$^2$, which is below the threshold for piloted ignition of wood products. The introduction of the water curtain decreased the heat flux readings on the target wall for all flow rates to below 4.6 kW/m$^2$. On the façade wall the heat flux was initially between 6.3 kW/m$^2$ and 13.3 kW/m$^2$. Use of the sprinklers at all flow rates caused the average heat flux to drop to below 3.2 kW/m$^2$, substantially lower than the initial values as well as the critical heat flux for piloted ignition of wood products.

The heat flux results for the 3-MW fire at 3 m are presented in Table 2. The average heat flux along the target wall was 13.6 kW/m$^2$ initially, greater than that required for piloted ignition of wood products. Reduction of the heat fluxes with the use of sprinklers increased with increasing sprinkler flow rate, lowering the initial heat flux below the critical value. On the facade wall the heat flux was initially between 10 kW/m$^2$ and 25 kW/m$^2$. As with the target wall, the front wall extension saw a similar decrease in the average heat flux with every increase in sprinkler flow rate.

The heat flux results for the 4-MW fire at 3 m are presented in Table 3. The initial average heat flux experienced by the target wall was 23.3 kW/m$^2$. The use of the water curtain showed a similar pattern to that seen for the 3-MW fire tests with each increase in flow rate having a larger impact on reducing the average heat flux. Heat fluxes were greater than the critical heat flux but it should be noted that the water curtain was wetting the target wall. During real fire events the water curtain is predicted to prevent sparks from reaching the adjacent properties. On the facade wall the heat flux was initially between

**Table 2.** Heat flux (kW/m$^2$) results for the 3-MW fire tests at a separation distance of 3 m.

| Sensor Location | 0 L/min | 57 L/min | % Reduction | 76 L/min | % Reduction | 91 L/min | % Reduction |
|---|---|---|---|---|---|---|---|
| 1 | 14.5 | 13 | 0.10 | 9.6 | 0.34 | 7.6 | 0.48 |
| 2 | 13.9 | 14.5 | −0.04 | 9.4 | 0.32 | 7 | 0.50 |
| 3 | 15.8 | 15.3 | 0.03 | 10.8 | 0.32 | 7.7 | 0.51 |
| 4 | 13.8 | 14.4 | −0.04 | 10.6 | 0.23 | 7.3 | 0.47 |
| 5 | 10.2 | 9.5 | 0.07 | 7.4 | 0.27 | 5.4 | 0.47 |
| Target Wall Average | 13.64 | 13.34 | 0.02 | 9.56 | 0.30 | 7.00 | 0.49 |
| 6 | 24.8 | 12 | 0.52 | 6.8 | 0.73 | 4.6 | 0.81 |
| 7 | 15.6 | 10.7 | 0.31 | 5.7 | 0.63 | 3.6 | 0.77 |
| 8 | 10.9 | 10.8 | 0.01 | 5.8 | 0.47 | 3.5 | 0.68 |
| Façade Wall Average | 17.10 | 11.17 | 0.28 | 6.10 | 0.61 | 3.90 | 0.75 |

**Table 3.** Heat flux (kW/m$^2$) results for the 4-MW fire tests at a separation distance of 3 m.

| Sensor Location | 0 L/min | 57 L/min | % Reduction | 76 L/min | % Reduction | 91 L/min | % Reduction |
|---|---|---|---|---|---|---|---|
| 1 | 25.1 | 18.8 | 0.25 | 15.5 | 0.38 | 13.8 | 0.45 |
| 2 | 23.8 | 20.8 | 0.13 | 15.7 | 0.34 | 13.2 | 0.45 |
| 3 | 26.6 | 20.1 | 0.24 | 17 | 0.36 | 14 | 0.47 |
| 4 | 23.4 | 18 | 0.23 | 15.8 | 0.32 | 13.1 | 0.44 |
| 5 | 17.4 | 12.6 | 0.28 | 11.2 | 0.36 | 9.1 | 0.48 |
| Target Wall Average | 23.26 | 18.06 | 0.23 | 15.04 | 0.35 | 12.64 | 0.46 |
| 6 | 41.2 | 18.3 | 0.56 | 11.2 | 0.73 | 8.6 | 0.79 |
| 7 | 22.2 | 15.7 | 0.29 | 9.2 | 0.59 | 5.7 | 0.74 |
| 8 | 16.7 | 14.8 | 0.11 | 9.8 | 0.41 | 5.4 | 0.68 |
| Façade Wall Average | 26.70 | 16.27 | 0.32 | 10.07 | 0.58 | 6.57 | 0.74 |

**Table 4.** Heat flux (kW/m$^2$) results for the 2-MW fire tests at a separation distance of 4 m.

| Sensor Location | 0 L/min | 57 L/min | % Reduction | 76 L/min | % Reduction | 91 L/min | % Reduction |
|---|---|---|---|---|---|---|---|
| 1 | 7.3 | 6.7 | 0.08 | 6.3 | 0.14 | 5.9 | 0.19 |
| 2 | 6.7 | 6 | 0.10 | 6 | 0.10 | 5.4 | 0.19 |
| 3 | 6.8 | 6.3 | 0.07 | 6 | 0.12 | 5.5 | 0.19 |
| 4 | 6.6 | 5.5 | 0.17 | 5.4 | 0.18 | 5.1 | 0.23 |
| 5 | 4.7 | 4.3 | 0.09 | 4.1 | 0.13 | 3.8 | 0.19 |
| Target Wall Average | 6.42 | 5.76 | 0.10 | 5.56 | 0.13 | 5.14 | 0.20 |
| 6 | 7.7 | 4.6 | 0.40 | 4.1 | 0.47 | 2.1 | 0.73 |
| 7 | 7 | 4.9 | 0.30 | 2.3 | 0.67 | 1.2 | 0.83 |
| 8 | 6.3 | 5.1 | 0.19 | 2.5 | 0.60 | 0.24 | 0.96 |
| Façade Wall Average | 7.00 | 4.87 | 0.30 | 2.97 | 0.58 | 1.18 | 0.84 |

16 kW/m$^2$ and 42 kW/m$^2$. The use of the sprinklers caused greater drops in the average heat flux with increasing in sprinkler flow rates.

### 3.2 Second test series: 4 m separation distance

The heat flux results for the 2-MW fire at 4 m are presented in Table 4. The initial average heat flux experienced by the target wall when no sprinklers were active was 6.4 kW/m$^2$, which is below the threshold for piloted ignition of wood products as it was for the previous 2-MW fire tests. On the façade wall the heat flux was initially between 6.3 kW/m$^2$ and 7.7 kW/m$^2$. Use of the sprinklers at all flow rates caused the average heat flux to drop.

The heat flux results for the 3-MW fire at 4 m is presented in Table 5. The average heat flux along the target wall was 13.7 kW/m$^2$ initially, greater than that required for piloted ignition of wood products. On the facade wall the heat flux was initially between 13 kW/m$^2$ and 19 kW/m$^2$. The use of the water

**Table 5.** Heat flux ($kW/m^2$) results for the 3-MW fire tests at a separation distance of 4 m.

| Sensor Location | 0 L/min | 57 L/min | % Reduction | 76 L/min | % Reduction | 91 L/min | % Reduction |
|---|---|---|---|---|---|---|---|
| **1** | 14.8 | 9.8 | 0.34 | 10.3 | 0.30 | 9.9 | 0.33 |
| **2** | 14 | 9 | 0.36 | 9.6 | 0.31 | 9.2 | 0.34 |
| **3** | 15.3 | 9.8 | 0.36 | 10.2 | 0.33 | 9.6 | 0.37 |
| **4** | 14.2 | 9.1 | 0.36 | 9.9 | 0.30 | 9.6 | 0.32 |
| **5** | 10.3 | 7.1 | 0.31 | 7.6 | 0.26 | 7.4 | 0.28 |
| **Target Wall Average** | 13.72 | 8.96 | 0.34 | 9.52 | 0.30 | 9.14 | 0.33 |
| **6** | 18.4 | 6.9 | 0.63 | 5 | 0.73 | 3.9 | 0.79 |
| **7** | 16.7 | 7.9 | 0.53 | 5.1 | 0.69 | 4.3 | 0.74 |
| **8** | 13.1 | 7.8 | 0.40 | 4 | 0.69 | 3.2 | 0.76 |
| **Façade Wall Average** | 16.07 | 7.53 | 0.52 | 4.70 | 0.71 | 3.80 | 0.76 |

**Table 6.** Heat flux ($kW/m^2$) results for the 4-MW fire tests at a separation distance of 4 m.

| Sensor Location | 0 L/min | 57 L/min | % Reduction | 76 L/min | % Reduction | 91 L/min | % Reduction |
|---|---|---|---|---|---|---|---|
| **1** | 20.9 | 16.7 | 0.20 | 15.9 | 0.24 | 14.4 | 0.31 |
| **2** | 19.9 | 14.8 | 0.26 | 15.2 | 0.24 | 13.6 | 0.32 |
| **3** | 21.2 | 15.6 | 0.26 | 15.7 | 0.26 | 14.2 | 0.33 |
| **4** | 19.8 | 14.8 | 0.25 | 14.6 | 0.26 | 13.4 | 0.32 |
| **5** | 14.4 | 11.4 | 0.21 | 11.2 | 0.22 | 10.1 | 0.30 |
| **Target Wall Average** | 19.24 | 14.66 | 0.24 | 14.52 | 0.24 | 13.14 | 0.32 |
| **6** | 39.4 | 13.9 | 0.65 | 10.7 | 0.73 | 7.6 | 0.81 |
| **7** | 33.2 | 12 | 0.64 | 10.3 | 0.69 | 7.9 | 0.76 |
| **8** | 22.9 | 9.6 | 0.58 | 7.6 | 0.67 | 6 | 0.74 |
| **Façade Wall Average** | 31.83 | 11.83 | 0.62 | 9.53 | 31.83 | 11.83 | 0.62 |

curtain caused a decrease in heat flux readings for both the target wall and the façade wall to below critical values.

The heat flux results for the 4-MW fire at 4 m is presented in Table 6. The initial average heat flux experienced by the target wall was 19.2 $kW/m^2$. Similar to the 4-MW fire at 3 m separation, the heat fluxes were greater than the critical heat flux but no sparks were getting through the water curtain. On the facade wall the heat flux was initially between 22 $kW/m^2$ and 40 $kW/m^2$. Use of the sprinklers caused the heat flux readings to decrease.

### 3.3 Discussion

Based on the 2-MW fire tests for both separation distances, 3 m and 4 m, the initial heat fluxes along the target wall were found to be below the critical heat flux required for piloted ignition of wood products. With the sprinkler flow the heat flux reductions for both distances were found to be similar for all flow rates meaning that increasing the water flow rates had minimal impact on the attenuation after the minimum flow rate test. The results from the façade wall sensors showed that the heat fluxes above the opening were varied due to the hot gases exiting the compartment. With the introduction of the sprinklers all heat fluxes dropped to well below the critical heat flux.

Based on the 3-MW fire tests the initial heat fluxes were greater than the critical heat flux for piloted ignition of wood products. With the sprinkler flow all heat fluxes fell below the critical heat flux for the 4-m tests but required a larger flow rate to bring the 3-m separation distance tests to below the critical heat flux.

Along the façade wall the initial heat fluxes were greater than the critical heat flux but all fell below the critical heat flux with the sprinkler water flow. A significant increase in attenuation was seen with the increase in the sprinkler flow rates for both separation distances.

The initial heat flux results of the 4-MW fire tests were much greater than the critical heat flux for piloted ignition of wood products. The sprinkler water flow failed to bring the heat fluxes below the

**Table 7.** Average calculated compartment temperatures based on the baseline test results.

| Test Series | Average Compartment Temperature (K/°C) |
|---|---|
| 2-MW @ 3-m | 1175(902) |
| 2-MW @ 4-m | 1281(1008) |
| 3-MW @ 3-m | 1223(950) |
| 3-MW @ 4-m | 1389(1116) |
| 4-MW @ 3-m | 1349(1076) |
| 4-MW @ 4-m | 1463(1190) |

critical heat flux but they were brought low enough to avoid spontaneous combustion. Even though the heat fluxes were above the critical value, the critical value is conservative and requires a spark or flame to be in close proximity to the target wall. With the use of the sprinklers, sparks and flames were unable to penetrate the water curtain and during the 3-m separation tests the water curtain was wetting the wall. The use of sprinklers caused uniform reductions to the heat fluxes for both separation distances. Along the façade wall initial heat fluxes were greater than the critical heat flux but fell below the critical heat flux with the larger flow rate sprinklers. A significant increase in attenuation was seen with the increase in the sprinkler flow rates for the 3-m separation distance with a smaller increase for the 4-m separation distance.

The attenuation along the target wall was found to be less effective for the larger separation distance. The heat fluxes for all the experiments were lowered to acceptable ranges such that flame spread could be avoided, protecting adjacent property. The largest reductions were generally seen during the sprinklers tests with the highest flow rate. It was found that the baseline heat fluxes for the 4-m separation tests were similar to or even higher than the 3-m separation tests with the exception of the 2-MW tests. For the 2-MW fire the 3-m separation tests had clearly higher heat fluxes. This is believed to be caused by one or more reasons. The reasons include; the room used for the 4-m separation tests had better insulation resulting in more heat exiting from the opening, the separation distance between the compartment and the target wall may have affected the amount of air available to be entrained into the compartment, and variations in the propane flow rate.

The values of the façade wall heat fluxes were not stable initially due to the turbulence of the hot gases and flames escaping the compartment. The heat fluxes were found to be greater than the critical heat flux required for piloted ignition but were brought to acceptable levels with the introduction of the water curtain. The larger fires required larger flow rates to bring the heat fluxes down to acceptable levels. The use of sprinklers caused a decrease of the temperature of the hot gases and radiation heat fluxes. This can help prevent upward flame spread from a lower compartment to a higher one through an opening. The sensors closest to the opening were impacted more than the higher sensors due to being in closer proximity to the hot gases and flames that were exiting the compartment.

## 4. DISCUSSION OF RESULTS

In order to study the effect the water curtain had on attenuating radiation the individual fire tests were broken down and analyzed. For each test the configuration factor was determined using the height of the flames exiting the compartment as well as the sensors along the target wall that fell with the boundaries of the radiator area. The radiator was assumed to be rectangular and to be in plane with the window such that the window and flames exiting the compartment were a single rectangular radiator. The radiators for each test were broken down into quadrants and each configuration factor for the quadrants was determined. Using the individual quadrant configuration factors the total configuration factor for each test was determined. Once the configuration factors were determined the compartment temperatures for each test series was determined, shown in Table 7.

**Table 8.** Calculated theoretical and experimental reduction factors.

| | 57 L/min | | 76 L/min | | 91 L/min | |
|---|---|---|---|---|---|---|
| | **Calculated** | **Experimental** | **Calculated** | **Experimental** | **Calculated** | **Experimental** |
| Average Reduction | 21.82 | 23.23 | 29.91 | 31.18 | 37.59 | 38.55 |

The average compartment temperatures were used to determine the theoretical reduction factor for each sensor using the following equations:

$$q''_{red} = (1 - f_t)\,\varphi\varepsilon\sigma\left(T_c^4 - T_a^4\right) \tag{1}$$

$$f_t = 1 - \left(\frac{q''_r}{\varphi\varepsilon\sigma\left(T_c^4 - T_a^4\right)}\right) \tag{2}$$

$q''_{red}$ = Experimental Reduced Heat Flux

$f_t$ = Theoretical Reduction Factor

$\varphi$ = Configuration Factor

$\sigma$ = Stefan Boltzmann constant,$5.67 \times 10^{-8}\ \mathrm{W/m^2K^4}$

$T_e$ = Average Compartment Temperature, K

$T_a$ = Ambient Temperature, K

$\varepsilon$ = emissivity, $\varepsilon = \left(\frac{1}{\varepsilon_e} + \frac{1}{\varepsilon_r} - 1\right)^{-1}$

Using Eq. (1) the reduction factor was solved for all sensors, fires, and flow rates. This theoretical reduction, Eq. (2), was then compared to the actual experimental reduction, Eq. (3), of the experimental heat fluxes with the water curtain to the baseline heat fluxes without the water curtain. Comparison of the calculated theoretical heat flux reductions and the actual experimental heat flux reductions found that the theoretical reductions were within −8% to 4% of the actual experimental heat flux reductions.

$$f_e = 1 - \frac{q''_{red}}{q''_{base}} \tag{3}$$

$q''_{red}$ = Experimental Reduced Heat Flux

$q''_{base}$ = Experimental Baseline Heat Flux

$f_e$ = Experimental Reduction Factor.

The reductions were then averaged for each flow rate for both theoretical and experimental to determine the reduction factor for each particular flow rate, Table 8. The difference between the theoretical reduction factors and the experimental reduction factors is reasonably accurate with a difference of less than 1.5%, which can be attributed to the use of ambient temperature instead of the actual temperature of the receiving surface.

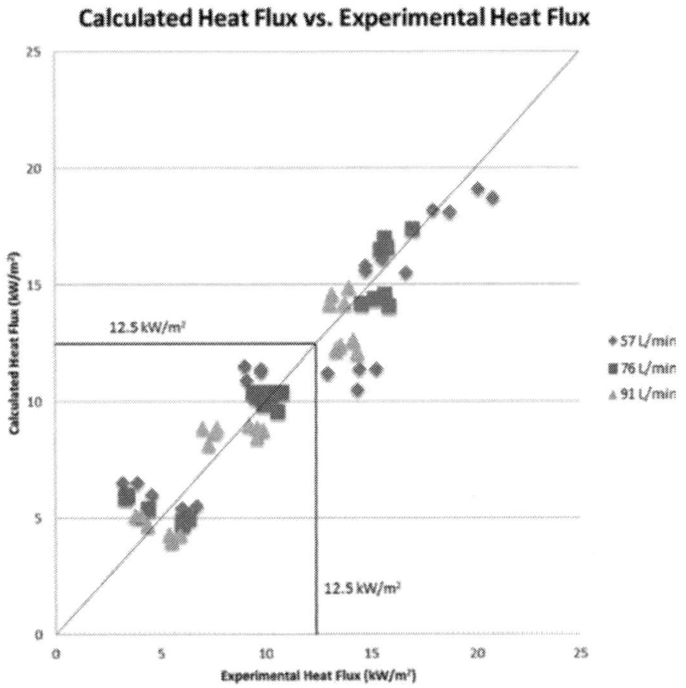


**Figure 3.** Calculated and experimental heat flux differences with critical heat flux displayed.

Using the average theoretical heat flux reductions the theoretical heat flux based on the reductions was calculated, using Eq. (4).

$$q_t'' = (1 - f_t)\,\varphi\varepsilon\sigma\left(T_c^4 - T_a^4\right) \tag{4}$$

$q_t''$ = Theoretical Reduced Heat Flux

$f_t$ = Theoretical Reduction Factor

$\varphi$ = Configuration Factor

$\sigma$ = Stefan Boltzmann constant, $5.67 \times 10^{-8}\,\mathrm{W/m^2K^4}$

$T_e$ = Average Compartment Temperature, K

$T_a$ = Ambient Temperature, K

$\varepsilon$ = emissivity, $\varepsilon = \left(\frac{1}{\varepsilon_e} + \frac{1}{\varepsilon_r} - 1\right)^{-1}$.

The theoretical reduced heat flux was calculated using a reduction factor of 21.9% for the 57 L/min flow rate, 29% for the 76 L/min flow rate, and 39% for the 91 L/min flow rate. The theoretical heat fluxes were then compared with the actual experimental heat fluxes for all sensors, fires, and flow rates. All the individual heat fluxes were compared and the average of the absolute differences was calculated. The average accuracy of the absolute differences was found to be within 18% for the 57 L/min flow rate, 12% for the 76 L/min flow rate, and 15% for the 91 L/min flow rate. The individual differences were to some extent accurate with a few tests showing large inaccuracies but the overall accuracy is reasonable, Fig. 3.

## 5. CONCLUSIONS

It was found that the reduction effect of the water curtain could be determined with reasonable accuracy such that it can be applied to the radiation equation to determine the reduced heat flux. When comparing the results of the theoretical and experimental reduced heat fluxes the average difference was less than 18% for the 2-MW tests and less than 12% for the 3-MW tests and 15% for the 4-MW tests. These findings can help establish guidelines for determining safe separation distances with the inclusion of a water curtain. With further study into the effectiveness of the water curtain the reduction factor can be modelled more effectively such that the reduction can be calculated using the flow rate and density of the water curtain.

Based on the experiments and research conducted in this thesis the following conclusions can be drawn:

1. For all tests the use of a water curtain was able to reduce the heat fluxes along the target wall to minimize the ignition of combustibles in adjacent properties.
2. For tests in which the heat fluxes were above the heat flux required for piloted ignition of wood the water curtain was wetting the target wall and preventing sparks from getting through to the target wall. This means that flame spread would not be possible without another flame source.
3. The water curtain was able to prevent upward flame spread by reducing the heat fluxes on the façade wall to within acceptable levels. The water curtain helps stabilize the turbulent flames and hot gases that were exiting the compartment and preventing the heat flux on the façade wall above the opening from reaching critical levels.
4. The piloted ignition of wood products criteria for the tests performed is a conservative value as the water curtain was preventing sparks from reaching the target wall. The water curtain was also wetting the target wall at close distances.
5. Heat flux and temperature reductions caused by the water curtain were found to be less effective as the separation distance increased.
6. The use of a water curtain lowered heat fluxes to allow for a reduced spatial separation requirements.
7. The effect of the water curtain on the attenuation of radiation can be modelled to within a reasonable accuracy based on the flow rate of the water curtain.

This research studied the ability of a water curtain placed over a compartment opening to attenuate radiation in order to protect adjacent property. The experiments conducted were on a single compartment with a single fixed opening. The factors that were tested included the flow rate of the water curtain, the fire sizes, and the separation distance. To ensure that the results of this research are used practically for assessing the performance of water curtains placed over compartment openings, further tests should be conducted using different compartment sizes and openings, as well as varying types of water curtains made with different sprinkler systems. In a building environment there are many possible parameters that could require different configurations that need to be studied.

In other areas of the world common construction materials may have different critical heat flux requirements. This research can be adjusted in order to compare reduced heat fluxes with critical heat fluxes of common construction materials, such as plastic cladding, in order to determine if a water curtain can be used as an effective fire protection system for other materials in those regions of the world. Comparison with building code requirements in those regions may allow for the installation of a water curtain to protect those materials.

Thanks to Ba Lamthien for assisting with construction and running the tests, and to the staff at the National Research Council of Canada's Fire Lab for their advice and help during these experiments.

## References

[1] Shorter, G. W., et al. *The St. Lawrence Burns*. Division of Building Research, National Research Council, 1960.

[2] Law, Margaret. *Heat radiation from fires and building separation*. London: HM Stationery Office, 1963.

[3] Law, Margaret. *Radiation from Fires in a Compartment*. HM Stationery Office, 1968.

[4] Lin, Ching-Yuan. "Study of exposure fire spread between buildings by radiation." *Journal of the Chinese institute of engineers* **23.4** (2000): 493–504.

[5] Chen, A., and J. Francis. "Radiant heat flux to external surfaces from escaping and extrusive flashover flames." *Proceedings of the Institution of Mechanical Engineers, Part C: Journal of Mechanical Engineering Science* **217.2** (2003): 247–256.

[6] Cheng, Hao, and George V. Hadjisophocleous. "Experimental study and modeling of radiation from compartment fires to adjacent buildings." *Fire Safety Journal* **53** (2012): 43–62.

[7] Keramida, E. P., et al. "Numerical modeling of radiant heat attenuation through water mist." *Combustion science and technology* **159.1** (2000): 351–371.

[8] Coppalle, A., D. Nedelka, and B. Bauer. "Fire protection: water curtains." *Fire Safety Journal* **20.3** (1993): 241–255.

[9] Parent, Gilles, et al. "Experimental investigation of radiation transmission through a water spray." *Journal of Quantitative Spectroscopy and Radiative Transfer* **97.1** (2006): 126–141.

[10] Hostikka, Simo, and Kevin McGrattan. "Numerical modeling of radiative heat transfer in water sprays." *Fire Safety Journal* **41.1** (2006): 76–86.

[11] Yang, Wenhua, et al. "The interaction of thermal radiation and water mist in fire suppression." *Fire Safety Journal* **39.1** (2004): 41–66.

[12] Boulet, Pascal, Anthony Collin, and Gilles Parent. "Heat transfer through a water spray curtain under the effect of a strong radiative source." *Fire Safety Journal* **41.1** (2006): 15–30.

[13] Tseng, C. C., and R. Viskanta. "Absorptance and transmittance of water spray/mist curtains." *Fire safety journal* **42.2** (2007): 106–114.

[14] Chow, Wan Ki, and Elaine YL Ma. "Experimental studies on thermal and smoke blockage by water curtains." *9th AIAA/ASME Joint Thermophysics and Heat Transfer Conference*. 2006.

[15] Choi, C.L. "RADIATION BLOCKAGE EFFECTS BY WATER CURTAIN." *International Journal on Engineering Performance-Based Fire Codes,* **6**, 4 (2004): P.248–254.

[16] Kim, Sung Chan, and Hong Sun Ryou. "An experimental and numerical study on fire suppression using a water mist in an enclosure." *Building and Environment* **38.11** (2003): 1309–1316.

[17] Grant, G., J. Brenton, and D. Drysdale. "Fire suppression by water sprays."*Progress in energy and combustion science* **26.2** (2000): 79–130.

[18] Buchlin, Jean-Marie. "Thermal shielding by water spray curtain." *Journal of loss prevention in the process industries* **18.4** (2005): 423–432.

MATEC Web of Conferences **46**, 04002 (2016)
DOI: 10.1051/matecconf/20164604002


# Fire safety engineering applied to high-rise building facades

Samuele Sassi[1], Paolo Setti[1], Giuseppe Amaro[2], Lamberto Mazziotti[3], Giuseppe Paduano[3], Piergiacomo Cancelliere[3] and Mauro Madeddu[1]

[1] FSC ENGINEERING SRL, Fire & Structure Consulting, via A. Zezon, 4, Milano, Italy
[2] GAE Engineering S.r.l. Corso G. Marconi, 20, 10125 Torino, Italy
[3] Ministry of Interior, Italian National Fire Services, Largo Santa Barbara, 2, Rome, Italy

**Abstract.** Fire safety of high-rise building facades is a complex problem, therefore the use of prescriptive fire codes could not be sufficient to ensure a proper building fire safety level. In new high rise buildings the fire safety of facades can be taken into account by means of performance based fire design that can help in the selection of the best technological solutions and material choices.

This article deals with a case study done on the Piedmont Region Headquarters (TORRE REGIONE PIEMONTE) which is one of the highest office buildings in Italy. The "Torre Regione Piemonte" is located in Turin and has 45 storey and a total height of 183,61 m. The building is characterized by the closed enclosure called "Grand Space" which is continuous for almost the entire height of the building and contains volumes such as offices/meeting rooms named as "Satellites".

In order to fulfil Italian fire safety requirements, the "Torre Regione Piemonte" building has been assessed using the performance base design and fire safety engineering approach. This paper deals with the design process selection of the most representative fire scenarios focusing on the fire resistance performance requirements of structural and glazed elements of the facades. Furthermore, all the simulation, calculations and performance fire resistant requirements of the high rise building have been supported by specific laboratory tests and experimental results, that are reported and discussed within paper.

## 1. INTRODUCTION

Fire safety of high-rise building facades is a complex problem; therefore the use of prescriptive fire codes could not be sufficient to ensure a proper building fire safety level. In new high rise buildings the fire safety of façades can be considered by means of performance based fire design that can help in the selection of the best technological solutions and material choices.

In this article, the façade of Piedmont Region Headquarters ("Torre Regione Piemonte", project of Arch. Massimiliano Fuksas, still under construction in Italy) is examined by fire safety engineering (FSE), as a case study due to its distinctive architectural and technological aspects.

The building is composed of a central part, used as offices (Office Area), which is next to a closed volume for the entire height named as "Grand Space". Inside this volume, there are box-shaped structures called "Satellite" used for meeting rooms connected with office areas.

The Office Area and the "Satellites" are built in reinforced concrete and pre-stressed concrete, enclosed by a aluminium/glass facade, on the other hand the "Grand Space" is enclosed by only a façade composed by a main steel frame and aluminum–glass cladding.

As the presence of "Grand Space" affects the fire behavior of the building, performance based design and FSE methods have been chosen to examine and conform the building "Torre Regione Piemonte" to the Italian fire safety requirements. More precisely, different fire scenarios have been considered both with respect to the effects on structures and to the occupants in order to optimize the choice of active and passive fire safety measures. In order to check the smoke and temperature distribution, 6 different fire scenarios have been performed by CFD. As a result, the passive fire protection of the critical points of steel structures are optimized and fire resistance characteristics of the façade are enhanced. In particular, according to the results of the scenarios, only Level -2 and level -1 of "Grand Space" steel structures had been protected with intumescent paint; glazed separations between Offices and "Grand Space" had been assessed for $EW60_{(i\rightarrow o)}$ fire rating instead of EI 120 (preliminary requirement); glazed separations between Offices and "Satellites" had been assessed for EI 60 fire rating instead of EI 120.

Furthermore, the facade types used in the building are selected with respect to the fire resistance requirements derived from performance fire design and CFD models, taking into account also the effects due to the installation and weather conditions.

In the article, the fire resistance performance requirements of structural and glazed elements of the facades and the results of the CFD models are reported. In addition, all the simulated calculations and performance fire resistant requirements assessment process, have been supported by specific laboratory tests. These fire resistant tests have been also carried out in accordance with EN 1364-3 to evaluate the fire behaviour of the chosen glazed façade system.

## 2. DESCRIPTION OF THE BUILDING

"Torre Regione Piemonte" building is composed of 2 basement levels and has 43 storeys above ground level (max. height: 183.61 m) with dimensions of 45 × 45 m. All space in the first three floor is allocated to entrance hall, consists of 2 mezzanine floors as exhibition areas. The upper floors are office areas. The 3-D view and a typical plan are show in Fig. 1.

The building is composed of a central part, used as offices (Office Area), which is next to a closed volume for the entire height named as "Grand Space". Inside this volume, there are box-shaped structures called "Satellite" used for meeting rooms connected with office areas.

The Office Area and the "Satellites" are built in reinforced concrete and pre-stressed concrete, enclosed by a aluminium/glass facade, on the other hand the "Grand Space" is enclosed by only a façade composed by a main steel frame and aluminum–glass cladding.

According to the Italian prescriptive fire codes, the project fire safety requirement should fulfil the following fire safety measures:

- Each floor of the offices (except the double/triple height zone - entrance hall) should ensure a fire compartment with R/REI/EI 120 fire resisting class and the installation of mechanical smoke and heat exhaust systems and sprinklers.
- Each Satellite should ensure a fire compartment with R/REI/EI 120 fire resisting class, the installation of sprinklers and water curtain on glazing.
- Grand Space is the only volume that requires R 60 fire resisting class; it is composed of aluminum façade structure supported by steel elements and is equipped with mechanical smoke and heat exhaust systems.
- The façade between two macro areas (including the separation between Offices and Satellites) should have EI 120 fire resisting class.

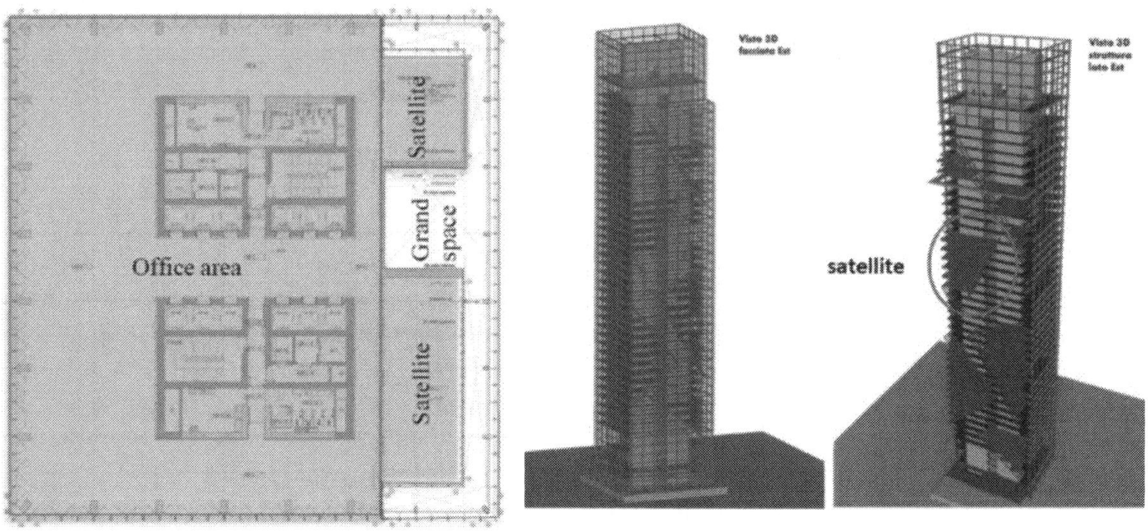


**Figure 1.** Typical plan of Tower Piedmont Region Headquarters where are highlighted Office Area, Satellites and Grand Space.

# 3. CONSTRUCTIONAL AND FUNCTIONAL PROBLEMS

During the construction phase of the building, there are three major problems also related with fire safety of the external façade.

- The first problem is the maintenance of fire proofing on the metal structure of Grand Space. Once the application of fire-proofing is completed there is no way of having an access to all parts of the structure.
- The second one is the possibility of having an alteration in the color of glazed elements of façade (fire resisting class EI 120) due to temperature changes in summer. In that case, it is difficult to replace the damaged glass panel with the new one.
- The last problem is the excessive additional load on load-bearing structure due to self weight of water curtain in the Satellite area, which can lead to further severe problems.

## 3.1 Objectives of the fire safety measures

In accordance to the Italian regulation the objectives of the taken measures to guarantee the fire safety of the façade are listed below by taking into account the problems mentioned in the previous sections:

- Avoiding the collapse of Grand Space structure for at least 60 minutes.
- Blocking fire propagation between different levels and Satellites.
- Guaranteeing the safety conditions for the occupants and fire rescue teams (fire brigades) in case of fire inside the Grand Space.

## 3.2 Technical solutions

In order to fulfill the specified fire safety objectives on each area the following solutions are implemented:

- Fire-proofing is applied only on accessible portion of the Grand Space structure and the structure is checked by FSE with partial fire-proofing to ensure its fire resistance.

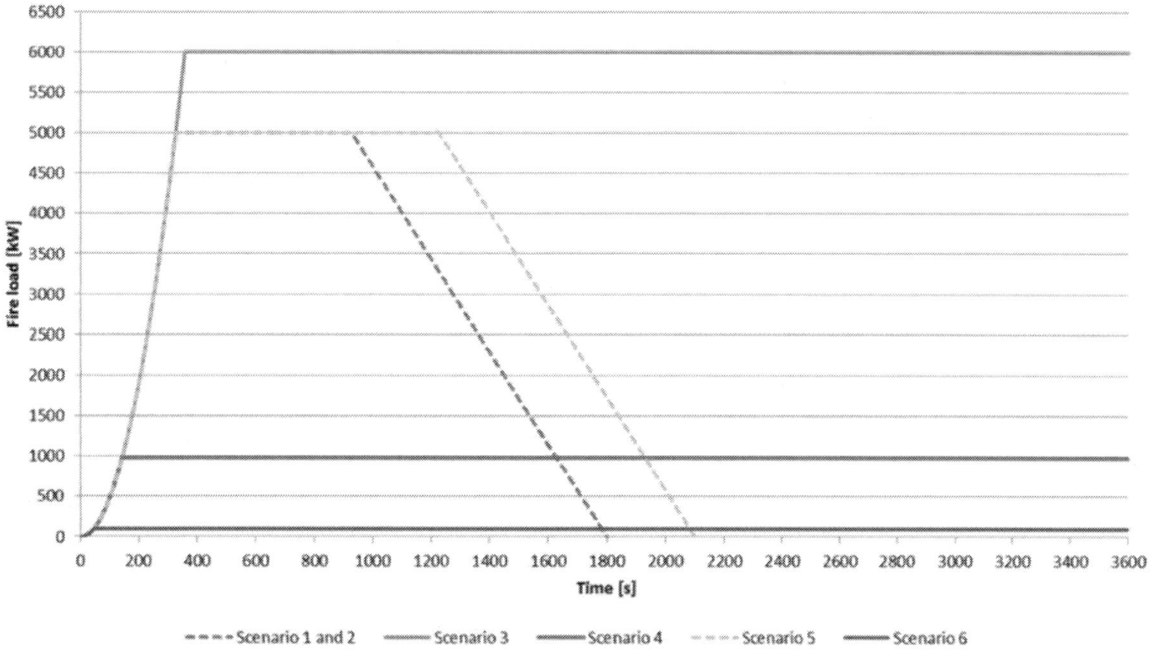


**Figure 2.** HRR define for the 6 selected scenarios.

- Performance class of the façades (among Offices – Satellites and Offices – Grand Space) is lowered from EI 120 to EI 60 for the areas where two sided fire is possible and where there is no excessive temperature rising during summer season.
- To eliminate the excessive additional load due to self weight of water curtain in the Satellite area water mist system is used instead of water curtain to suppress an outbreak of a fire in that area. Besides, Grand Space structure is also checked by FSE to ensure its fire resistance.
- Performance class of the façades (between Offices – Grand Space) is lowered from EI 120 to EW $60_{(i \rightarrow o)}$ and the Grand Space structure is checked by FSE to ensure its fire resistance.
- Fire safety conditions are checked by FSE inside the Grand Space taking into account natural smoke and heat exhaust systems.

## 4. CFD MODELS

The geometry for the CFD models is based on 2D AutoCad drawings of the building. The CAD models is then transferred and used in the CFD software where a computational mesh is constructed. Mesh refinements are applied in and around the fire areas to ensure correct flame temperatures. The simulations are run on a large computer cluster to speed up the solution process.

Four models are assembled to simulate six different fire locations. In the following the fire scenarios are listed:

1) Fire in second basement level inside the Grand Space (scenario 1)
2) Fire in ground floor level inside the Grand Space (scenario 2)
3) Fire inside Offices (scenario 3 e 4)
4) Fire inside a Satellite (scenario 5 e 6).

Figure 2 report the Heat of Heat Released (HRR) for the 6 selected scenarios.

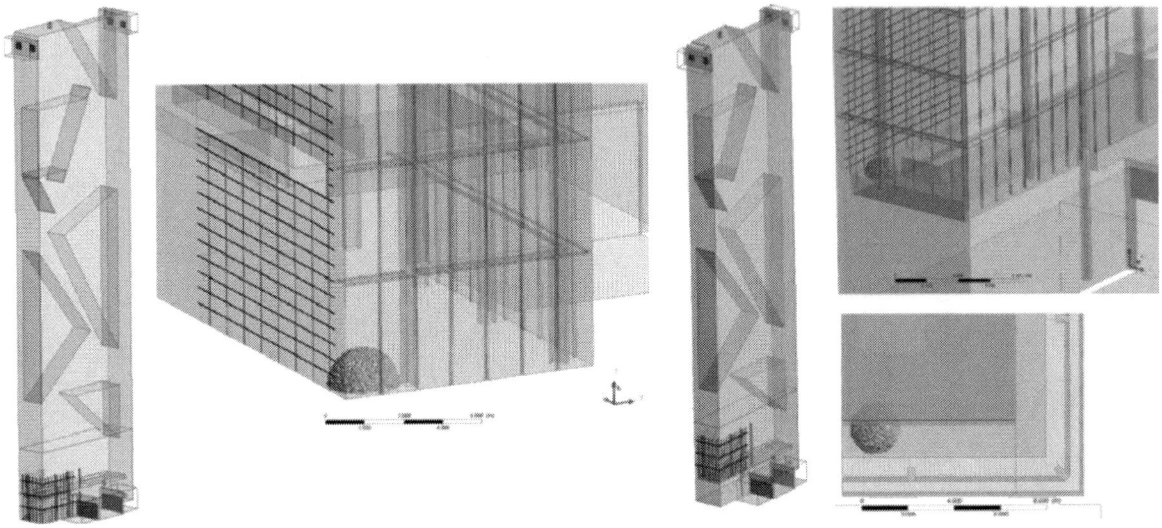

**Figure 3.** CFD models for the scenario 1 and scenario 2.

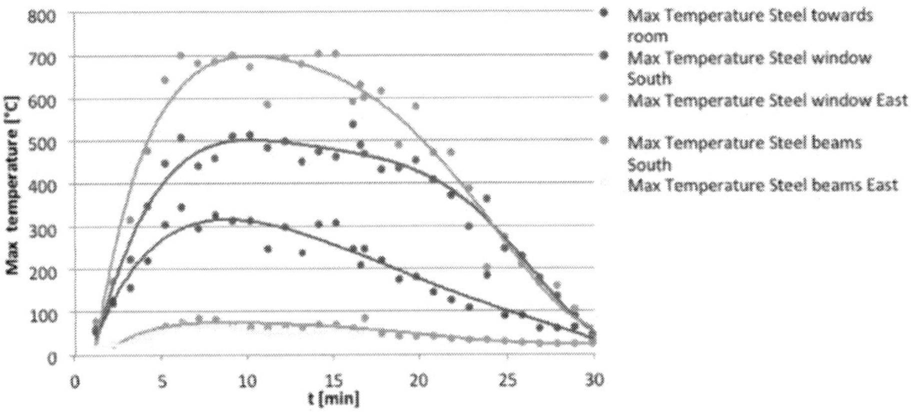


**Figure 4.** Maximum temperature on the steel at the eastern and southern part of the façade and the steel towards the adjacent room for the scenario 1.

The present CFD calculations have been performed by CFD program ANSYS CFX in collaboration with Ramboll group.

## 4.1 Scenario 1 and 2 – Fire inside the Grand Space

The fire is located at level -2 (scenario 1) and at level 0 (scenario 2) inside the atrium close to the glass façade and steel structure. The fire development coefficient is 0.047 kW/s$^2$ and the fire effect is 5 MW. The fire thereby reaches maximum HRR after 326 s. The HRR is kept constant for 10 minutes after it has reach maximum and then it starts to decay linearly. The openings for fresh air in the bottom and the openings for smoke ventilation in the top are open after 60 s. Figure 3 depicts CFD Models for the scenario 1 and scenario 2.

For the fire location at level -2, the flame will stick very close to the southern steel structure in the façade and results in high temperature levels. This impingement of the flames results in high temperature level that can be maintain over a period of 20–25 min for this fire location as it can be seen on Fig. 4. The main problem is that the flame is very close or touching the steel structure.

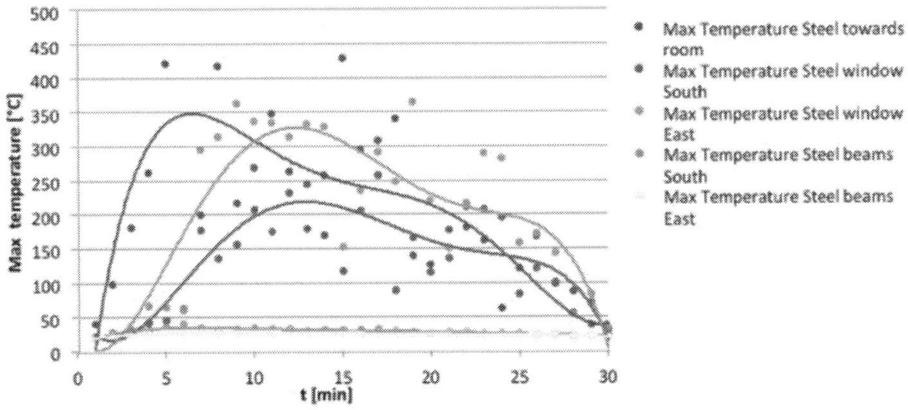


**Figure 5.** Maximum temperature on the steel at the eastern and southern part of the façade and the steel towards the adjacent room for the scenario 2.

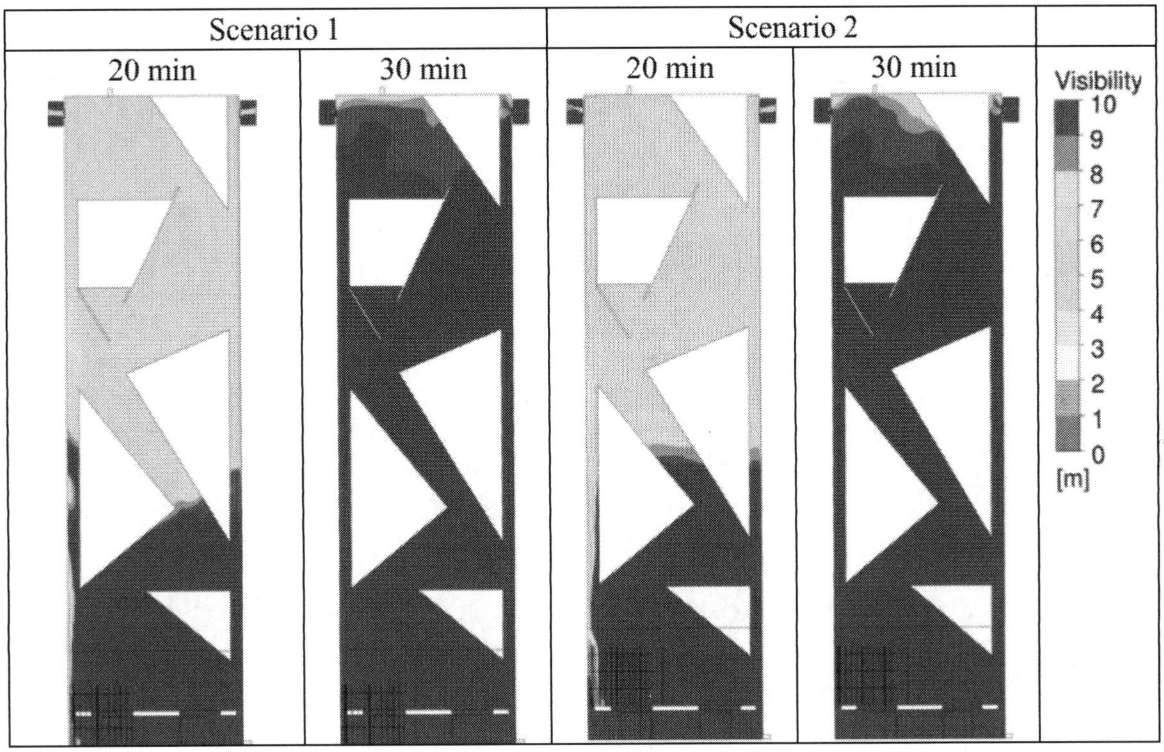


**Figure 6.** Visibility inside the Grand Space for the scenario 1 and scenario 2.

For the fire location at level 0, characterized by a 1 m of distance form Grand Space facades, the flame are not able to increase the temperature of the steel structure over the critical level.

The peak value for the maximum surface temperature of the steel towards south and the adjacent room is around 500 °C – 600 °C, as reported in Fig. 5.

The visibility inside the Grand Space is shown on figure 6 at different times at 20 min to 30 min for Scenario 1 and Scenario 2. From Fig. 6, it can be seen that the visibility will not be critical in the occupied zone at level -2 and 0. From the figure it can be seen that the atrium is filled with smoke since the visibility is reduced. This however occurs above the occupied zone on level -2 and 0 and is therefore not critical.

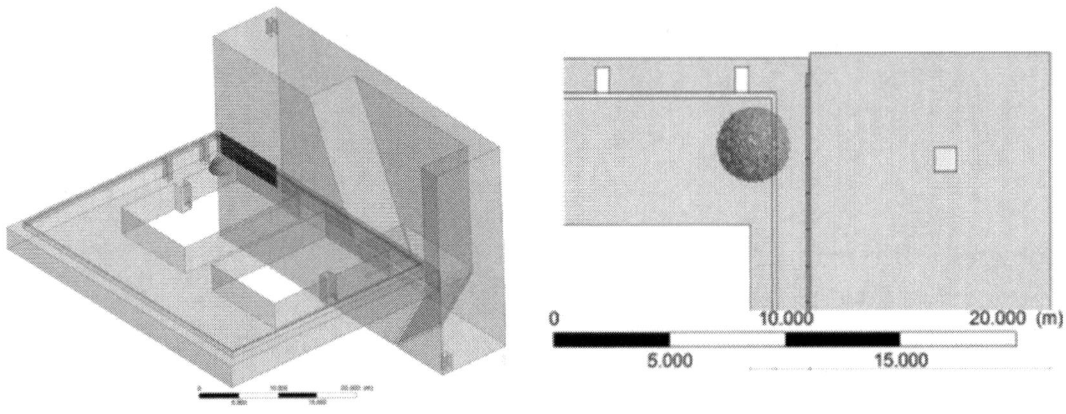


**Figure 7.** CFD model for the scenario 3 and 4.

As a result of the scenarios 1 and 2, it is necessary to protect steel structure of the Grand Space between level -2 and 0 to guarantee the R 60 fire resistance class. In addition, the presence of natural smoke and heat exhaust system makes certain the adequate level of safety inside the Grand Space in case of fire for evacuation of occupants and intervention of Fire Brigades.

## 4.2 Scenario 3 and 4 – Fire inside office area

The office is an open space office with a panorama view out to the atrium (Grand Space). The glass between the office area and the atrium is a fire resistance glass EW $60_{(i\rightarrow o)}$. This means that the glass maximum allows an incident radiation less than15 $kW/m^2$. In the model only the glass and steel in close proximity of the fire are modelled. The distance from the centrum of the fire to the atrium surface is around 2 meters. Figure 7 reports the CFD models for scenario 3 and scenario 4.

Including the adjacent atrium into the model makes possible to see the effects of the temperature and radiation from the fire on the steel structure of the atrium.

In the scenario 3 the sprinklers in the office area fail to operate and the fire develops to up to an HRR value of 6 MW. It reaches its maximum effect after 326 s and remains constant for 1 hour. The mechanical smoke ventilation starts after 40 s and reaches its full capacity after 100 s. The incident radiation is shown in Fig. 8. The effect of the fire resistance glass (EW $60_{(i\rightarrow o)}$ ) is clear and reduces the radiation incident to the steel structure at the exterior façade. After 20 minutes the fire continues at the same fire load and so does the thermal radiation that not change much after this period. This can be seen on Fig. 8 where the incident radiation on the northern and eastern façade is given as a function of time.

The temperature at the external façade will increase with time after 20 min due to the accumulation of the thermal radiation in the atrium. However, after some time the air convection mechanism due to the airflow helps limiting the temperature levels of the external façade. The temperature development on the façade is given on the right side of Fig. 8.

In the scenario 4, made for checking the tenability condition of the occupants inside the office, the sprinklers are activated after 148 s decreasing the HRR from 6 MW to 0.98 MW. The fire development coefficient is 0.047 $kW/s^2$. The smoke ventilation is activated after 40 s and reaches its full capacity after 100 s. The visibility is reduced in 5 minutes and the temperature would be critical in only 8 minutes.

The results of the scenarios 3 and 4 show that it is not necessary to protect the steel elements of the Grand Space since the façade EW $60_{(i\rightarrow o)}$ is able to reduce their temperature. Therefore, the presence of EW $60_{(i\rightarrow o)}$ façade hinders vertical propagation of fire from "office area".

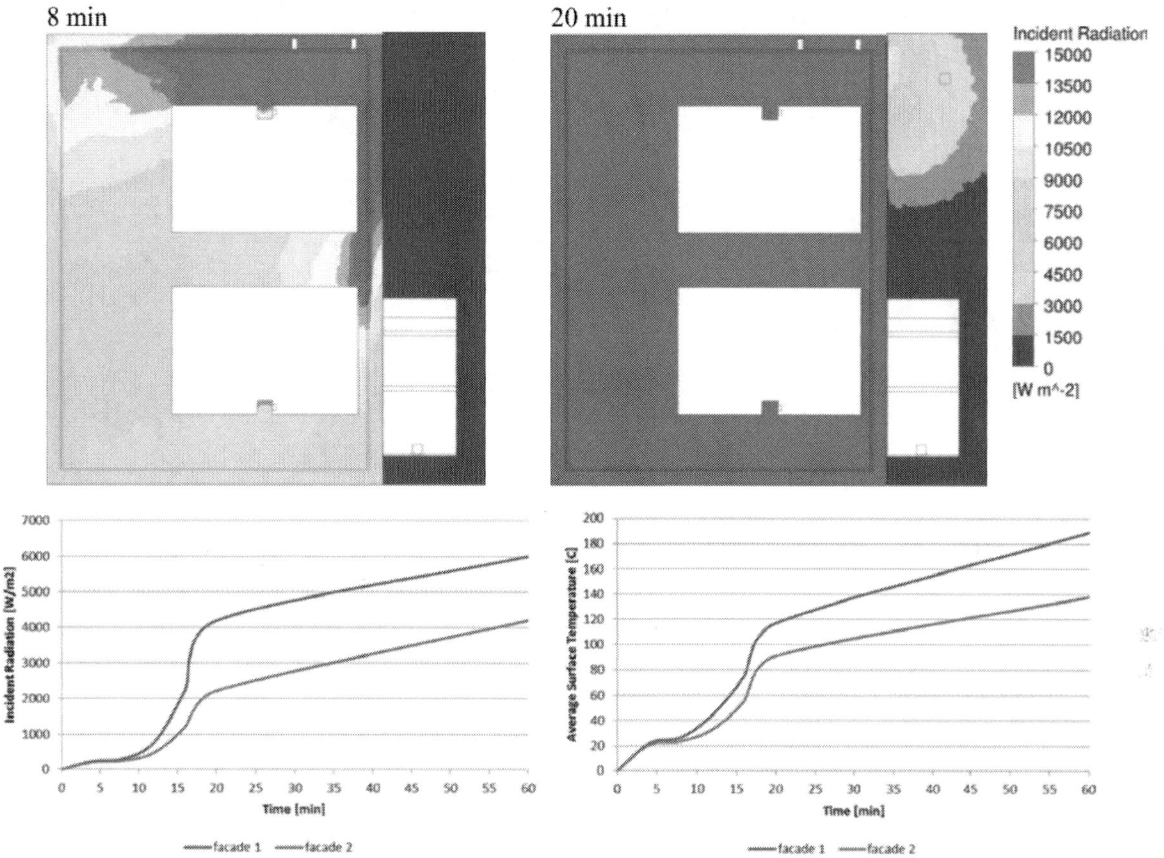


**Figure 8.** Incident radiation and temperature at façade 1 and 2 – scenario 3.

## 4.3 Scenario 5 and 6 – Fire inside Satellite

In the Satellites there are doors which connect the meeting room to the office area. In the meeting room the glass façade are mounted on a steel frame through the atrium and the office area. The glass of the meeting room is normal glass without any kind of fire resistance properties.. The problem is sketched on Fig. 9.

The main focus for a fire in the meeting room with failure of the water mist system is to see the temperature effects on the steel structure at the external glass façade. Therefore the model only consists of the air volume outside the meeting room including the steel structure. The heat from the fire is then applied on the surface of the glass in the meeting room through the atrium space.

In the scenario 5 the water mist system fails in the meeting room and the HRR load reaches 5 MW and remains constant for 15 minutes and then starts to decay.

The average surface temperature on the steel has a peak value around at 80 °C for a period of around 15 minutes. The peak value for the maximum surface temperature on the steel structure is 140 °C for around 15 minutes. For the steel structure at the southern façade the peak value for average surface temperature is 110 °C for 15 minutes and the peak value for maximum surface temperature is 200 °C for 15 minutes period. No temperatures above 300 °C are registered for any of the steel structure surfaces under consideration. Figure 10 shows the maximum surface temperature on steel at the façade for the scenario 5.

In the scenario 6, made for checking the tenability condition of the occupants inside the satellite, it is assumed there are no fire ventilation in the meeting room and the water mist activates and thereby reducing the HRR from 5 MW to 0.1 MW. The fire growth coefficient is 0.047 kW/s$^2$.

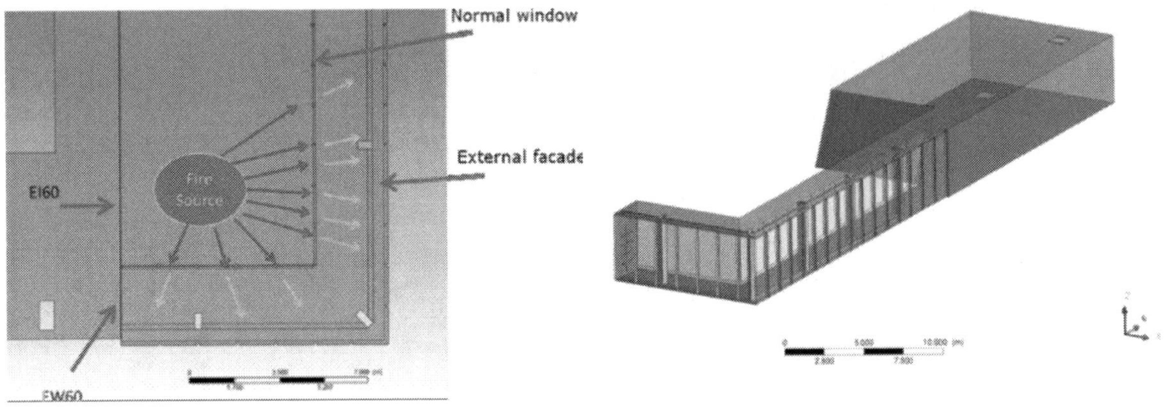


**Figure 9.** CFD model for the scenario 5 and 6.

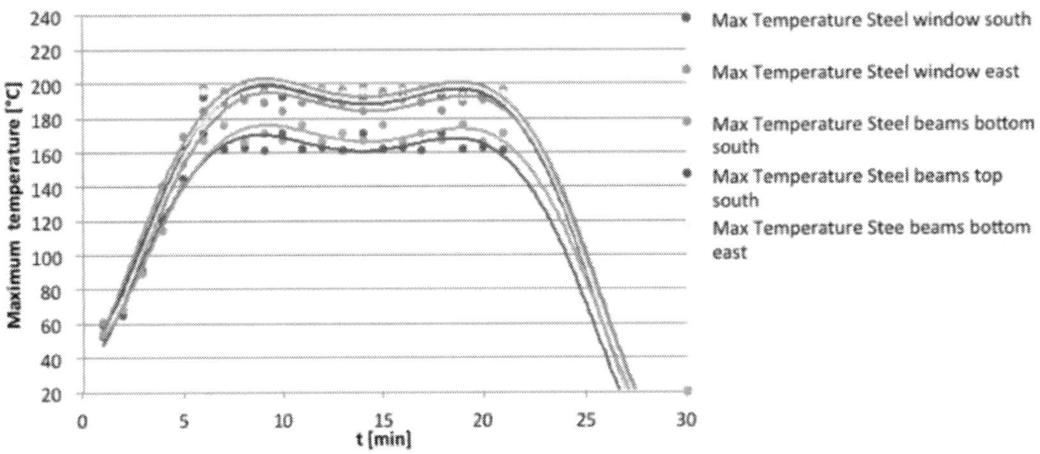


**Figure 10.** Maximum surface temperature on steel at the façade for the scenario 5.

The visibility is critical in the meeting room after 2 minutes and the temperature would be critical in most of the meeting room after 4 minutes.

The results of the scenario 5 and 6 show that temperatures of steel members of the Grand Space does not exceed the critical limits in case of a fire inside a Satellite. Therefore, it is not necessary to protect the steel elements as it guarantees fire resistance class R 60 without any fire protection. The presence of water mist system enhances the actual fire safety conditions to prevent fire propagation at higher levels or inside the Grand Space.

## 5. SPECIFIC LABORATORY TESTS

In order to guarantee the performance class EW $60_{(i \rightarrow o)}$ in the glazed façades between Offices and Grand Space, the fire resistant glazed panels (without intumescent gel coat) with a special composition to hinder possible color variations in summer conditions is taken into consideration.

A specific test in accordance to EN 1364-3 is performed to check fire performance of panels with respect to the CFD models. In Fig. 11 a typical node of façade, the lay-out of the test and the radiation curve obtained during the fire test are shown.

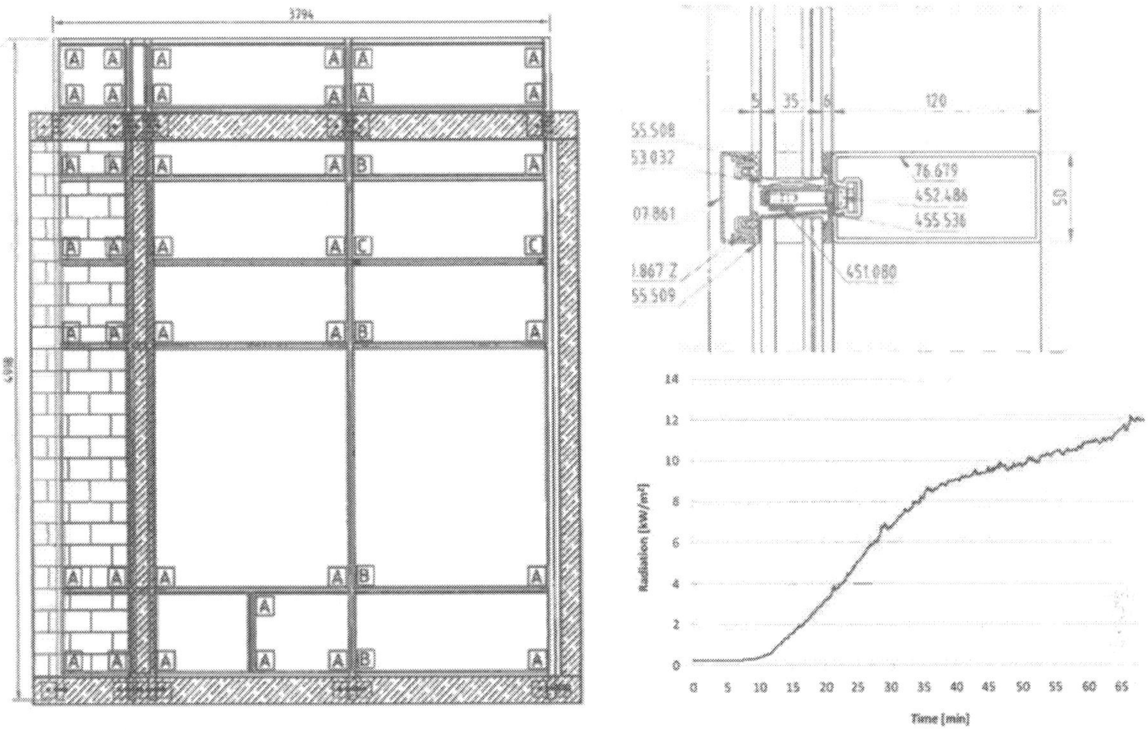


**Figure 11.** Lay-out of the test (left), typical node of façade (right-top), radiation curve (right-down).

## 6. CONCLUSION

Fire safety of high-rise building façades is a complex problem, therefore the use of prescriptive fire codes could not be sufficient to ensure a proper building fire safety level. This article emphasizes that FSE and CFD fire modeling are used in a real case as useful tools to solve fire safety problems of façades addressing the fire safety level of the overall building.

The case analyzed in the "Torre Regione Piemonte" gives some opportunities to optimize construction technology and fire safety measures in order to fulfill the required safety objectives in accordance to the Italian national fire codes.

The results of the fire scenarios show that fire protection of the steel elements can be made most efficient. It has be seen that there is need of fire protection on Grand Space façade only from level -2 to 0. Furthermore, the presence of EW $60_{(i \to o)}$ façade and water mist system provide the required fire safety conditions in case of an outbreak of a fire for higher building levels.

In addiction, CFD models are performed to optimize the smoke and heat exhaust systems inside the Grand Space and hence, the fire resistance class of the separating façade between Offices and Grand Space can be reduced from EI 120 to EW $60_{(i \to o)}$.

This case study confirms that performance based approach makes possible to analyze façades complexly, to guarantee the fulfillment of the required fire safety level and to optimize cost and time of construction, reducing the overall maintenance cost and maintenance time of the building.

### References

[1] Fuksas, `http://www.fuksas.it/home.htm#/progetti/0401/`, last checked 14 January 2014.
[2] User Manuel Ansys CFX14.0, Ansys CFX-International, 2012.

[3] Mathematical Model and Solution Algorithms Manual, Ansys 14.0, Ansys CFX-International, 2010.
[4] Drysdale, D., Fire Dynamics, John Wiley & Sons, Chichester, 1985.
[5] Mulholland G.W.: Smoke production and properties in DiNenno P.J., et al, Editors, SPPE Handbook of Fire Protection Engineering, 2. Edition, Chapter 15, Section 2. pp. 217–227, 2002.
[6] Leif Staffanson, Selecting design fires, Department of Fire Safety Engineering and Sys-tems Safety, Lund University, 2010.
[7] Boverket, Boverkets allmänna råd 2011 – Vägledning i analytisk dimensionering av byggnaders brandskydd, 2011.
[8] Calculating smoke potential for use in Argos, Danish Institute of Fire and Security Tech-nology, Bjarne Husted, 2004.

MATEC Web of Conferences **46**, 04003 (2016)
DOI: 10.1051/matecconf/20164604003


# Vertical safe separation distance between openings in multi-storey buildings having a fire resistant spandrel

Michael A Delichatsios, Joe Ryan, Ning Tian and Jianping Zhang

Fire Safety Engineering Research and Technology Centre, University of Ulster, UK

**Abstract.** A novel comprehensive methodology is presented for the determination of the vertical separation distance between floor glazing openings in buildings having a fire resistant non-flammable spandrel. The spandrel height is specified by requiring that the heat flux is less than a critical heat flux for glazing failure at the base of the opening of the floor above the floor of fire origin. For a specified room, opening geometries and fuel type, the heat flux at the base of this opening is calculated by using correlations for the flame height and heat flux distribution of the façade flames for under ventilated fire conditions in the compartment of fire origin. As the compartment opening increases a transition occurs from under ventilated to over ventilated conditions reducing the façade fire spread hazard. The methodology and results can also be applied for wooden and lightweight façade walls to predict fire spread behavior on the wall and towards the openings to floors above the floor of fire origin.

## 1. INTRODUCTION

The main mechanisms of façade flame spread after an initiating fire at the floor of fire origin are summarized as:

- Fire spread to the interior of level above via openings such as windows causing secondary interior fires on levels above resulting in level to level fire spread.
- Flame spread over the external surface of the wall.
- Flame spread within an interval vertical cavity /air gap.
- Heat flux impacts cause degradation/separation of non-combustible external skin resulting on flame spread on internal core.
- Secondary external fires to lower (ground) levels arising from falling burning debris.

In this work we examine flame spread from floor to floor through glazing openings when the spandrel is fire resistant. Radiant heat to adjacent buildings, combustibility of external wall claddings and fire-stopping internally between floors are dealt with in great detail in fire safety codes in the British Isles as well as internationally. The possibility of fire spread externally from floor to floor is generally not considered in detail

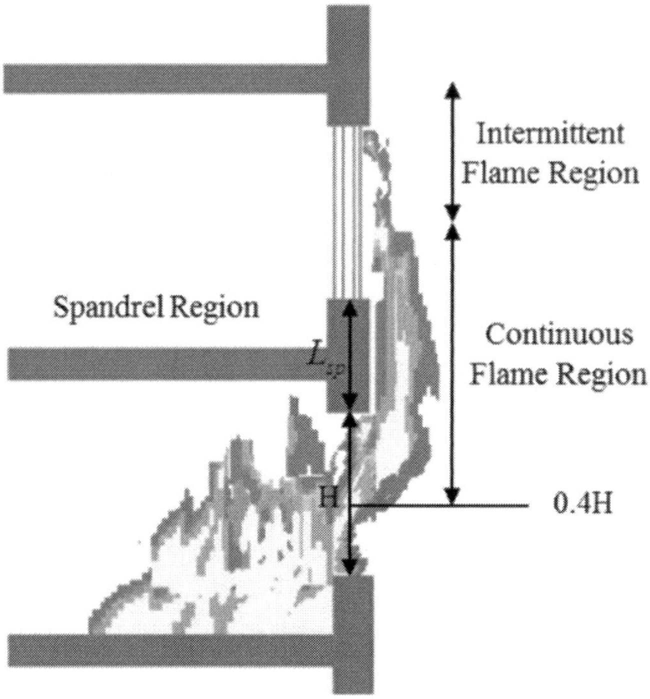


**Figure 1.** Floor to Floor External Fire Spread.

In under-ventilated fires the unburned fuel will flow out and burn at the exterior of the building after the window in the enclosure breaks. Because the unburned fuel has been sufficiently heated and ignited, it is possible to have external flame heights with associated heat flux levels applied on the floor above the floor of fire origin, which could cause fire spread between floors through windows. This is also possible for well-ventilated fires, but as the fuel has enough air to burn inside the compartment exterior flames would be much smaller. We demonstrate in this work that considerable flame height and heat flux values can be reached when specific opening sizes are associated with specific room sizes and fuel loads, which could lead to fire breakthrough on the floor above the floor of fire origin through windows even for the case that fire resistant standard block or concrete construction is used for the outside wall.

Currently, there are no test methods for assessing fire resistance of wall systems in UK in any of the jurisdictions. There is also no specification or guideline in the UK for required external vertical separation distance between floors with regard to the threat of external fire spread. The general view taken is that once fire-stopping is brought to the external weather line of the building, then external fire spread is not an issue. American codes require protection between floor openings where buildings are more than three stories high not sprinklered Specifically, e.g. NFPA 5000 37.1.4 requires a fire resistant spandrel when openings in the floor above are within 1.5 m of the floor below. In addition, the Taiwanese building regulations require a spandrel separation distance of greater than 0.90 m between floor openings. There is also an ASTM test standard E2307 for determining the fire resistance of perimeter fire barrier systems i.e. the joint between an exterior wall assembly and a floor system. The test method determines the ability of wall systems to prevent fire spread as they deflect and deform under intensive heat.

In this work we consider that the area between floor openings is fire resistant, as it would be in standard block or concrete construction by emphasizing that fire spread from a floor to the floor above will occur due to glazing failure and not because of failure of a spandrel region of a curtain wall system.

Figure 1 illustrates how fire spread from floor to floor occurs. The extent of the spandrel is defined as the length from the top of the opening in the room of fire origin to the bottom of the opening in the next floor as Fig. 1 indicates. The worst condition for fire spread from floor to floor develops when fires in the floor of fire origin are under-ventilated. In a well-ventilated fire scenario all, or nearly all, of the fuel is burned inside the compartment of fire origin so that any heat fluxes from short exterior elongated flames are not as high as in under-ventilated fires.

In under-ventilated fires unburned fuel escapes from the compartment ignited and burning outside the compartment generating a façade flame as in Fig. 1. If there is sufficient unburned fuel outside the compartment this will burn with ambient air and can cause fire breakthrough from the floor of fire origin to the floor level above. The focus of this work is to calculate the amount of unburned fuel that could be created in a given situation and to establish the parameters for which this could cause fire breakthrough and spread through the upper floor openings. We are primarily interested in the vertical separation distance required between floors in a given situation. There are a number of factors that will affect this, including: Room size, Room Type, Window Opening Area, Window Opening Dimensions, Room/Store Height, and Window Opening Location.

We focus in this work on window opening size, compartment dimensions and fuel load. We first describe the methodology to find out the specific range of the opening size in relation to the room size to generate façade flames that can cause fire breakthrough. Then we show that breakthrough occurs when the opening dimensions have certain relations to the fuel load available (for given room size and type) in order to produce external flame heights and heat fluxes.

## 2. THEORY AND METHODOLOGY (UNDER-VENTILATED FIRES)

The theory and methodology is based on the extensive work done in FireSERT and SKLFS China for enclosure and façade fires [1–6]. In reference [6] a review of previous works on façade fires are included.

The results from [1–5] used in this paper are next summarized for under-ventilated fires:

1. The air inflow for the opening:

$$\dot{m}_a = 0.5AH^{0.5}\,(\text{kg/s}) \tag{1}$$

   Here $A$ = area of opening ($m^2$), $H$ = height of the opening (m) and the Coefficient 0.5 has a dimensional value kg/s.$m^{5/2}$.

2. The total mass pyrolysis rate inside the enclosure:

$$\dot{m}_T = 0.1AH^{0.5} = 0.2\dot{m}_a\,(\text{kg/s}) \tag{2}$$

3. The fuel burned inside the enclosure is at stoichiometric proportions to the incoming air, $S$ is air to fuel mass stoichiometric ratio:

$$\dot{m}_{ins} = \frac{\dot{m}_a}{S}\,(\text{kg/s}) \tag{3}$$

4. It follows that the excess mass pyrolysate rate burning outside on the façade is :

$$\dot{m}_{ext} = \dot{m}_T - \frac{\dot{m}_a}{S} = \dot{m}_a\left(0.2 - \frac{1}{S}\right) \tag{4}$$

   and the heat released externally on the façade is

$$\dot{Q}_{ext} = \dot{m}_{ext}\Delta H_C = \dot{m}_a\left(0.2 - \frac{1}{S}\right)\Delta H_C = \left(\dot{m}_T - \frac{\dot{m}_a}{S}\right)\Delta H_C \tag{5}$$

   where $\Delta H_C$ is the het of combustion of the fuel.

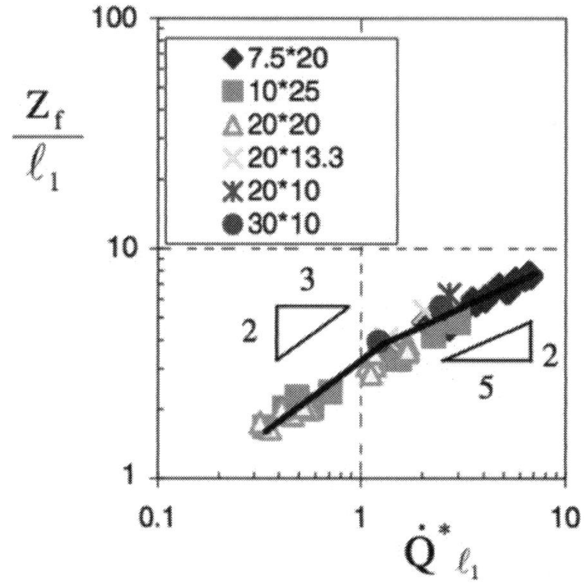


**Figure 2.** Flame height correlation with normalized external heat release rate.

5. Figure 2 shows the relation for the façade flame heights to the normalized external heat release rate.

The flame height generated by the external heat release rate is:

$$Z_f = 3.2\dot{Q}_{\ell 1}^{*2/3} \text{ for } \dot{Q}^* < 1.3 \tag{6a}$$

$$Z_f = 3.2\dot{Q}_{\ell_1}^{*2/5} \text{ for } \dot{Q}^* > 1.3 \tag{6b}$$

where the length scale $\ell_1 = (AH^{0.5})^{2/5}$ and the normalized heat release rate is defined as :

$$\dot{Q}^*_{\ell_1} = \frac{\dot{Q}_{ext}}{\rho_\infty c_p T_\infty \sqrt{g}\,\ell_1^{5/2}} = \frac{\Delta H_C}{c_p T_\infty}\left(\frac{0.5}{\sqrt{g}\rho_\infty}\right)\left(0.2 - \frac{1}{S}\right) \tag{6c}$$

Equation (6c) has been derived by using Eqs. ((1)–(5)).
where
$c_p$ is specific heat capacity of air equal to 1kJ/kg K.
$T_\infty$ is ambient air temperature (K) equal to 293 K.
$g$ is acceleration due to gravity equal to 9.81 m/s$^2$.
$\rho_\infty$ is ambient air density equal to 1.2 kg/m$^3$.
$S$ is stoichiometric air-fuel mass ratio.

We note that the normalized external heat release rate in Eq. (6c) is independent of the opening or room geometry. Moreover, the origin of the flame height is located at 0.4H, which corresponds to the neutral plane for the flow at the opening. For enclosure under-ventilated fires the normalized heat release rate is less than one so that Eq. (6a) applies.

6. The normalized heat flux distribution along the façade wall for the external flames is shown in Fig. 3.

For subsequent use, we make the following observations regarding the heat flux to the façade wall:

a. The continuous flame height is at $0.8Z_f$ of the mean flame height ($Z_f$).
b. The heat flux over the continuous flame height (regime I) takes a maximum constant value given by the following equation after using Eq. (6a):

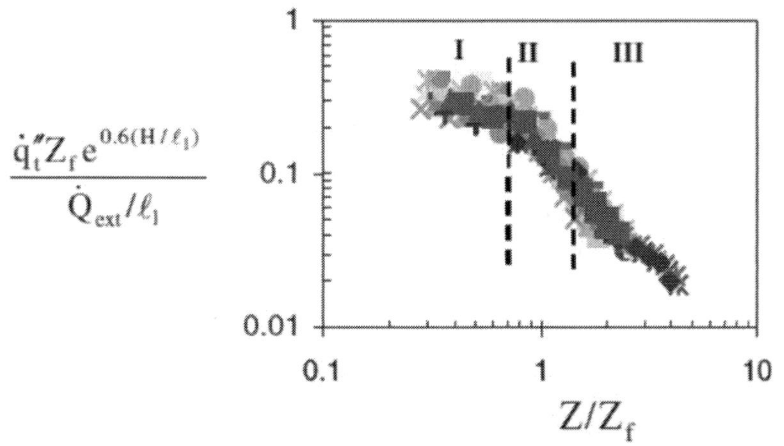


**Figure 3.** Normalized heat flux distribution along the façade against vertical location normalized by the flame height. The origin of the flame height is at 0.4H and there regimes of the heat flux distribution are shown, continuous (I), intermittent (II) and thermal (III).

$$q''_{t,max} = 0.35\frac{\dot{Q}_{ext}}{Z_f e^{0.6(\frac{H}{l_1})}\ell_1} = 0.35\frac{\dot{Q}^{*\ 1/3}_{\ell_1}\rho_\infty c_p T_\infty \sqrt{g}\,\ell_1^{1/2}}{3.2e^{0.6(\frac{H}{l_1})}} = 0.1\frac{\dot{Q}^{*\ 1/3}_{\ell_1}\rho_\infty c_p T_\infty \sqrt{g}\,\ell_1^{1/2}}{e^{0.6(\frac{H}{\ell_1})}} \quad (7)$$

c. The continuous flame regime is followed by the intermittent regime II and then the buoyant thermal flow regime III above the flame extent. The heat flux distribution over the first two regimes is given as

$$q''_t = q''_{t,max} \text{ for } Z < 0.8Z_f \quad (8a)$$

$$\frac{\dot{q}''_t}{\dot{q}''_{t,max}} = 0.64\left(\frac{Z}{Z_f}\right)^{-2} \text{for} Z > 0.8Z_f. \quad (8b)$$

## 3. DETERMINATION OF A SAFE SPANDREL HEIGHT

Flame spread from the façade flames to the floor above the floor of fire origin through its window will not occur if the heat flux at the bottom of this window is less than a critical heat flux for glass breakage, $\dot{q}''_{cr}$,which depends on the glazing system. The location of the bottom of the window from the origin of the flames is:

$$Z_{sp} = 0.6H + L_{sp} \quad (9a)$$

where $L_{sp}$ is the length of the spandrel section (see Fig. 1).

If the critical heat flux is less or equal to the maximum flame heat flux (Eq. (7)) the required spandrel length for no flame spread is given by:

$$\frac{Z_{sp}(= 0.6H + L_{sp})}{Z_f} = 0.8\left(\frac{\dot{q}''_{t,max}}{\dot{q}''_{cr}}\right)^{1/2} \text{ if } \dot{q}''_{cr} \le \dot{q}''_{t,max}. \quad (9b)$$

If the critical heat flux is greater than the maximum heat flux

$$\dot{q}''_{cr} > \dot{q}''_{t,\max} \ \underline{\text{there is no restriction on the spandrel height}} \tag{9c}$$

so that the minimum height can be chosen based only on construction requirements.

We note that the maximum heat flux for the façade flames in Eq. (7) depends primarily on the geometry of the opening namely the length scale $l_1 = (AH^{0.5})^{2/5}$ and the ratio $H/l_1$. It follows that the condition $\dot{q}''_{cr} = \dot{q}''_{t,\max}$ can be satisfied for certain geometric characteristics of the opening.

## 4. CROSSOVER FROM UNDERVENTILATED TO OVERVENTILATED CONDITIONS AS OPENING SIZE INCREASES

Simple inspection for the flame height Eq. (6) and heat flux relations Eqs. (7) and (8) shows that the conditions for spandrel separation heights become more restrictive as the opening factor ($AH^{0.5}$) or equivalently the length scale ($l_1 = (AH^{0.5})^{2/5}$) increase. As the opening dimensions increase more air flows into the compartment leading to the change of conditions from under-ventilated to over-ventilated. This cross over occurs when the total mass pyrolysis rate becomes less than the stoichiometric air requirements [1–5]:

$$\dot{m}_T < \frac{\dot{m}_a}{S} = \frac{0.5AH^{1/2}}{S}. \tag{10a}$$

The maximum mass pyrolysis for over ventilated conditions in an enclosure is equal to mass pyrolysis rate per unit area multiplied by the effective fuel area $\dot{m}_T = \dot{m}''_F \cdot A_F$ so that Eq. (10a) provides the minimum opening factor above which over ventilated conditions prevail:

$$AH^{1/2} = \ell_1^{5/2} \geq 2S \cdot \dot{m}''_F \cdot A_F \tag{10b}$$

$$\ell_1 \geq (2S \cdot \dot{m}''_F \cdot A_F)^{2/5}. \tag{10c}$$

As the conditions are over-ventilated when Eq. (10) is satisfied all combustion occurs nearly inside the enclosure and the conditions are much less critical than when conditions were just under-ventilated.

## 5. ILLUSTRATION OF THE METHODOLOGY USING EQUATIONS (1)–(10)

We apply the methodology for determining spandrel height separation for a typical office and a typical shop as shown in Figs. 4 and 5. In both cases the floor area is 4 m by 5 m. The heat of combustion and stoichiometric ratio for the office is $\Delta H_C = 19000$ kJ/kg and $S = 6.3$ whereas the corresponding values for the office are $\Delta H_C = 22500$ kJ/Kg and $S = 7.4$. These values represent an average fuel being primarily paper for office and primarily polymer such as PMMA for the shop. In Fig. 4 we keep the opening width fixed for the office at 1 m and vary the opening height. In Fig. 5 we keep the opening height fixed and vary the opening width. Figures 4 and 5 plot the variation of the maximum heat flux (Eq. (7)) and the continuous flame height ($0.8Z_f$ from Eq. (6a)) as the opening height and width changes respectively.

In addition, Figs. 4 and 5 include two critical heat fluxes as horizontal lines corresponding to failure of the glazing. In our analysis, we will use the values, 70 kW/m$^2$ and 20 kW/m$^2$ [13], as our cut-off points for fire breakthrough for tempered and non-tempered glass. While this is an obvious simplification, and

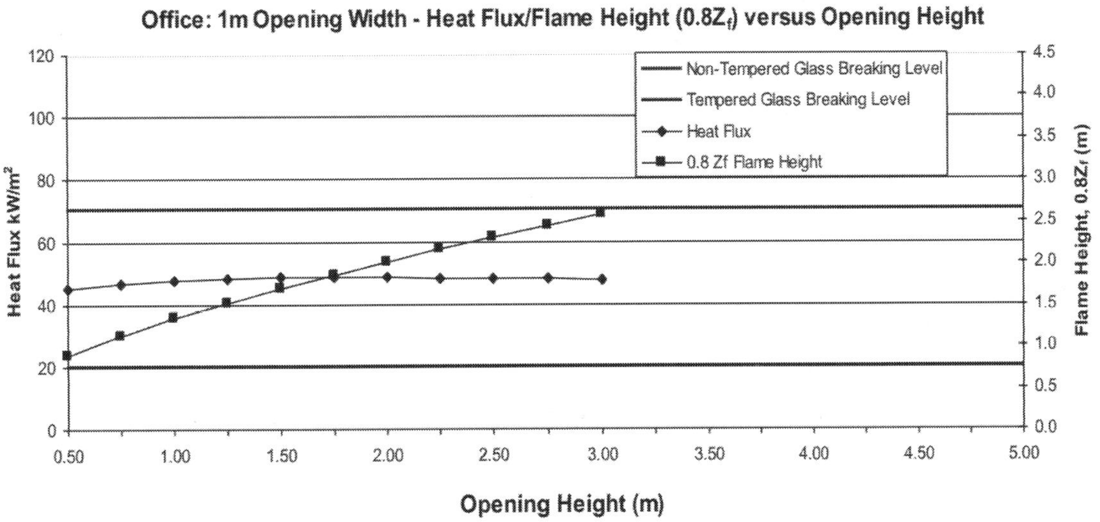


**Figure 4.** Office: maximum heat flux and continuous flame height calculations.

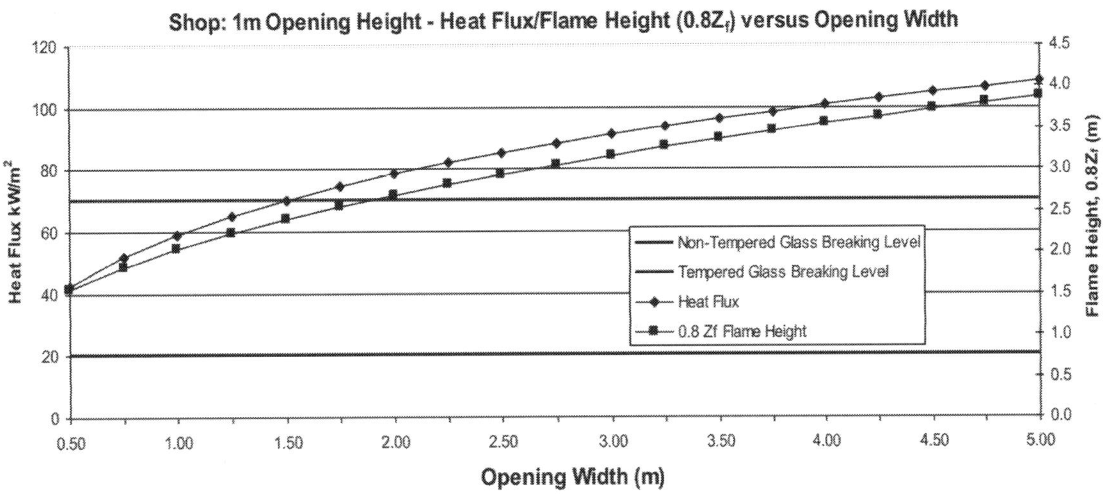


**Figure 5.** Shop: maximum heat Flux and continuous flame height calculations.

various different studies have given a range of different values for all types of glazing, these values still give us a good general indication for analysis purposes.

Finally, we note in Figs. 4 and 5 that calculations stop at certain height for office (Fig. 4 for office) and a certain width for shop (Fig. 5 for shop) at which point conditions change from under-ventilated to over-ventilated as the opening factor takes an appropriate value. The crossover condition provides the length scale $\ell_1$ as given by Eq. (10) where we have used the value $\dot{m}''_F = 0.02\,\text{kg/m}^2$ [7] for the mass pyrolysis rate at over-ventilated conditions. We also have taken the whole floor area $(4 \times 5 = 20\,\text{m}^2)$ as the effective fuel area AF in Eq. (10). Using the so determined value of the length scale $\ell_1$ the maximum flame height and heat flux are determined from Eqs. (6a) and (7). Beyond these extreme values (continuous flame height 2.62 m for office and 4.05 m for shop, see Figs. 4 and 5) the fire in the enclosure becomes over ventilated and the extent of the flame height is minimal primarily determined by the ceiling area and length of the enclosure. Figs. 6 and 7 illustrate the crossover from under-ventilated to over-ventilated fires for the case of the office. The spandrel height can be determined for given opening dimensions in Figs. 4 and 5 using Eqs. (9a), (9b) and (9c).

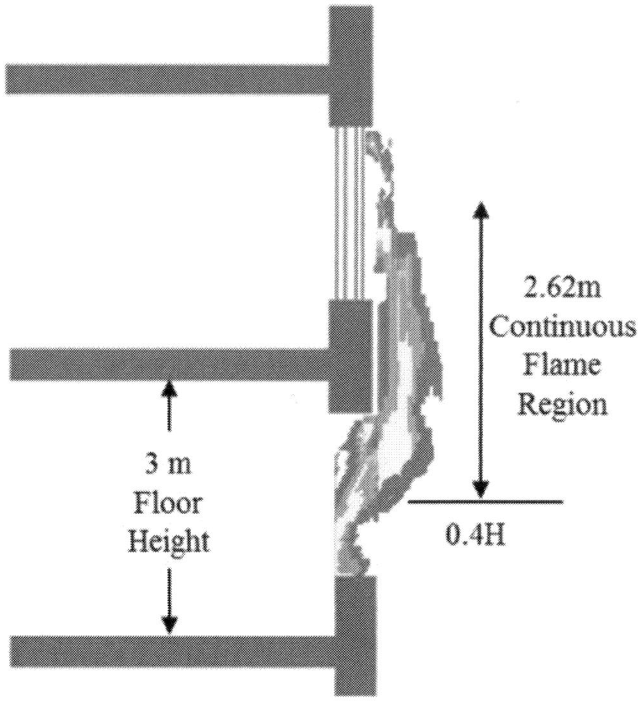


**Figure 6.** Office: 2.62 m Flame Height – Under-Ventilated Crossover Fire Measured from 0.4H Level.

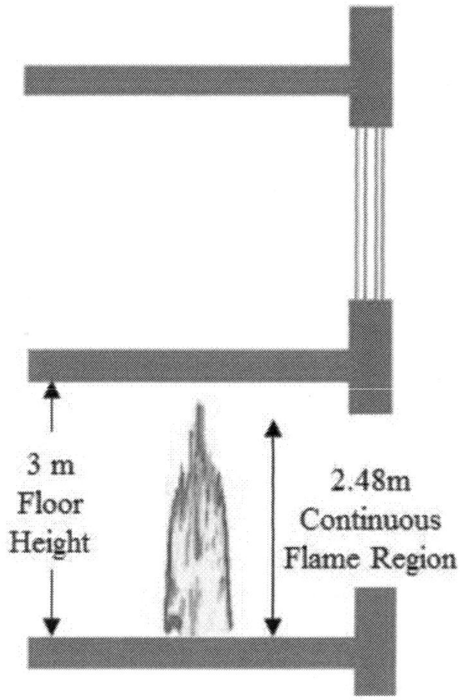


**Figure 7.** Office: Flame Height 2.48 m – Just Well-Ventilated Fire Calculation from Floor Level.

# 6. CONCLUSIONS

The major results of this work for specifying spandrel height to prevent façade fire spread from floor to floor through glazing (see Fig. 1) are applied in Figs. 4–7 and summarized as follows:

1. For given room and opening sizes and fuel type, the flame height and heat flux distributions are shown in Figs. 2 and 3 and calculated using Eqs. (6) and (8) for under ventilated conditions.
2. Glazing failure occurs when the façade flame heat flux is greater than a critical heat flux for glazing failure at the base of the opening at the floor above the floor of fire origin. The critical heat flux is taken to be 70 kW/m$^2$ for tempered glass and 20 kW/m$^2$ for normal glass.
3. The spandrel height to prevent fire spared is then determined using Eqs. (9a), (9b) and (9c).
4. As the opening size at the room of fire origin increases, the conditions cross over from under ventilated to over ventilated conditions (see Eq. (10)) thus reducing the façade fire spread hazard.
5. The results for flame height, heat flux distribution and cross over can also be applied to assess the façade fire spread hazard owing to other failure mechanisms listed at the beginning of the introduction including lightweight wall façades and wooden façades.

## References


[1] Delichatsios, M.A., "Enclosure and Façade Fires: Physics and Applications", *Fire Safety Science* **11**, 3–27, 2014.

[2] Lee, YP., Delichatsios, M.A., and Ohmiya, Y., "The Physics of the Outflow from the Opening of an Enclosure Fire and Re-examination of Yokoi's Correlation", *Fire Safety Journal*, **49**, 82–88, 2013.

[3] Lee, YP., Delichatsios, M.A., Ohmiya, Y., Wakatsuki, K., Yanagisawa, A., and Goto, D., "Heat Fluxes on Opposite Building Wall by Flames Emerging from an Enclosure", *Proceedings of the Combustion Institute*, **32** (2), 2551–2558, 2009.

[4] Lee, YP., Delichatsios, M.A., and Silcock , G.W.H., "Heat Fluxes and Flame Heights in Façades from Fires in Enclosures of Varying Geometry", *Proceedings of the Combustion Institute*, **31** (2), 2521–2528, 2007.

[5] Lu, K., Hua, L., Delichatsios, M.A., Tang, F., Qiu, Z., and He, L., "Merging Behavior of Façade Flames Ejected from Two Windows of an Under-ventilated Compartment Fire", *Proceedings of the Combustion Institute*, **35** (3), 2615–2622, 2015.

[6] White, N., and Delichatsios, M.A, "Fire Hazards of Exterior Wall Assemblies Containing Combustible Components", Fire Protection Research Foundation, Document Number: EP142293, Quincy, Massachusetts, U.S.A, 2014.

[7] Quintiere, J.G., *Principles of Fire Behaviour*, Delmar Publishers, 1998.

[8] Drysdale, D., *An Introduction to Fire Dynamics*, John Wiley & Sons, 1985.

[9] Steckler, K.D., Quintiere, J.G., and Rinkinen, W.J., "Flow Induced by a Fire in a Compartment", 19th Symposium on Combustion, 913–920, Haifa, Israel, 8–13 August, 1982.

[10] Kawagoe, K., "Estimation of Fire Temperature-Time Curve in Rooms", Building Research Institute, Third Report BRI Research Paper No.29, Tokyo, Japan, 1967.

[11] "Application of Fire Safety Engineering Principles to the Design of Buildings - Initiation and Development of Fire within the Enclosure of Origin", PD 7974-1, British Standards Institute, 2003.

[12] Fire Engineering CIBSE Guide E, 2003.

[13] Babrauskas, V., "Glass Breakage in Fires", The Fire Place (Washington Chapter IAAI Newsletter), 15-18 (Feb/Mar 1998). Reprinted in: Hotline (Missouri Chapter IAAI Newsletter) 14, 18–21 (Summer 1998).

MATEC Web of Conferences **46**, 04004 (2016)
DOI: 10.1051/matecconf/20164604004


# Experimental study on vertical fire spread along a wooden façade

Mathieu Duny[1,2], Dhionis Dhima[1], Jean-Pierre Garo[2] and Hui-Ying Wang[2]

[1] Centre Scientifique et Technique du Bâtiment, Université Paris Est, Marne-la-Vallée, France
[2] Institut P', UPR CNRS 3346, Département Fluides, Thermique, Combustion, ENSMA, BP. 40109, 86961 Futuroscope Chasseneuil, France

## 1. CONTEXT

The main purpose of these tests is to get some information on fire spread along a combustible façade in order to assess the ability of numerical codes [1] to simulate flame propagation. Many authors have already studied compartment fire with external flames for non-combustible façades [2–4] and also combustible façades [5, 6]. Studies on non-combustible façade allow to describe flames (flame depth, flame height, thermal actions, etc.) and to explain the influence of geometrical parameters (opening dimensions, "U" configuration, etc.) on flame shape and flame behaviour. These studies highlight the importance of some of these parameters on vertical spread and correlations are issued from these experimental studies. If many researches on combustible façades exist, it is difficult to find academic configurations with a large instrumentation and generally, these studies focus on a facade system involving complex combustible materials (insulation or composite). That's why it was decided to perform some large scale tests to study fire spread along a vertical wood wall.

## 2. EXPERIMENTAL SET-UP

A principle drawing of the experimental set-up is presented in Fig. 1. The fire room is made of lightweight concrete. As a façade, a fire resistant board is set on a steel frame structure. The total area of the façade is 2.5 m × 4.9 m. The compartment is 1.5 m on a side and 1.15 m high. One side of the compartment is completely open on the façade in order to ensure well-ventilated condition inside the compartment. This test rig is really similar to the British standard façade test which is a real fire propagation test. But in the present test there is no side wall along the façade. As you can see in Table 1, three different tests are carried out. The first test is made without any combustible cladding on the calcium silicate board in order to measure the temperature and the calorific on inert façade. In the second and third tests a cladding of 18 mm thick plywood is mounted directly onto the calcium silicate board, i.e. no ventilation cavity behind plywood. In these two tests, two different type of wood are used, respectively Birch wood and Okoumé (African wood), in order to observe the influence of the essence of the wood on flame propagation and on calorific values.

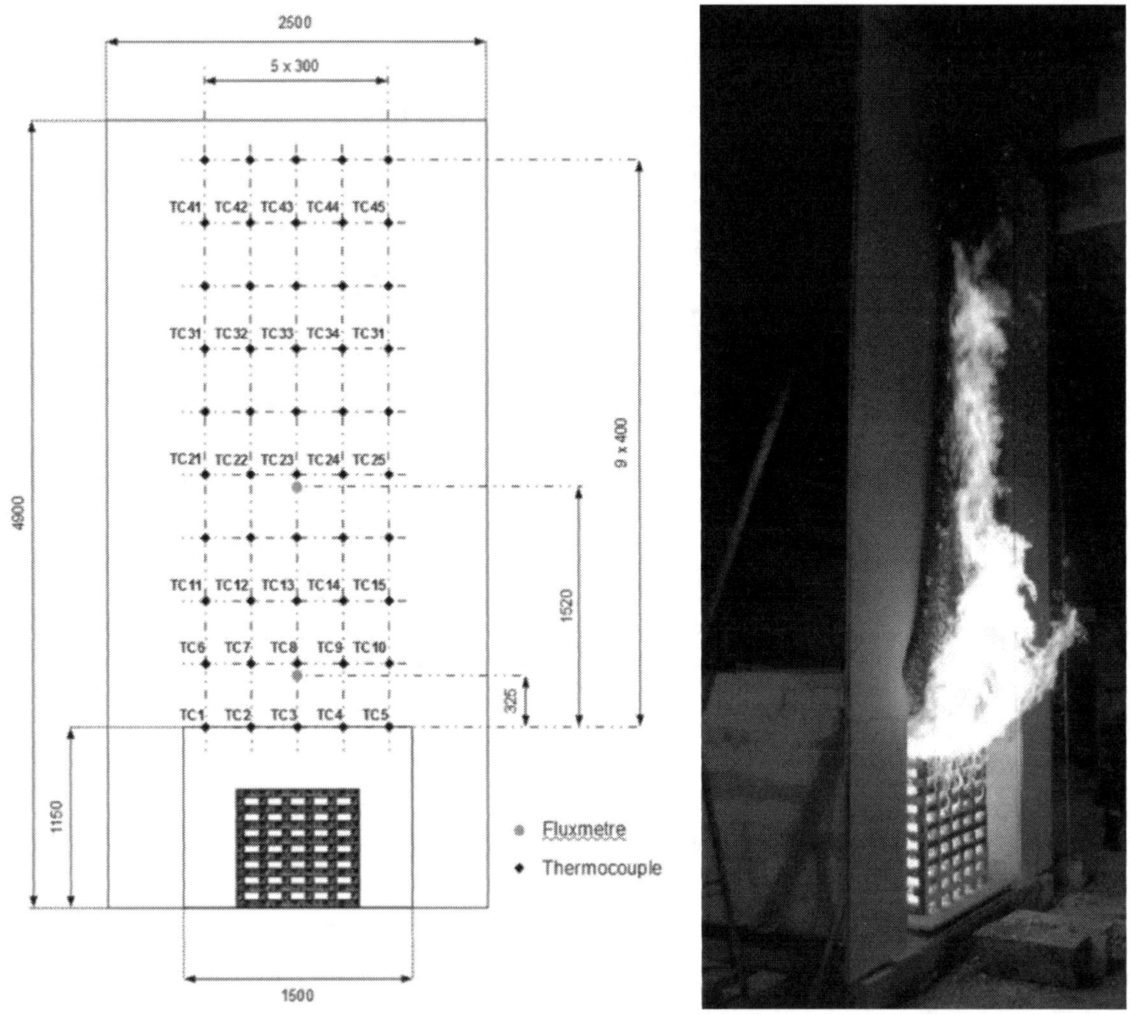


**Figure 1.** Experimental set-up.

**Table 1.** Description of tests.

| | Test_ 1 & Test_ 2 | Test_ 3 et Test_ 5 | Test_ 4 et Test_ 6 |
|---|---|---|---|
| Type of façade | Calcium silicate | Birch | Okoumé |
| Thickness (mm) | 13 | 18 | 18 |
| Density (kg/m$^3$) | 870 | 678 | 527 |

## 2.1 Scaling of the fire source

The heat release rate (HRR) is deduced as follows:

$$HRR = \dot{m} \cdot \Delta H_c \tag{1}$$

Where ṁ is the mass loss rate of the wood crib measured during the test. In this study, it was supposed that the heat of combustion ($\Delta H_c$) of the fire source is equal to 17.5 MJ/kg. The theoretical HRR can be dimensioned before the test using a simple equation from the SFPE Handbook [7]. The equation of the mass loss rate commonly used for this type of fire is as follows:

$$\dot{m} = \frac{4}{D} m_0 \nu_p \left(1 - \frac{2\nu_p t}{D}\right). \tag{2}$$

**Figure 2.** Picture of the wood crib used in the tests.

In this equation, $D$ is the stick thickness, $m_0$ is the initial mass, t is the time since ignition and $v_p$ is the fuel regression. Stick thickness (D) is dimensioned in order to obtain a maximum value of the HRR (t = 0) of about 1 500 kW. This value is high enough to get external flames simulating a post-flashover fire scenario and to obtain sufficient heat flux and temperature to ignite wood on the façade. In consequence, a 0.8 m × 0.8 m × 0.8 m wood crib constitutes the designed fire source (Fig. 2). Oil-soaked rags are used to ignite the fire.

The heat release rate (HRR) is obtained with two different methods:

- By measuring the mass loss during the test (Eq. (1))
- By collecting exhaust gases in a hood above the façade test rig (Eq. (3)).

$$\mathrm{HRR} = \rho_{O_2}\dot{m}_e\Delta H_{O_2}\frac{\frac{X_{O_2}^{\infty}}{100} - \frac{X_{O_2}}{100}\frac{X_{N_2}^{\infty}}{X_{N_2}}}{1 + (1.1 - 1)\left(1 - \frac{X_{O_2}X_{N_2}^{\infty}}{X_{N_2}X_{O_2}^{\infty}}\right)}. \tag{3}$$

In this equation, $\rho_{O_2}$ and $\dot{m}_e$ are respectively the oxygen density (temperature dependent) and the exhaust volume flow. The quantity $\Delta H_{O_2}$ (MJ/kg) is the amount of energy released per unit mass of oxygen consumed, generally taken equal to 13.1 MJ/kg. Quantities $X_{O_2}^{\infty}$ and $X_{O_2}$ are the ambient and exhaust mass fractions of oxygen. $X_{N_2}^{\infty}$ and $X_{N_2}$ are the ambient and exhaust mass fractions of nitrogen.

### 2.2 Measurement

Tests are performed with a deep instrumentation. A series of 2 mm K-type thermocouples is placed at a distance of 25 mm from the façade. Two other series of 10 thermocouples are located respectively 30 cm and 60 cm from the façade. In addition, two total radiometers are located at two different levels. Figure 1 shows the location of the measuring devices on the surface of the specimens.

## 3. RESULTS AND DISCUSSIONS

### 3.1 Heat release rate

HRR deduced from mass loss rate (Eq. (1)) is plotted in Fig. 3 for each test. A technical issue occurred during Test_ 1 (blue curve) resulted in an inaccurate measurement of the mass. In consequence, the curve issued from this test diverges from the others. Test duration is approximately between 18 and 19 minutes. HRR evolution is similar in tests 2 to 4. Fire grows linearly up to its maximum value. Moreover, maximum HRR is almost the same, close to the theoretical value (1567 kW). Also, measured mass loss rates traduce a good repeatability of the fire source behaviour.

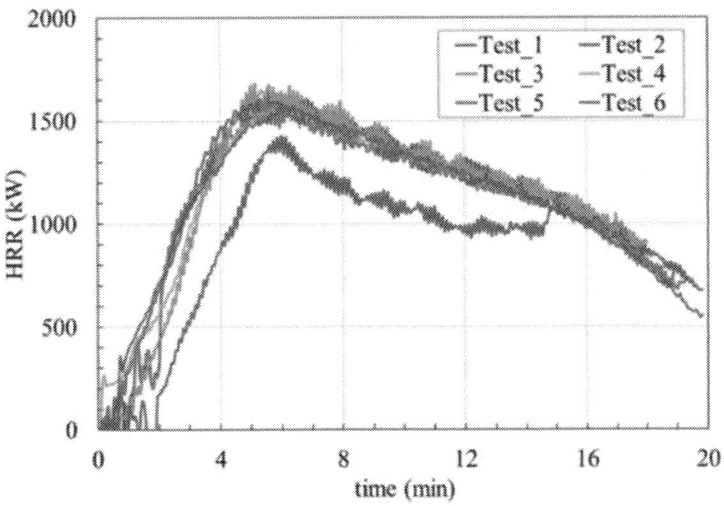


**Figure 3.** HRR deduced from mass loss rate.

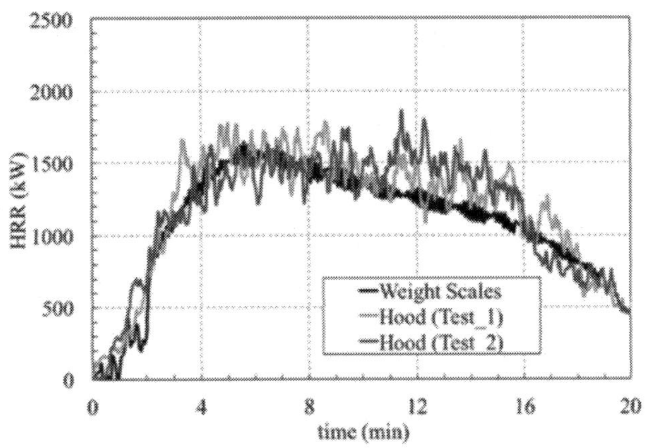


**Figure 4.** Comparison of HRR obtained using the two methods (Test_ 1 & Test_ 2).

Note that HRR estimation from fuel mass loss rate is a useful method to verify test repeatability and to heat released by the wood crib. However, this method does not allow to estimate façade contribution i.e. heat released by the combustible façade after fire propagation.

Figure 4 represents a comparison of the two methods used to determine the HRR for Test_ 1 and Test_ 2. It shows similar results between the two methods for both tests. Indeed, growth phases (between 0 and 6 minutes) are really close and maximum values are almost equal. The steady state is quite similar for the three curves. While HRR obtained from the mass loss rate decreases slowly, HRR obtained from oxygen consumption into the hood seems to be constant before falling sharply and reaches the black curve. Moreover, we can see that results obtained with the second method (hood) are really close highlighting the repeatability of these tests.

Figure 5 shows the heat release rate obtained for Test_ 2 (inert), Test_ 3 (Birch) and Test_ 4 (Okoumé) using the gas analysis. We can clearly see the contribution of the combustible claddings for the three tests. It can be noted that the initial phase of the tests, up to 2 minutes is similar. After 3 minutes, for the combustible façade, the HRR increases of about 500 kW compared with the inert façade, for both Test_ 3 (Borch) and Test_ 4 (Okoumé). This contribution slowly decreases to the end of the test. It is really interesting to observe that the two combustible façades have the same contribution during the same time interval. These data are very important in order to validate numerical simulation.

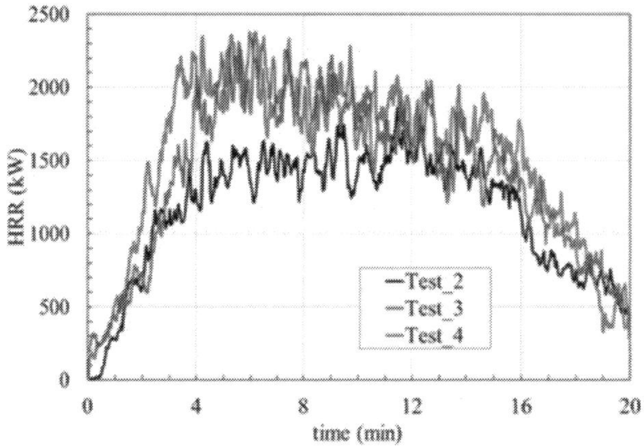


**Figure 5.** HRR measured in the hood during Test_ 2, Test_ 3 and Test_ 5.

## 3.2 Gas temperatures along the façade

Thermocouples are used in order to compare thermal actions for the three tests. On Fig. 6, temperatures for Test_ 2 (inert façade), Test_ 3 (Birch façade) and Test_ 4 (Okoumé) are compared. For the three tests, the flame ejected through the opening is symmetric, and maximum temperatures are always located in the centreline of the façade. That's why only centreline temperatures are presented here.

During the three tests, temperatures rapidly increase and we can observe a steady state on temperature between 3 and 16 minutes. The lowest temperatures are obtained with the inert façade in the higher part of the façade due to the presence of flames in this area. However, in the lower part (from TC3 to TC13), temperatures are higher for the inert façade. This can be due to the accumulation of pyrolysis gases that reduces the combustion efficiency and move up the combustion area leading to a decrease of temperature at these locations for combustible façades.

We can observe that thermal actions are not equivalent for the tests with combustible claddings. Indeed, temperatures are higher with Birch wood. Moreover, vertical fire spread seems to be more important for this essence. During Test_ 3, temperatures at location TC 33 to TC 43 show that the flame spreads all along the façade and reaches the upper part of the test rig. Temperatures are near 800°C 4 minutes after ignition 3.2 m above the opening. During Test_ 4, there is no flame at this location. It is important to see that thermal actions are lower during this test because heat release rates are similar for these two tests. This shows that the type of wood has a great influence on the flame spread.

## 3.3 Fire spread over the façade surface

As you can see on Fig. 7 and Fig. 8, flame shape is really symmetric. Visual observations after the test confirm that there is fire spread and charring up to the top of test rig for Birch wood (Test_ 3). Moreover, a large part of plywood burns or falls down directly above the opening. After Test_ 4, we can see that a smaller area burned. However, as observed on Fig. 5, the energy released by the combustion of this façade is exactly the same for both tests (Approximately 500 kW). Thus, this may be due the heat of combustion. The Okoumé essence seems to have a higher value. Additional tests with a bomb calorimeter are planned in order to determine the heat of combustion.

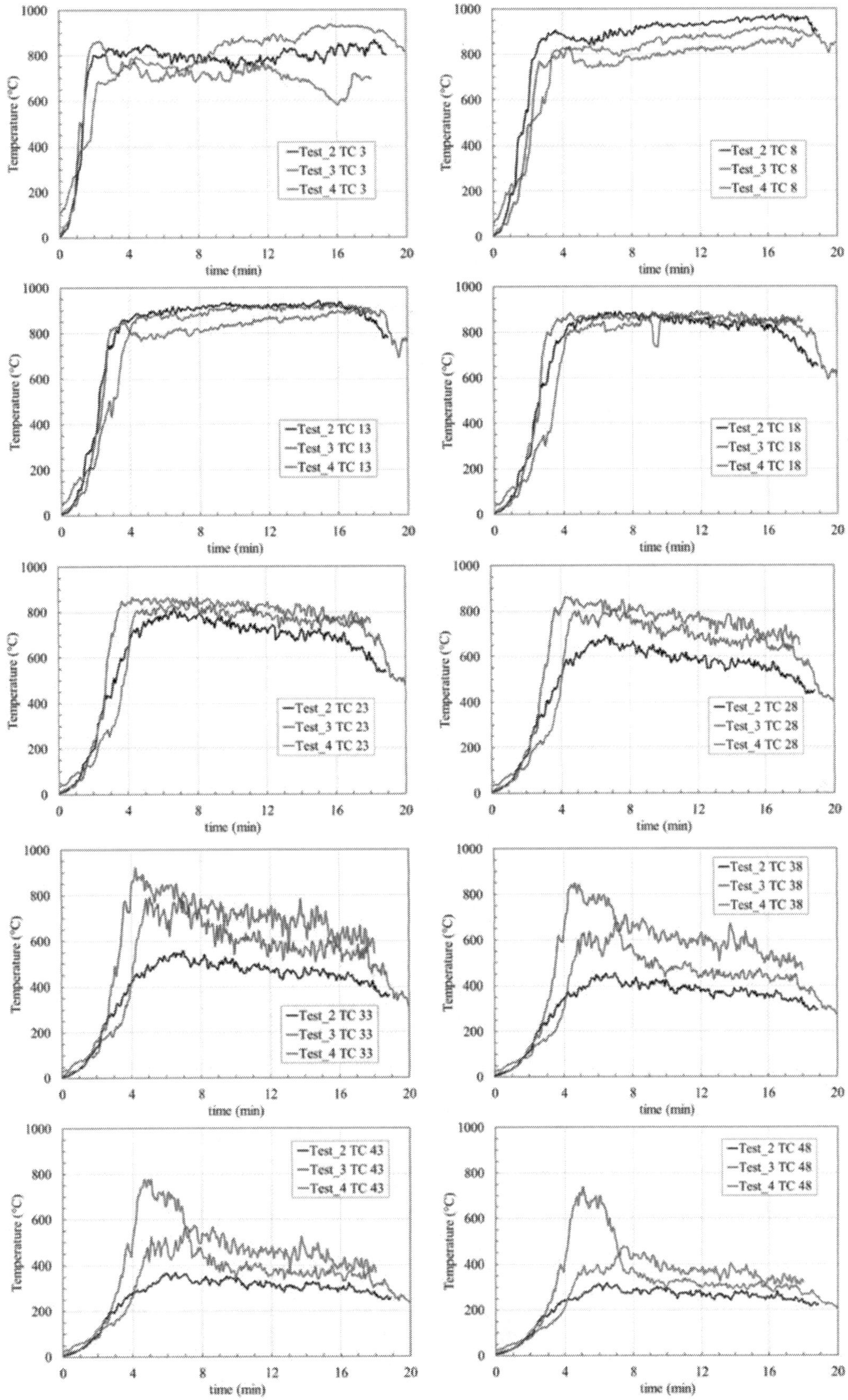


**Figure 6.** Comparison of temperatures in the centerline of the façade.

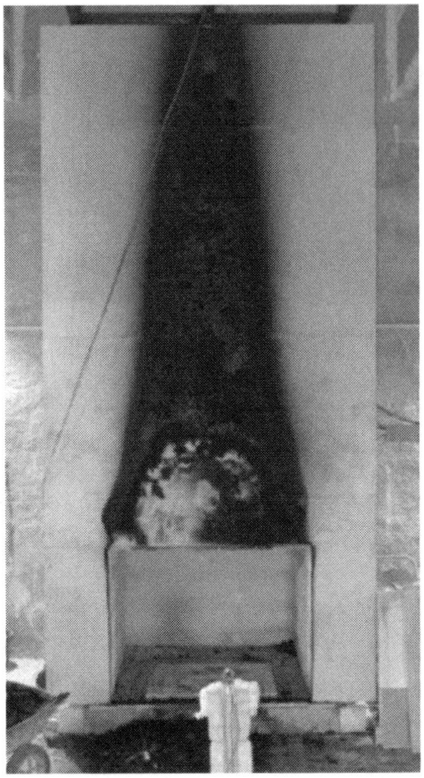

**Figure 7.** After Test_3.

**Figure 8.** After Test_4.

## 4. CONCLUSION

The aim of these tests was to perform an academic test in order to obtain a lot of measurements in order to validate fire models. These results give the possibility to compare experimental data with numerical predictions. Gas temperatures along the façade and heat release rates are the main results. A good repeatability was observed with the fire source behaviour. The use of two essences of wood was relevant. Whereas heat releases rates were identical during the two tests with combustible claddings, gas temperatures and observations post fire show that the vertical fire spread was really different.

### References

[1] McGrattan, K., McDermott, R., Hostikka, S., Floyd, J., "*Fire Dynamics Simulator (Version 6), user's guide*", NIST Special Publication, 2013.

[2] Lu, K.H., Hu, L.H, Tang, F., Delichatsios, M.A., Zhang, X., He, L., "*Facade flame heights from enclosure fires with side wall at the opening*", Procedia Engineering, The 9th Asia-Oceania Symposium on Fire Science and Technology, 2013.

[3] Tang, F., Lu, K.H., Hu, L.H, Delichatsios, M., "*Experimental study on flame height and temperature profile of buoyant window spill plume from an under-ventilated compartment fire*", Journal of Heat and Mass Transfer, 2011.

[4] Lu, K.H., Hu, L.H, Tang, F., Zhang, X., He, L., "*Experimental investigation on window ejected facade flame heights with different constraint side wall lengths and global correlation*", Journal of Heat and Mass Transfer, **78**, 17–24, 2014.

[5] Yoshioka, H., "*Study of test method for evaluation of fire propagation along façade wall with exterior thermal insulation*", Fire Science and Technology, **30**, 27–44, 2011.

[6] Nishio, Y., "*Experimental study on fire propagation over combustible exterior facades in Japan*", MATEC Web of Conferences **9**, 04001, 2013

[7] Babrauskas, V., "Heat Release Rate". In *The SFPE Handbook of Fire Protection Engineering Third Edition*, Chapter 3–1, 2002.

MATEC Web of Conferences **46**, 04005 (2016)
DOI: 10.1051/matecconf/20164604005


# Performance profiles of exterior fire protective building envelopes

Tobias Jarnskjold, Geir Jensen and Atle Knudsen

COWI AS Grenseveien 86, Norway

**Abstract.** The fire protective envelope of any building consists of multiple elements with widely differing properties relating to a fire, such as glass, roof tiles and sheathings, wood cladding, gaps and openings. Where resistance to an exterior fire is required, all elements should be verified to provide a comparable risk of burn-through.

Elements are rated by either the material response to fire or fire resistance. In Europe, cladding sheets and wall membranes can be rated by K classes, which effectively include a measure of the time to burn through. A determination of burn-through time of each element of a specific building envelope should be obtained.

A design tool to verify the performance of a building's fire protective envelope has been developed. In this paper, a general description of passive elements of the envelope, which should be included in a risk assessment tool such as an index method, is presented. An illustrative approach to visualise the profiles for areas densely spaced structures where an exterior fire may trigger building-to-building fire spread is also included.

The research is based on the hypothesis that a relatively subtle and pointed upgrading of an exterior building envelope will severely reduce the speed of building-to-building fire spread, thus allowing firefighting efforts to get on top of the situation. For a burning structure to expose other buildings to fire, the fire has to settle, which leads to a burn-through to the inside. Once inside, an enclosure fire needs to develop and burn through the roof or break one or more large window panes. It is estimated that a 5–10 min delay for a structure to expose other structures to fire can be sufficient to avoid loss of multiple structures. A 10–50 min burn-through time allows for an extended intervention by the fire brigade, which is significant in rural areas. A fire protective envelope may prevent an exterior fire from penetrating the protective envelope at all and the structure can be saved.

## 1. INTRODUCTION

Exterior fires and wildland urban interface (WUI) fires are global challenges and subject to increasing interest due to the emerging risk from climate change. This type of fire is hard to prevent and threatens communities that are vulnerable to exterior fires. Economic losses are mainly related to structural losses. A general approach is to reduce the potential for ignition by hardening the exterior of structures [1]. The US and Australia are familiar with such fires and have established regulations to prevent structural losses due to exterior fire events, referred to as WUI fires. In these countries, the standards NFPA 1144 "Reducing structure ignition hazards from wildland fire" [2] and AS 3959 "Construction of buildings in

**Figure 1.** An exterior fire led to an indoor fire at Ringve Museum.

bush fire prone areas" [3] include requirements for the surrounding area and the building structure. The Nordic countries have less focus on this type of fire, since such events have been rare there.

However, in 2014, three fires with WUI characteristics occurred. Two of these occurred in Norway during the winter, namely the fires of Lærdal [4, 5] and Flatanger [6]. The third fire occurred in Sweden during July and August. These fires have increased the focus on exterior fire protection and motivated the work of a master thesis in 2015 [7]. The result is a first-hand risk assessment tool, which is described in general in this paper and from now on it is referred to as FireScore.

In the fall of 2015, Ringve Musical Instruments Museum was damaged by a weed burner near its facade (Fig. 1) [8]. The fire took hold in the air gap in the rain screen and spread within the gap to two floor voids, and the rescue team had to break up floors and the rain screen in multiple locations. A full-coverage fixed extinguishing system inside could not have prevented the damage caused by such an exterior fire entering through wall cavities and floor voids. Such fires can be prevented by identifying risks and systematically devising adequate, simple and strategic passive protection measures.

Mitigating the effects of exterior fire exposure on buildings in the Nordic countries has traditionally always focused on radiant heat transmission and direct flame contact. However, a report by SPFR [4] and a report on the Witch and Guejito fires in the US [9] indicate that a fire spread by firebrands is underrated in regard to exterior fires since firebrands are often the source of the fire spreading to other buildings and structures.

Adapting regulations and risk tools in Nordic countries from the US and Australia is not necessarily a good approach, since vegetation fuels, the climate and the occurrences of such events are different. Still, wildland fires are an increasing problem all over the world and protective measures are needed. The Norwegian Institute of Bioeconomy Research (NIBIO) has highlighted the problem and requested tools and regulations to protect built areas with dense natural vegetation [10]. Regarding forest fires, the focus in Norway has been on loss of forest, though the economic loss to buildings can be as large. Forest fires are natural events in an ecosystem. However, when built environments including infrastructures are threatened, adequate protection is required.

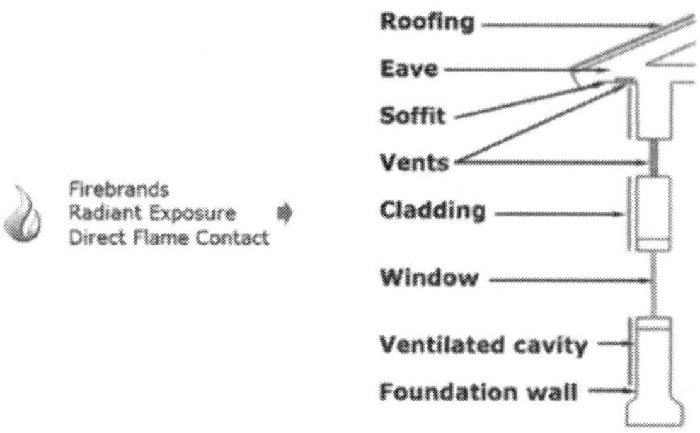


**Figure 2.** The fire protective envelope.

Active measures to prevent the spread of forest fires have been the common approach for decades. However, passive measures to harden structures might be more cost-effective and robust while allowing natural loss of vegetation.

Over the years, several types of risk assessment tools have been developed. Examples are given in [11]. However, none of these tools takes into account a building's response to exterior fire exposure. It must be mentioned, as indicated by Cohen [12], that risk scores resulting from these tools rate the potential loss of buildings rather than predicting the loss. It is assumed that lowering the risk will lead to improved chances of survival, but a guarantee can never be achieved.

Based on an index method, the developed risk and design tool FireScore is aimed at assessing the performance of single or multiple structures to resist exterior exposure to fire. The output report identifies and visualises weak spots of any single structure or of an area with multiple structures, which subsequently can be improved and upgraded.

A better understanding of exterior fire behaviour in areas of interest and an overview of the high-risk areas will in many cases eliminate the need for highly accurate real event simulations, such as by computational fluid dynamics (CFD) modelling. Such simulations demand a high level of precision and a large amount of input data and they tend to be expensive and time-consuming.

Critical elements of a given structure envelope can be graded and implemented in an index method such as by the FireScore. In this way, comparable risk value can be obtained for a structure. The method itself provides a transparent and systematic approach, which can be communicated to involved parties. Furthermore, the output is mapped and visualised. Tools like Google Maps are suited for making heatmaps to highlight occurrences. New handheld technology allows for maps to be created in the field, prior to the implementation of risk-reducing measures based on statistical analyses.

## 2. THE FIRE PROTECTIVE ENVELOPE

The risk assessment requires a set of elements that are critical in the event of an exterior fire. Main exposures and building elements are shown in Fig. 2. Hardening the fire protective envelope to extend burn-through time will increase the probability of effective fire fighting efforts in the case of a passing conflagration or other exterior exposure.

The protective envelope comprises multiple elements with widely different properties relating to a fire, such as glass, roof tiles and sheathings, wood cladding, gaps and openings. When resistance to an exterior fire is required, all elements should be verified to see if they have the same performance. In practice, some elements are classified by the material response to fire and others by a fire-resistance rating. The K-class rating takes into account surface and backing-sheet properties. Typically, one attempts to determine the burn-through times for all elements to ensure they all meet or exceed a minimum time.

The building elements of the fire protective envelope are listed below by importance in descending order. Combustibility and burn-through time are of interest for all the building elements.

## 2.1 Building elements of the fire protective envelope

The fire protective envelope is divided into six elements: (1) roof, (2) eaves and gables, (3) walls, (4) foundation walls, (5) windows and (6) doors. Passive designs to harden the structure should increase the resistance against exterior fires. A protective envelope is evaluated by its effectiveness in respect to firebrand showers, spot fires, direct flames and radiant heat.

**Roofs** are the most important element of a building envelope in a wildfire [7, 13, 14]. They are vulnerable to firebrand penetration at the eaves, ridges or other vulnerable locations. The type of roofing is also important [15]. When assessing a roof, combustibility and burn-through of the roofing material should be included; e.g., Howard [13] has observed and linked the extensive loss of buildings to wooden shingles used as a roofing material.

Secondly, **eaves and gables** are vulnerable to penetration of firebrands through the eave vents and to fire at the exterior wall. Ventilated air gaps in the eaves or gable should be assessed. Ventilated roofs with eaves made of wood have a considerable risk of a fire spreading [16]. This is also the case in cold attics with openings for natural air ventilation. Ventilated roofs take advantage of the high pressure on the windward side and the low pressure on the leeward side in the removal of moisture and heat, and therefore, they are especially vulnerable. Protection measures are considered, for instance, in the California Building Code [17] and the Australian standard [3] in which a non-combustible mesh of maximum sizes 3.1 mm and 2 mm, respectively, is implemented.

Ventilated air gaps in the rain screen of **exterior walls** are common in the Nordic countries of Europe. The gaps are vulnerable to firebrand penetration and direct flame contact. Flames in wall cavities may be elongated 5–10 times compared to exterior flames due to combustion and the flow in the cavities [18]. A fire established at a wall assembly has a high risk of spreading to the roof and eaves, subsequently to the attic, which often leads to total loss of the structure. Openings to rain-screen air gaps should be assessed and measures taken to prevent the ingress and spread of fire.

Crawl spaces and openings in the **foundation walls** must be assessed. Manzello [15] observed that firebrands accumulate in front of wall structures, igniting mulch beds and other combustible materials to threaten the further spread of the fire to the exterior of the building. Therefore, the possibility of the fire spreading into crawl spaces and the average height of the foundation wall up to ordinary wall elements should be considered.

**Windows** should be assessed by the type of glazing in the protective envelope, including the number of layers, glass and frame quality. Glazing is vulnerable to local heating, which results in rapid failure. Window qualities are to some extent addressed by the California Building Code [17]. The transition between the wall and window frames should also be assessed with respect to gaps and firebrand penetration.

**Doors** should be assessed in regards to the door leaf, the gaps between the door leaf and doorframe, and the transition between the frame and wall. Gaps of 2 mm or more [15] are critical and improvements are necessary.

Figure 3 shows the combination of highly fire-resistive and preventive measures for the fire protective envelope of an invaluable historic structure. It is to be protected from any conflagration of the surrounding residential buildings. Modifications of elements of the building envelope are aimed at making them fire resistive for 50 min.

The thickness of some panels was increased from 30 to 35 mm, and overlaps were protected by intumescent seals. Windows were protected by fire-resistant glass in the frame behind the outer glass and the door sets were upgraded. Ventilating cavity barriers rated for fire resistance were installed in the rain-screen cladding, at the eaves, at the foundation, in the walls and in the tower constructions. Hidden

**Figure 3.** Pilot project [19].

battens were used to harden the eaves to extend their burn-through time. The most exposed wall has a hidden 6-mm fire-rated $Ca_2SiO_4$ membrane carefully located to avoid upsetting the function of the rain screen. The fire protective envelope is designed to resist exterior exposure, which is fundamentally different to exposure from room enclosure fires.

In addition to the above mentioned elements, the type of exposure varies as well. Any element of an envelope may offer different resistance to different types of exposure. Interactions between elements may also cause the overall performance to vary when exposed. For example, a roof assembly that performs well against sustained radiation and flame attack from a fire of a nearby structure may easily fail when subjected to firebrand showers from a vegetation fire further away. Glass may prevent the ingress of firebrands, but easily break in a high temperature gradient even when due to small flames. In general, unprotected openings allow the penetration of firebrands, flames and radiation, which may ignite an indoor fire in seconds.

To estimate and implement measures for a building's fire protective envelope an analysis of the exposure is required. Seven types of exposures need to be assessed and they are listed below by importance in descending order.

## 2.2 Exposure elements

The most important exposure element when assessing the spread of an exterior fire is the **climate**. This element can be divided in two: first, by assessing the severity and frequency of forest and wildland fires, e.g. by looking at forest or wildland fire risk indexes like the Canadian Forest Fire Weather Index [20], and second, by assessing the probability of challenging wind speeds. If the wind speed exceeds 7 m/s, which is necessary for firebrands to be lofted [15], the probability of a fire spreading over bigger distances increases.

**Natural vegetation** close to buildings has been shown to contribute to and increase the risk of fire spreading between buildings [7, 11, 21, 22]. How a fire spreads due to vegetation is related to the vegetation's distance to the building. The space between is commonly divided into zones [23]. The amount of (natural) vegetation, the space between bushes or trees and the flammability of vegetation species should be evaluated throughout the zones. There is evidence of ignition to structures at 40 m from forest crown fire fronts [12]. The principles of fuel treatment are described in [21].

**Nearby buildings** also need to be considered regarding direct flame contact, radiant heat and firebrand generation. This includes sheds and annexes that are easily ignited due to their low or no fire protection. As with vegetation, the risk decreases with increasing distance between buildings.

A vegetated **slope** downhill of a building by a gradient exceeding 20% is associated with increased damage to the building [24]. The risk of a fire spreading also increases if the distance between the top of the slope and the building is short. In addition, this element should be considered together with wind direction and fuel.

**Bridging fuels** are surrounding fuels that allow a fire to spread to a building, e.g. vehicles, construction materials, hedges, wood piles, wooden porches, terraces, furniture etc., and are close or directly attached to the building. Connecting fuels affect fire exposure (treatment 19 for connecting fences [24]). This parameter should include assessing plant beds and bushes that act as a link between the ground beside the building wall and the combustible parts of the building envelope. Plant beds covered with mulch have been proven to increase base wall exposure [15].

**Fire breaks** are areas with incombustible surfaces that can act as a barrier to hinder or limit fire spread. Treatment 7 in [24] indicates such breaks are parking spaces, driveways, paths, roads or trails. Fire breaks do not necessarily stop a conflagration but limit the damage to buildings as they may reduce the grade of exposure. Yet, spot fires are known to occur as far as 20 kilometres from the fire front [25], also known as spotting.

**Gardening and tidiness** refer to the condition of orchards, flowerbeds etc. surrounding the building. By trimming hedges, bushes and trees and keeping grass short, the amount of fuel is reduced. Dry fuels are a particular concern as they may support ignition that can spread to the surroundings [23].

## 3. A SYSTEMATIC RISK EVALUATION TOOL

These elements greatly affect the risk of fire spreading and should be evaluated systematically and methodically. CFD modelling is a simulation tool, which can be used to predict the spread of an exterior fire. Fires in built environments are complex, involving many physical mechanisms. To analyse the absolute risk level of such fires in a probabilistic setting can be cumbersome, and it is often not feasible as input data are too scarce or uncertain. Furthermore, the cost of laborious risk analyses may prevent a risk assessment from actually being performed. Methods combining simple use with sufficient information output and providing an adequate risk assessment and risk improvement options are therefore valuable. Risk ranking produces estimates of relative fire risk and aggregate rankings into a single index [26].

A simple tool to consider the risk of a fire spreading has been developed. FireScore is based on an index method and the result of a master thesis [7]. Elements, which are of importance to exterior fire spread were mapped by an interview and literature study. The elements were identified, sorted, ranked and weighed. The literature for the data input was mainly from the US [17] and Australia [3], complemented by a study of the Lærdal fire [7]. Literature published in Norway and interviews of the firefighting personnel involved during the Lærdal fire provided useful material for the development of FireScore.

The index method is a risk assessment tool known in the Nordic countries for its contribution to evaluating the risk of indoor room fires, i.e. "Fire Risk Index Method – Multistorey Apartment Buildings" by Björn Karlsson [27]. The same principles for risk assessment were used to establish the FireScore tool. It addresses exposure to an exterior fire and the response of the building envelope. An index method is a semi-quantitative risk assessment method, in that it assigns grades to influential elements based on a qualitative assessment. It further combines the grades of the elements by an arithmetic process where they are assigned relative weights. The weighing of the elements can easily be adjusted depending on the area of interest. The adjustment can use the analytical hierarchy process

(AHP) [28]. The index is a comparative value for assessing risk levels. Such an index method provides transparency and documentation, which is useful for decision makers.

FireScore is divided into two distinct parts: the first gives an indication of the exposure severity and the second indicates the robustness of the fire protective envelope. The ratio of resistance to exposure gives a total risk score to a building. A low score indicates a reduced probability of loss.

# 4. MAPPING AND VISUALISATION

A risk characterised by a number is pointless in itself. It has to be communicated in an understandable and meaningful way. Visualisation of risk using colours and maps is a common way to communicate risk, e.g. as California Fire Hazard Severity Zone Map [29]. Such maps indicate the hazard in areas based on potential natural vegetation fuels over a 30–50-year horizon and their associated expected fire behaviour and exposure to buildings. Several tools have been developed to visualise and understand the risk of wild fires. Tools such as FIREHARM [30] and a method to characterise and map wildland urban interface (WUI) fires at large scale [31] all contribute to managing the problem of wildland fires.

## 4.1 Tools to facilitate registration and visualisation

In addition to the FireScore risk tool, a tool to register exposures and building elements has been developed to collect a dataset for calculating and visualising the risk score by heatmap using Google Maps. The registration tool provides an online representation of the FireScore. It takes advantage of new technology, i.e. phones, tablets, the internet etc. It is useful in the field and enables fast and cost-effective registration of larger communities. Results are stored in a database, providing vast possibilities for statistics, visualisation, reports etc.

The data collected and registered can provide an overview of an area as well as detailed identification of crucial elements for single buildings by allowing the user to examine the details further. The registration tool can be accessed in the field or in the office. If used in the field combined with the right tool, a building can be geo-located for further visualisation on a map.

The user interface has two levels of detail. The first level has general information about an area and the second has information about individual buildings.

A heatmap have been made to see if it improves understanding and visualisation of the risk calculated by FireScore. It is a visualisation layer on top of a map where areas of higher fire risk (high intensity) are coloured red and areas of lower risk appear in green [32].

Figure 4 shows how a heatmap can be applied for visualisation by implementing the risk score calculated by FireScore and registered for further input. This allows multiple scores to be gathered in a single map giving an overview of the risk in an area. By zooming in, the points spread out and the score for a specific address becomes visible.

Furthermore, drones can be useful for higher-quality inspections of tall buildings. ASTM has established a task group to develop a guide for the visual inspection of building facades using drones [33]. High-quality cameras allow for close-up views of parts of facades that otherwise cannot be seen.

This type of knowledge can be useful for area planning processes as well, i.e. to map built-up areas in need of special attention by conservationists or others.

# 5. DISCUSSION

Fire dynamics are never straightforward and in most cases simulated results that forecast the exact route of a fire are never achievable. But we do have tools that make us better able to implement risk-reducing measures and to a greater extent reduce potential losses.

**Figure 4.** Application and illustration of risk visualisation by a heatmap.

An index method involves weighing elements. For exterior fire scenarios, locations and areas may differ in several respects and adjustments of the weights may, therefore, be required to obtain an adequate risk estimate. Mathematical adjustments of the element weights are simple but require professional insight, which can be obtained by a Delphi panel.

FireScore provides a risk score as a mathematical value. Changing the score changes the theoretical probability of a structure's survival. However, by mapping the index results of multiple structures in a specific area, a more accurate estimate of the risk to the community can be evaluated.

CFD tools, which simulate exterior fires, are available and useful, although they wholly depend on the input parameters. The number of parameters is hardly manageable. They are time-consuming with respect to exterior fire development and the fire spreading to the exterior envelopes of buildings. FireScore accounts for direct flame contact and heat radiation and the risk of fire spreading by firebrands linked to the exterior design of the building envelope.

The method described combined with a registration tool permits visualisation of the risk due to exposure to an exterior fire in built areas. It is not meant to simulate real events and scenarios, but to give a probabilistic picture of the situation by pinpointing critical areas, a building's details and the exposure elements.

One of the advantages of an index method is its transparency, which improves the evaluation of the risk of an exterior fire between professionals of all disciplines and owners. It also ensures that important elements are properly evaluated so that necessary measures can be taken.

Applying new technology to map built areas provides an opportunity to make statistical calculations and compare similar areas to each other. Experience thus gained can be stored in a virtual bank for later use or statistical evaluations. When events occur later in mapped areas, it is possible to learn even more about the risk by evaluating the measures that were implemented. Technology such as drones, allows for higher-quality inspections of building elements that are not visible from the ground. Nevertheless, inspections should be carried out by personnel competent in the field of exterior fires.

## 6. CONCLUSION

Decisive crucial elements that govern the development of a conflagration and WUI fires have been identified and ranked according to information based on fire-resistive performance, a literature study and interviews. Subsequently, an index method has been developed as a fire protective design tool by applying the findings. To simplify use and to add transparency, a mapping method was integrated.

A design tool was developed to provide a map of exterior fire risk in a given wild land area of multiple structures. The tool was evaluated based on exercises in pilot projects. For a user, it is a suitable tool for systematic risk assessment of the loss due to the spread of an exterior fire. The user may be challenged by ranking elements where site-specific conditions are not covered by the current tool. The tool identifies weak points and encourages the user to address them. Other regions may not fit the current FireScore weighing and it is recommended that expert panels with relevant knowledge specify default element weights.

## References

[1] Mell, W.E., et al., *The wildland–urban interface fire problem–current approaches and research needs.* International Journal of Wildland Fire, 2010. **19**(2): p. 238–251.

[2] NFPA, *NFPA 1144 Standard for reducing structure ignition hazards from wildland fire.* 2013: National Fire Protection Association (NFPA).

[3] Australian_Standard, *Construction of Buildings in Bush Fire Prone Areas*, in *Standards Australia.* 2009, Australian Standard: Australia.

[4] Steen-Hansen, A., et al., *What can we learn from the Lærdal fire in January 2014? Evaluation of the fire spread. (in Norwegian)*, in *SPFR-Rapport*, S.F. Research, Editor. 2014, SP Fire Research AS: Norway.

[5] Steen-Hansen, A., et al., *Evaluation of fire spread in the large Lærdal fire, January 2014*, in *Fire and Materials 2015, 14th International Conference.* 2015, Interscience communications: San Francisco, USA. p. 1014–1024.

[6] PWC, *Evaluation of the fires: Lærdal, Flatanger og Frøya.* 2014, PricewaterhouseCoopers (PWC).

[7] Jarnskjold, N.M.T., *Brannskallsikring i tett trehusmiljø; English: Securing fire-protective envelopes in dense wooden built environments*, in *Department of Civil and Transport Engineering.* 2015, Norwegian University of Science and Technology (NTNU): Norway.

[8] Ulgenes, T. *Ringve Museum Fire.* 2015; Available from: `https://prezi.com/vdz_khotnbth/ringve-3-august-2015`.

[9] Maranghides, A. and W. Mell, *A case study of a community affected by the Witch and Guejito wildland fires.* Fire technology, 2011. **47**(2): p. 379–420.

[10] Per Holm Nygaard, S.S., *Skogbrann og hytter, - risiko og forebygging*, in *Byggeindustrien.* 2015: `http://www.bygg.no/article/1255632`.

[11] Gollner, M.J., et al., *Pathways for Building Fire Spread at the Wildland Urban Interface*, F.P.R. Foundation, Editor. 2015, Fire Protection Research Foundation; National Fire Protection Association NFPA: QUINCY, MASSACHUSETTS, U.S.A.

[12] Cohen, J.D. *Reducing the wildland fire threat to homes: where and how much?* in *Proceedings of the symposium on fire economics, planning, and policy: bottom lines.* 1999. San Diego, California, U.S.A.

[13] Howard, R.A., *Decision Analysis of Fire Protection Strategy for the Santa Monica Mountains: An Initial Assessment: Preliminary Report.* 1973: Stanford Research Institute.

[14] Maranghides, A. and W.E. Mell, *Framework for Addressing the National Wildland Urban Interface Fire Problem-Determining Fire and Ember Exposure Zones Using a WUI Hazard Scale.* 2012: US Department of Commerce, National Institute of Standards and Technology.

[15] Manzello, S.L., S. Suzuki, and Y. Hayashi, *Enabling the study of structure vulnerabilities to ignition from wind driven firebrand showers: A summary of experimental results.* Fire Safety Journal, 2012. **54**: p. 181–196.

[16] Hakkarainen, T. and T. Oksanen, *Fire safety assessment of wooden facades.* Fire and Materials, 2002. **26**(1): p. 7–27.

[17] CBSC, *California Building Code, Title 24, Part 2, Chapter 7A*, I.C.C. California, Editor. 2013, California Building Standard Commission.

[18] Colwell, S. and T. Baker, *Fire performance of external thermal insulation for walls of multistorey buildings: (BR 135) Third edition*. Building Research Establishment Report BR135, ed. B. Trust. 2013: BRE publications.
[19] Puntabyrgi, P., *Pilot project-passive exterior fire protection of small church*, P. 1, Editor. 2015, Lervike AS.
[20] Van Wagner, C.E., *Development and Structure of the Canadian Forest Fire Weather Index System*. Forestry Technical Report. 1987, Ottawa, Canada: Canadian Forestry Service.
[21] Vicars, M. and J. Luckhurst, *Firesmart: Protecting your community from wildfire*. Second edition ed. 2003, Edmonton, Alberta, Canada: Partners in Protection.
[22] NFPA, *Firewise Toolkit. A guide to Firewise principles*, N.f.p.a.N. Firewise communities, Editor. Undated.
[23] Vicars, M. and J. Luckhurst, *Firesmart: protecting your community from wildfire*. Partners in Protection. Edmonton, Alberta, 1999.
[24] Maranghides, A., et al., *A Case Study of a Community Affected by the Witch and Guejito Fires: Report #2–Evaluating the Effects of Hazard Mitigation Actions on Structure Ignitions*. 2013, Natianal Institute of Standards and Technology (NIST).
[25] Teague, B., R. McLeod, and S. Pascoe, *Final report, 2009 Victorian bushfires royal commission*. Parliament of Victoria, Melbourne Victoria, Australia, 2010.
[26] Watts, J. *Criteria for fire risk ranking*. in *Fire Safety Science-Proceedings of the Second International Symposium*. 1991. Tokyo, Japan: Hemisphere Publishing Corporation.
[27] Karlsson, B., *Fire Risk Index Method - Multistorey Apartment Buildings, FRIM-MAB Version 2.0*. Report P0212053, Trätek AB, Stockholm, 2002.
[28] Saaty, T.L., *How to make a decision: the analytic hierarchy process*. European journal of operational research, 1990. **48**(1): p. 9–26.
[29] FIRE, C., *California Fire Hazard Severity Zone Map Update Project*, T.C.D.o.F.a.F. Protection, Editor. 2007, CAL FIRE: `http://www.ca.gov`.
[30] Keane, R.E., et al., *A method for mapping fire hazard and risk across multiple scales and its application in fire management*. Ecological Modelling, 2010. **221**(1): p. 2–18.
[31] Lampin-Maillet, C., et al., *Mapping wildland-urban interfaces at large scales integrating housing density and vegetation aggregation for fire prevention in the South of France*. Journal of Environmental Management, 2010. **91**(3): p. 732–741.
[32] Developers, G., *Heatmap Layer*. Google: `https://developers.google.com/maps/documentation/javascript/heatmaplayer`.
[33] ASTM. *New ASTM Standards Aim to Help with Building Facade Inspections, Including Drone Use*. 2016 [cited 2016 08.01.2016].

MATEC Web of Conferences **46**, 05001 (2016)
DOI: 10.1051/matecconf/20164605001


# Fire risk related to the use of PV systems in building facades

Lamberto Mazziotti[1], Piergiacomo Cancelliere[1], Giuseppe Paduano[1], Paolo Setti[2] and Samuele Sassi[2]

[1] Ministry of Interior Italian National Fire Services Largo Santa Barbara 2, Rome, Italy
[2] FSC ENGINEERING SRL, Fire & Structure Consulting, via A. Zenon 4, Milano, Italy

**Abstract.** Nowadays the use of photovoltaic (PV) systems in buildings is not only related to the solar energy conversion into electrical one, but these PV modules or panels could also be used with aesthetic features or, even more, as thermal protection systems in building facades.

Thanks to the technical development of the photovoltaic industry, PV system can easily be architectonically integrated into building construction elements such as roofs, vertical façade components, both with opaque or transparent surfaces. Furthermore, PV construction facades elements could also be provided by openings like doors or windows.

Accident analysis show that the use of PV systems as construction elements could increase the risk of fire in buildings. In fact, international and National data report a growing number of fire caused by PV system applied or integrated in buildings. The Italian National Fire Service, that is the Authority having jurisdiction for fire safety in buildings (in Italy), in 2012 has released a Guideline in order to asses and mitigate the risk of fire when a PV system is put in place on a building as a façade or as a roof. The Guideline addresses not only the reduction of the PV fire ignitions causes and the aspects related to the fire spread due to the combustible parts that constitute PV modules or panels, but also take into account the safety of both the maintenance personnel and the rescue teams.

This paper focuses on the fire safety aspects related to the use of fire PV panels and systems in building facades, showing some interesting experimental data related to the fire behaviour of these components and underlining the factors that promote the spread of fire, like the high operating temperature of the PV system itself.

## 1. INTRODUCTION

Photovoltaic (PV) Systems exploit the photovoltaic effect to convert the solar radiation into electrical energy. Environmental incentive policies worldwide are supporting the use of solar PV electrical energy converters. Over the last decade, even during difficult periods of industry consolidation and economic crises, a considerable increase in PV plant installations has been recorded. In fact, the global PV market has marked a record growth in 2011, relenting in 2012 and growing again in 2013. In 2013 alone, PV plants have produced at least 38.4 GW of newly-added capacity, and almost 11 GW only in Europe, reaching more than 138 GW of installed electrical power around the globe [1].

PV systems for buildings are designed and built not only to perform the function of electric energy production, but also to be put in place as construction elements in order to protect the building by the

**Figure 1.** PV façade of the airport terminal of Bari Palese (Italy).

action of water, snow, mechanical shocks or to meet requirements related to the energy classification of the building and the thermal comfort.

The PVs offer enormous potential for designers of buildings, but they must however be watched very carefully in terms of fire risk.

PVs can influence the building's orientation, footprint, layout and form; they will affect the building fabric and will be an important element of the environmental and building systems. They need to be considered as an integral part of the energy strategy of the building and of its functioning. The integration of PVs with the other building elements is critical to success and appearance and aesthetics are, as ever, especially important. The PV building products today can be important constructive elements of the building construction envelopes, and often they are part of its facades.

Much PV claddings can be considered to be panes of glass to which PV cells are applied and so the extensive experience of glazed facades can be built upon. In addition, modules can be easily incorporated into other proven systems such as rain-screen cladding.

Curtain walling systems are a well-established technology used in numerous prestige projects such as city centre offices. The mullion/transom stick system is the most common. Vision areas are normally double-glazed and non-vision areas are either opaque glass or insulated metal panels. PV modules can be incorporated easily as factory-assembled double-glazed units. The outer pane might be laminated glass on which in applied PV-resin glass; then we have the inner pane (glass). A sealed air gap is also placed between the two glasses. The overall thickness of the module would typically be under 30 mm. As an example, next figure reports a PV façade of the airport terminal of Bari Palese (Italy).

Since the first objective of the energy conversion systems is efficiency, the design and installation of PV systems focus mainly on efficiency aspects in order to exploit the maximum amount of the solar irradiance. Therefore, the fire risk due to an installation of a photovoltaic system is generally not well addressed. According to a recent survey of the National Fire Brigade (Italy) , around 1600 fire-related accidents have occurred in the nearly 560,000 PV plants currently installed in Italy. Some severe fires involving buildings with PV systems have lately occurred in many European countries and in the USA. Even though PV solar energy converters are not the cause of ignition, in case of an outbreak of a fire in hosting buildings, they will enhance fire spread and trigger flame propagation mechanisms. Hence, fire safety issues of a photovoltaic solar energy converter must be addressed both as a cause and as a victim of a fire. In fact they may be causing the fire that suffer it and contribute to its propagation.

The article opens with a section where the main causes and faulty mechanisms of PV fire ignition are illustrated, and continues with a section dedicates to the PV fire safety guideline issued by the Italian National Fire Services. The third section deals with the importance of testing procedures for the fire rating characterization of PV modules.

## 2. PV MAIN CAUSES AND MECHANISMS OF FIRE IGNITION

Experimental data demonstrated that special attention should be paid to the technological aspects of PV systems and, above all, to the possible reduction of some critical production defects, that should be avoided in order to reduce the risk of fire in buildings.

Since PV electrical plants are a subset of the family of electrical plants, they show the two main causes of electrical fire ignition: overload and short circuit. Moreover, if a PV electrical plant is designed and put in place according to national, European or international electrical code recommendations (i.e., NFPA 70, IEC 60364-7-712, CEI 64-8) the fire risk due to overload and short circuit is well addressed and mitigated. Furthermore, PV cell outputs characteristics shows a short circuit current a little greater than the nominal one.

Unfortunately, PV power plants show a set of proper causes of electrical fire ignition. There are two major considerations for fire safety of photovoltaic systems [2]: Fires could be caused by PV components and systems – “origin fire scenario”; Fires could be caused by external sources – “victim fire scenario”.

Arcing (Series Arcing, Parallel Arcing, Grounding Arcing) is one of the main causes of fire ignition [3]. Another important cause of fire ignition is related to the mechanism called in literature as “hot spot” [4]. An Hot spot on a PV module produces localized heating on the back sheet that could ignite and start a proper fire. Flowing of reverse currents in a module due to shadows or PV cell misalignment is responsible for the mechanism of fire ignition called “mismatch” [5]. Lack or bad connection among modules, strings (a series connection of two or more modules) and inverter could generate an enough amount of heat energy that could start a fire (as represented in Fig. 2).

Furthermore, during a PV fire event it is not possible to turn off the whole power system in order to guarantee that all the components are de-energized, while PV module and panels are exposed to solar or artificial light [6]. Whenever sufficient voltage is developed by a number of PV modules in series, there is a risk of electric shocks. Higher direct current (DC) voltage may be used in PV systems because it can lead to reduced wiring size and higher inverter efficiency. Hence, electrical hazards to firefighters involved in the mitigation of a fire concerning photovoltaic modules [7] ought to be also considered and addressed in a PV building.

## 3. THE PV FIRE SAFETY GUIDELINE

The presence of photovoltaic systems on buildings, both applied (BIPV) and integrated (BIPV), could worsen the existing level of risk of fire because the photovoltaic modules, panels and components could contribute significantly to the spread of fire outside or through the construction, interferes with smoke and vent system of the combustion products, obstacles to the operations of firefighting and introduce an additional danger for electric shock to the firefighters and rescue workers to the presence of elements of the circuit energized.

The Italian National Fire Services have issued a Guideline [8], [9] and [10] with the aim to reduce the outbreak of a PV fire related incident and their consequences on host buildings and construction works. The guideline has been developed with a non-prescriptive approach; in fact, according to the Regulation of the Construction Product – CPR Regulation [11] (EU) n.305/2011 of 9 March 2011, the

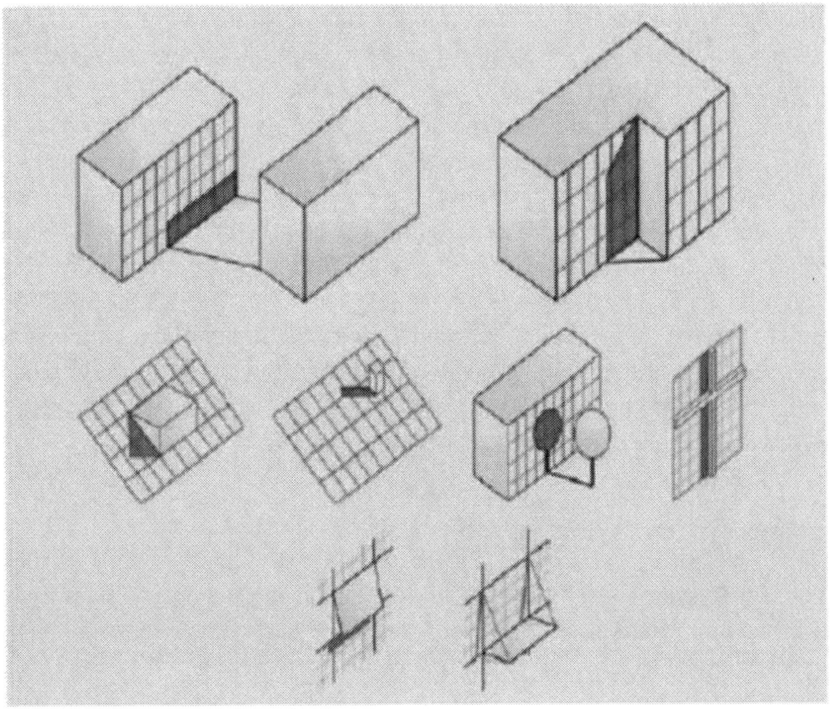

**Figure 2.** PV Self-shading failure mechanisms.

assessment and mitigation of the fire risk as a result of a PV plant installation on a building should be performed in order to meet the basic requirement n. 2 of construction work: "Safety in case of fire". The basic requirement n. 2 – Safety in case of fire – claims that a construction work must be designed and built in such a way that in the event of an outbreak of a fire,

(1) The load-bearing capacity of the construction can be assumed for a specific period of time;
(2) The generation and spread of fire and smoke within the construction works are limited;
(3) The spread of fire to neighbouring construction work is limited;
(4) Occupants can leave the construction work or be rescued by other means;
(5) The safety of rescue teams is taken into consideration.

The design and the installation of a PV power plant must be carried out in order to meet the basic requirement n. 2 listed above, addressing the following issues:

1. Avoiding interference with the ventilation system of the combustion products, obstruction of skylights and impediment of natural smoke and heat exhaust ventilators;
2. Avoiding the spread of an outbreak fire from the photovoltaic generator to the building and /or between its compartments in which its incorporated;
3. Evaluating the influence on the fire spread due to PV cables, switchboards and modules on the roof and or the façade of the host building/construction work;
4. Safety of maintenance personnel
5. Safety of the rescue team.

A complete analysis and discussion of the Italian Guideline for the fire safety of PV Systems applied or integrated in buildings is reported in [12].

One of the main fire safety goal for a building design is to restrict the vertical fire spread so that the smoke and flames are limited to the fire origin floor. The new building façade and curtain wall topologies could overwhelms concerns for fire safety, therefore the Italian National fire service has released a Fire Code Guideline in order to address the fire safety design for an high rise building façade [13].

As well known, the most used façades are built whit a series of layers of materials that increase insulating and sound absorbing performances without any considerations of combustibility and participation to fire tendency. For this reason, often, an ignition of a façade becomes a severe problem for the protection of people and property. Furthermore, if a fire starts at road level – i.e. a burning car or a waste container –, and ignite a nearby building façade covered or completely built up by combustible materials, taking into account the vertical position of the fuel and the unlimited presence of oxidizer quantity (the oxygen in the atmosphere), the fire façade could become a real huge torch. In order to reduce the vulnerability of the overall cover building system, the lightweight, high insulation and sound absorbing materials, and also PV cladding and covers used in the construction of a facade must have a proper reaction to fire performance. The next section reports the fire rating characterisations of PV module and panels used as applied or integrated system in building.

## 4. FIRE RATING OF PV MODULES

Fire spread and flame propagation on a roof or a building façade are strongly influenced by the Reaction-to-Fire of roof and building products and materials. In fact, Reaction-to-Fire of a construction product deals with characteristics such as ignition, flame spread, heat release rate, smoke and gas production and the occurrence of burning droplets.

The fire rating of PV modules in the EU market is not mandatory; in fact when the PV module international standard IEC 61730-2 "Photovoltaic (PV) module safety qualification – Part 2: Requirements for testing" becomes the European standard EN 61730-2, the Fire Test MTS 23 could be performed only if requested by the manufacturer of the PV modules. The MTS 23 test is carried out according to the ANSI UL/790 standard ("Standard for Standard Test Methods for Fire Tests of Roof Coverings"), and the fire performance classification of PV module ranges from Class C – "*fundamental fire rating*" – , to Class B to Class A – "*highest fire rating*" – ; the standard also requires a minimum fire resistance rating of Class C for any building-mounted module. Furthermore, UL 790 Standard addresses two fire protection concerns: flame spread along the roof and fire penetration.

The EN 61730-2 standard also states that the MTS 23 test specifies fundamental requirements and may not be sufficient to satisfy the needs for a module intended for building applications according to local or national building fire code requirements.

Following the last sentence, the Italian National Fire Services, that is the Authority Having Jurisdiction for fire safety in Italy, according to the experimental activities carried out by the Passive Fire Protection Area of the Central Directorate for Fire Prevention and Technical Safety has issued a national Resolution in order to test and classify the Reaction to fire rating of PV modules.

Since the classification of the fire behaviour and the certification of materials for the purpose of fire prevention are regulated by the Decree of the Ministry of the Interior issued on June 26, 1984 [14], the PV Resolution states that a PV module can be classified following the Italian Reaction to fire Classification provided by the standards, test methods and classification procedures stated in the aforementioned Decree.

The national reaction-to-fire performance classifications for combustible materials are determined by using the results from a combination of the three tests shown below:

- UNI 9174, Reaction to fire of products subjected to a flame in the presence of radiant heat [15];
- UNI 8456, Combustible products which can be hit by flames on both surfaces – Small flame test [16];
- UNI 8457, Combustible products which can be hit by flames on one surface – Small flame test [17];

**Table 1.** Italian classification of reaction to fire for materials.

| Test Method | Test Results | | | | | | Reaction to fire Rating (CLASSE) |
|---|---|---|---|---|---|---|---|
| UNI 9174<br>UNI 8457 | I Category<br>I Category | | | | | | 1 |
| UNI 9174<br>UNI 8457 | II Category<br>I Category | or I<br>II | | | | | 2 |
| UNI 9174<br>UNI 8457 | III Category<br>I Category | or II<br>III | or I<br>III | or II<br>II | | | 3 |
| UNI 9174<br>UNI 8457 | IV Category<br>III Category | or III<br>IV | or III<br>III | or IV<br>II | or II<br>IV | or I<br>IV | 4 |
| UNI 9174<br>UNI 8457 | IV Category<br>IV Category | | | | | | 5 |

Table 1 illustrates how the outputs results of each test are combined to produce the overall national classifications: combustible materials and products are divided into 5 categories, 1, 2, 3, 4 and 5, with 1 the best level of performance and 5 the worst. The table is derived from UNI 9177 "Reaction to fire – combustible products classification" [18].

The UNI 8457 test applies a little ignition flame to the lower border of the specimen under test for 30 seconds, observing and measuring the flame propagation speed, the damaged zone of the specimen, the post glowing time and the flaming droplet/particles.

The UNI 9174 deals with the reaction to fire performance of products that can be hit by a flame and a radiant heating; the specimen under test is placed on an incombustible support exposed to a radiant panel and hit by a flame to the lower side of the specimen. During the test, the heat emitted by the radial panel is equal to 6,2 W/cm$^2$. During the radiant panel test, the following parameters are observed and recorded:

- ignition velocity;
- flame propagation speed;
- damaged zone extension;
- Post glowing time;
- Flaming droplet/particles.

Next figure shows some fire test result on a façade PV module during the test with a flame in the presence of radiant heat: the left image of the figure highlights the layer separation due to the PV backsheet combustion, while the centre of the figure shows the mechanics of PV backsheet blistering. The right part of the figure depicts the height of the flame during the PV combustion.

The class 1 reaction to fire rating according to the Italian test procedure illustrated above can be easily achieved for PV module that are "glass to glass" type. In the most common PV module technology, the glass is only used on the side exposed to sun, while the backsheet is generally constitute by combustible materials. The flame spread behavior as well as the flame speed propagation are strictly related to the materials chosen to built up a PV module backsheet. In order to comply to the Italian Reaction to Fire code requirements, a PV module manufacturer should asses the material chosen for building up a proper reaction to fire PV module. A detailed explanation of the Italian reaction to fire method applied to PV module characterisation is reported in [19].

## 5. PV MODULE AS CONSTRUCTION PRODUCTS

In construction terms, Building integrated PV systems should play the same role as the traditional wall and roofing cladding elements they replace. Consequently, they must address all the normal issues of a

*Backsheet layer separation* *Backsheet blistering* *Flame height*

**Figure 3.** Façade PV module subjected to Italian radiant panel reaction to fire test.

construction product, i.e. appearance, weather tightness and protection from the elements, wind loading, lifetime of materials and risks and consequences of failure and all the safety aspect (construction, fire, electrical, etc.).

PV modules and systems are not yet regulated by any European harmonized standard. There is an European project performed by a working group and recently submitted to CEN ad CENELEC entitled as "prEN 50583 Photovoltaics in Buildings:2012" [20]. This prEN introduces for the first time the definition of the terms BIPV and BAPV.

BIPV are Building Integrated Photovoltaic modules providing a function as defined in the European Construction Product Directive CPD 89/106/CE [11]. Therefore, a PV module can be considered as a BIPV if it take parts for the integrity of the building's functionality. In other words, if an integrated PV module is dismounted, it would have to be replaced by an appropriate building component. This means that the BIPV modules must be able to perform on or more of the following functions:

- mechanical rigidity and structural integrity;
- primary weather impact protection: rain, snow, wind, hail;
- energy economy, such as shading, daylighting, thermal isulation;
- fire protection;
- noise protection.

BAPV are considered Building Attached photovoltaic, if the PV modules are mounted on a building envelope and do not fulfil the above criteria for building integration. This means that the integrity of the building functionality is independent of the existence of a building-attached photovoltaic module.

Furthermore, the European draft standard introduces several application categories and subdivides these into the essential requirements of the European Construction Product Directive. Façade elements fall into Category C and Category D as reported in Fig. 4.

The prEN 50583 assigns application – specific requirements to PV modules – divided into the main categories "containing" and "not containing glass panes".

| | | |
|---|---|---|
| Category C: | **Non-sloped (vertically) mounted not accessible from within the building** | 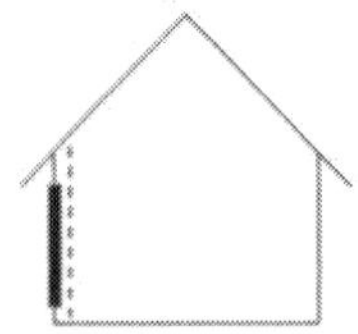 |
| | The PV modules are mounted in the building envelope at an angle between and including both 75° and 90° (see Fig. 1) with a barrier behind preventing large pieces of glass or persons falling to an adjacent lower area inside the building | |
| | | |
| Category D: | **Non-sloped (vertically) mounted accessible from within the building** |  |
| | The PV modules are mounted in the building envelope at an angle of between and including both 75° and 90° (see Fig. 1) | |

**Figure 4.** PV Façade elements application categories according to prEN 50583 "Photovoltaic in Buildings".

For the safety in case of fire, the general requirements for all BIPV categories refers to the EN 13501-1 standard, having a minimum requirement of European reaction to fire class E. For Category C façade PV elements fire classification standard should be performed according to EN 13501-2 and EN 13501-5, while Category D requires only EN 13501-2 Fire classification. For PV flexible plastic and rubber sheets, the fire performance could require Broof T1, T2, T3 or T4 classification according to EN TS 1187:2012.

## 6. CONCLUSION

Safety is a prime concern for the photovoltaics (PV) industry. As a technology deployed on residential and commercial buildings, it is critical that PV not cause damage to the buildings nor harm the occupants. The issue of an European harmonised standard could be useful for making façade PV modules inherently safer especially against fire risk.

The main conclusion is that the reduction of fire risk, fire spread and proper ignition mechanisms caused by such systems are strongly related to a proper and accurate risk assessment of the PV building installations and to the testing, qualification and certification aspects of the fire behaviour of the PV façade elements. Combining safer PV modules with better system designs can solve many of the safety problems observed with today's façade PV systems.

### References


[1] EPIA (2014–2018) Global market outlook for photovoltaic. www.epia.org/news/publications/global-market-outlook-for-photovoltaics-2014-2018/].

[2] Zhao Y, de Palma J, Mosesian J, Lyons R, Lehman B. Line–Line Fault Analysis and Protection Challenges in Solar Photovoltaic Arrays. IEEE Transactions On Industrial Electronics 2013; 60(9). DOI: 10.1109/TIE.2012.2205355.

[3] International Photovoltaic Reliability Workshop II PV Fire: Experience and Studies Experience Studies Liang Ji Underwriters Laboratories Inc (UL) July 31, 2009 T A i Copyright© 1995–2007 Underwriters Laboratories Inc. Tempe, Arizona, USA.

[4] Qiu YN. Electrical mismatch within single junction amorphous silicon and micromorph tandem thin film PV modules, 34th IEEE Photovoltaic Specialists Conference (PVSC), 2009 7–12 June 2009].

[5] Spertino F, Akilimali JS. Are manufacturing I–V mismatch and reverse currents key factors in large photovoltaic arrays. IEEE Transactions on Industrial Electronics 2009; **56**(11):4520–4531.

[6] Behavior of the Electrical Parameters of PV Modules Subject to a Flame Ignition, P. Cancelliere, V. Puccia, G. Longobardo, C. Liciotti, M. Cardinali, 28th European Photovoltaic Solar Energy Conference and Exhibition Session.

[7] Risk of electrocution during fire suppression activities involving photovoltaic systems, Riccardo Tommasini, Enrico Pons, Federica Palamara, Calogero Turturici, Pietro Colella – Fire Safety Journal, **67**, July 2014, Pages 35–41.

[8] Note n. 1324 February 7, 2012 "Guideline for PV plant installations", 2012 Edition ("Guida per l'installazione degli impianti fotovoltaici Edizione", Anno 2012).

[9] Note. n. 6334 May 4, 2012 – Explanation to the requirements of the Note n. 1324 February 7, 2012 (in Italian).

[10] Note PROTEM 622/867 February 18, 2011 "Procedures in case of intervention in the presence of photovoltaic panels – firefighters safety" (in Italian).

[11] REGULATION (EU) No 305/2011 OF THE EUROPEAN PARLIAMENT AND OF THE COUNCIL. Laying down harmonised conditions for the marketing of construction products and repealing Council Directive 89/106,EEC. Official Journal of the European Union 2011; L88/5, **54**, 4 April 2011.

[12] PV electrical plants fire risk assessment and mitigation according to the Italian national fire services guidelines, Piergiacomo Cancelliere, FIRE AND MATERIALS Article first published online: 5 DEC 2014, DOI: 10.1002/fam.2290.

[13] The Italian National Guidelines for the fire safety of facades, Mazziotti Lamberto and Piergiacomo Cancelliere, 1st International Seminar for Fire Safety of Facades Paris, France, November 14–15, 2013 Published online: 29 November 2013, DOI: `http://dx.doi.org/10.1051/matecconf/20130901005`.

[14] Decree of the Ministry of the Interior June 26, 1984 "Classification of reaction to fire and approval of materials for the fire prevention" (in Italian).

[15] UNI 9174, Reaction to fire of products subjected to a flame in the presence of radiant heat.

[16] UNI 8456, Combustible products which can be hit by flames on both surfaces – Small flame test.

[17] UNI 8457, Combustible products which can be hit by flames on one surface – Small flame test.

[18] UNI 9177 " Reaction to fire – combustible products classification".

[19] Fire Behaviour and Performance of Photovoltaic Module Backsheets, Piergiacomo Cancelliere, Claudio Liciotti, Fire Technology pp 1–16 First online: 11 June 2015.

[20] Testing and Standards for new BIPV products, Pellegrino, M.; ENEA (Agenzia Naz. per le Nuove Tecnol., l'Energia e lo Sviluppo Sostenibile), Portici, Italy; Flaminio, G.; Graditi, G., Industrial Electronics Society, IECON 2013 – 39th Annual Conference of the IEEE.

MATEC Web of Conferences **46**, 05002 (2016)
DOI: 10.1051/matecconf/20164605002


# Means of regulating combustible materials and products in external walls

Esko Mikkola

KK-Fireconsult Ltd., FI-02240 Espoo, Finland

**Abstract.** This report presents proposals for defining means of regulating the use of combustible materials and products in external walls. Required protections are based on the quantities of fire loads and their contribution to fire development. The study is based on life safety and protection of property priorities taking into account reaction to fire classes related to different types of fire loads and fire compartmentation requirements of the adjacent spaces of concern.

The proposals include the following main principles in relation to fire-separation requirements: In case of internal fire exposure the protective structure for combustible building parts needs to meet at least half of the fire-separating requirement for the compartment of concern. In case of external fire exposure the protection time requirement can be 15 minutes less than for the internal protection. The proposals are applicable for residential buildings and offices. In case of buildings with longer evacuation times more stringent requirement levels may be considered.

For verification of protection performance of fire loads it is proposed to use existing standardized test methods (fire protection ability (K classes) and fire-separating function (EI classes) validated methods of calculation and/or large scale fire testing.

## 1. INTRODUCTION

More stringent energy efficiency requirements for buildings have increased insulation layer thicknesses and the use of combustible insulation materials in external walls. Other changes during the last decades have been at least the following: The quantity and quality (more flammable items) of movable fire loads have changed, new construction products have entered to the market and new structural and architectural solutions have been developed. The main topic of this paper is to assess the contribution of thermal insulation materials (and other building materials within external walls) to the fire spread and required means of protection.

In regulations restrictions for use of combustible materials may be used or when such materials and products are allowed, minimum requirements of protection can be defined or fire performance may be verified by using large scale fire testing.

A prerequisite for the assumption that combustible building materials/products need not to be taken into account in fire development (and in consequences of fire spread), is that the fire load of concern will not contribute to fire in a certain time (e.g. time needed for safe evacuation). This means that combustible products usually need to be covered by protective layers and the protective performance is related to the assumed fire exposures of concern.

## 2. FIRE LOADS AND THEIR CONTRIBUTION TO FIRE

Fixed fire loads within building structures need not to be assumed to contribute to fire in all cases (e.g. within a defined time related to use of the building) or it can be assumed that the fire load burns only partly, if the fire load can be protected in a reliable manner. There are different protection methods available. Coverings (=products/boards having fire protection ability) are aimed to protect materials behind the covering from fire for specified times, such as 10 or 30 minutes. Similar protective effects can be achieved by using fire-separating building elements (e.g., EI15, EI30, EI60, etc.). Fire loads can be protected from fire also by preventing or limiting the fire by using automatic extinguishing systems.

It can be assumed that combustible building materials/products need not to be taken into account in any case in fire, when the fire load of concern will not contribute to fire development during the whole fire duration (including cooling phase). In practice, this can lead e.g. to the following interpretations:

- If the fire load is at maximum 600 $MJ/m^2$, then 60 min protection time (according to standard fire exposure curve) is always sufficient (even when there is no automatic extinguishing).
- The protection can also consist of a combination of passive (protection board) and active (automatic extinguishing) methods.

On the other hand, as an example the Finnish fire regulations [1] use 10 minutes fire protection time in 1–2 floor buildings assuming that a sufficient delay in the contribution to fire development of the protected fire load is reached. In this case, the emphasis is in life safety and property protection is having a lower priority (because total losses and risks to the neighborhood are lower than in case of higher buildings).

The higher the buildings are, the more important it is to limit the amount of fixed fire load contributing to fire. This is not only because of life safety (long evacuation times) but also because total effective fire loads which are higher than the design values can lead to collapse of the building.

Fire load consists of moveable property and fixed fire load in building structures. From regulatory point of view these two parts can be described as follows:

- Movable property as a fire load:
  - Furniture, fixtures, equipment, storage of goods, etc.
  - Mostly not regulated; room size and heat of combustion values of materials give the limits.

- Fixed fire loads within building materials and structures:
  - Surface linings, thermal insulations and other building products.
  - Load-bearing structures.
  - Building regulations give restrictions regarding combustibility (fire load) in load-bearing structures and thermal insulations depending on fire class of the building, number of storeys and use of the building.

## 3. SIGNIFICANCE OF THERMAL INSULATIONS IN FIRE LOADS

Figure 1 provides a summary of fire load changes concerning thermal insulations in external walls when normalized per apartment floor area in Finland [2]. The illustration shows the extreme limiting cases where either one of the apartment walls is an outer wall or three walls are outer walls.

In EPS and PIR insulated buildings the increase in total fire load (per floor area) caused by exterior walls between years 1976–2010 was $10 \pm 3\%$ assuming that all of the fire load could contribute to fire. Corresponding increase in case of stone wool insulation has been less than 1.7%.

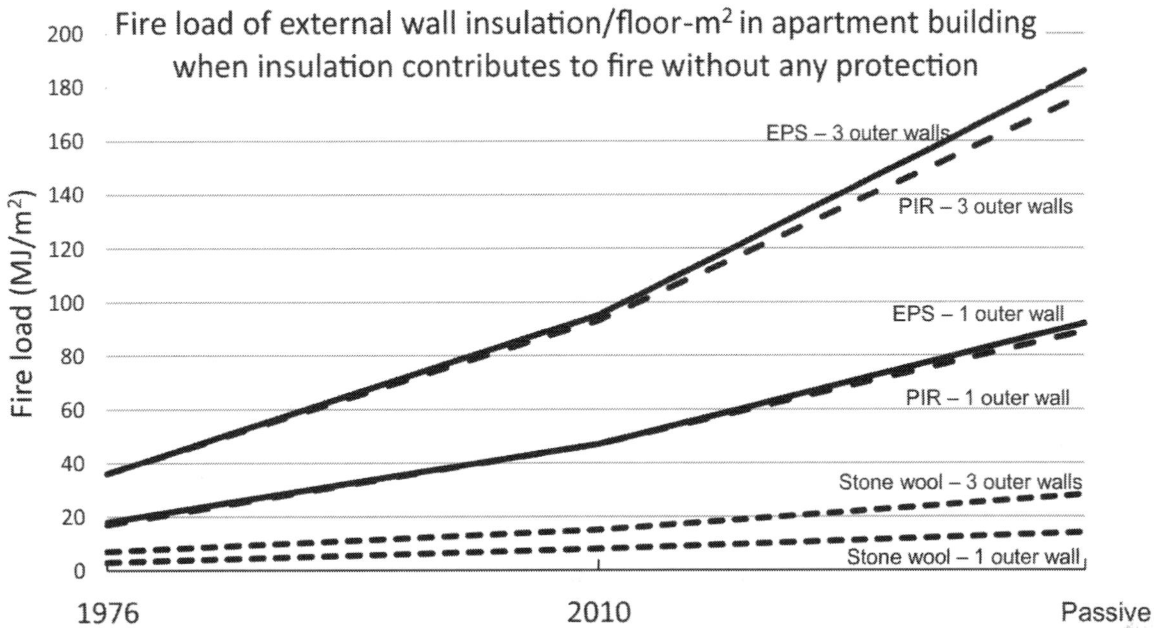


**Figure 1.** Changes in fire loads of external wall thermal insulations normalized per floor area in cases where one or three of walls of an apartment are outer walls (Case Finland; Passive refers to possible future requirements).

**Table 1.** Estimated values of fire load and rate of heat release for 200 mm thick thermal insulation materials without any protection in different reaction to fire classes.

| Fire class and Density of insulation | Fire load (MJ/m$^2$) | Rate of heat release (kW/m$^2$) 900 s mean value | Heat of combustion estimates |
|---|---|---|---|
| A2-s1, d0; 50–120 kg/m$^3$ | 30–70 | < 20 | A2 limit = 3 MJ/kg |
| B-s1, d0; 30–50 kg/m$^3$ | 60–200 | < 100 [3] | 6–20 MJ/kg |
| D-s2, d2; 25–50 kg/m$^3$ | 100–200 | < 250 [3] | 10–20 MJ/kg |
| E–F; 15–20 kg/m$^3$ | 100–150 [4] | 200–500 [4] | 30–40 MJ/kg |

# 4. FIRE PERFORMANCE AND PREVENTION OF SPREAD OF FIRE

## 4.1 Reaction to fire classes of materials and their contribution to fire

Thermal insulations are installed on the whole area of external walls with the exception of windows, doors and other openings. In Table 1 fire load and heat release values per building envelope area are given in the case when the 200 mm thick thermal insulation product is without any protection. These estimated values are based on relevant parameter limit values of each reaction to fire class.

In addition to the characteristics described in Table 1, significant differences between fire performance of products arise on whether the products char or melt: in charring materials spread of fire is usually slow and the propagation of char front can be predicted, whereas in melting products voids can be created which may cause additional spread of fire.

## 4.2 Protection of combustibles within building structures

For regulatory purposes it is proposed that protection of combustible products within external walls is defined in relation to the fire exposure of concern. The assumed fire exposure can be based e.g. on internal partitioning requirement (which is related to assumed fire load) or on external fire exposure based on large flames from a window (flashover fire).

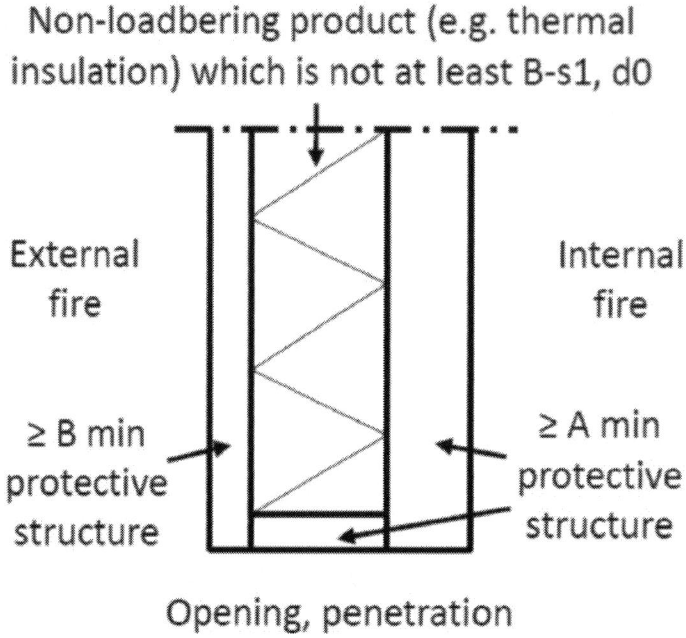


**Figure 2.** Protection against internal and external fire.

Main principles of the proposals in relation to internal compartmentation requirements (see Fig. 2) are:

Internal fire exposure (subject to inner surfaces and edges around openings):

- Protective structure fulfils at least half of the requirement for fire separating building elements in the space of concern.
- Applies also to protection of any ventilation gaps.

External fire exposure (subject to outer surfaces):

- Requirement for interior surfaces reduced by 15 minutes (except in case of EI30 requirement for interior spaces, reaction to fire requirements should be applied).

Note that the protection requirement against the internal fire is applicable also to edges of all windows and doors (including possible openings to ventilation gaps) as well as penetrations. Requirements for external surfaces are usually lower than for internal surfaces because of lower total fire exposures (open space and faster cooling).

The proposals are applicable for residential buildings and offices. In case of hotels, schools, day care centres and hospitals there is a need to take into account possible longer evacuation times and as a consequence more stringent requirement levels may be used.

## 5. VERIFICATION OF PROTECTION METHODS OF FIRE LOADS

For verification of protection performance levels of combustible materials and products within external walls the following options are being proposed:

A. Application of existing European test methods (fire protection ability (K classes) [5] and fire-separating function (EI classes) [6]) and calculation methods (e.g. Eurocode).
B. Large-scale facade testing – need for internationally accepted method (one test method capable of differentiating performance levels (classes) according to national needs).

C. Case by cases assessment using performance based fire safety design.
   - E.g. analysis based on in-depth temperature measurements of a wall construction in a fire resistance test.

Concerning fire-separating requirement, the protective structure shown in Fig. 2 shall fulfil the criteria given in the test method and in the classification standard either tested as a protective layer alone, or alternatively, as part of an end-use structure. In the latter case, verification of performance of the protection ability is based on measurement of insulation performance (thermocouples on the unexposed face of the protective structure), and on integrity (observation of cracks and holes in the protective structure).

### 5.1 Calculation methods

EN 1995-1-2 (Eurocode 5. Design of timber structures. Part 1-2: General. Structural fire design) [7] includes calculation methods for insulation performance for wooden boards, gypsum boards, stone wool and glass wool insulation materials. The calculation results must be associated with necessary guidance on mounting methods and sealing of joints.

Also other suitable (validated) calculation methods can also be used. The calculation method must be able to demonstrate that fire performance levels corresponding to the covering classes or fire-separating classes of concern are reached.

## 6. DISCUSSION AND CONCLUSIONS

As far as it concerns fire safety the most simple way to regulate the use of combustibles in external walls is to forbid their use according to type the building, use of the building, etc. However, this may not always lead to the best overall performance of the building. Thus use of combustibles can be allowed, but defined protective layers need to be used according to required fire performance levels taking into account both safety of life and property protection.

From the point of view of regulations, it would be desirable to use a fire load definition, which usually (in most cases) does not lead to a specific assessment of the fire load, but assumed values for fire load categories can be used. Two types of fire loads can be determined: Internal fire load of a fire compartment including movable fire load, fixed furniture and fire load within surface layers of structural elements, and protected fire load within building elements (protective performance determined according to the type of the building and fire-separating requirements).

For protection levels of fire loads within structures the following is proposed in relation to internal fire compartmentation:

Internal fire exposure (subject to inner surfaces and edges around openings):
- Protective structure fulfils at least half of the requirement (and not more than the requirement) for fire-separating building elements in the space of concern.

External fire exposure (subject to outer surfaces):
- Requirement for internal surfaces reduced by 15 minutes (except in case of EI30 requirement for interior spaces, reaction to fire requirements should be applied).

For verification of protection performance of fire loads it is proposed to use existing standardized test methods (fire protection ability (K classes) and fire-separating function (EI classes)), validated calculation methods or large-scale fire testing.

Tuomo Rinne and Peter Grönberg are thanked on fire load analysis. The financial support of Finnisolry is gratefully acknowledged.

## References

[1] E1. The national building code of Finland. Fire safety of buildings. Regulations and guide-lines 2011. Helsinki. Ministry of the Environment (In Finnish).
[2] Mikkola, Esko. Fire loads and fire safety in buildings. Part 2: Protection of fire loads within building structures. Research report. KK-Palokonsultti Oy, Espoo, 2015.
[3] Östman, Birgit; Tsantaridis, Lazaros; Mikkola, Esko; Hakkarainen, Tuula; Belloni, Kaisa; Brumer, Harry; Piispanen, Peter. 2006. Innovative eco-efficient high fire performance wood products for demanding applications. Final report for Vinnova-Tekes project InnoFIreWood. Stockholm, SP Wood Technology. 98 s. SP Swedish National Testing and Research Institute, Report 2006: 30.
[4] Mikkola, Esko; Hakkarainen, Tuula; Matala, Anna. 2013. Fire safety of EPS ETICS in resi-dential multi-storey buildings, VTT. 37 p. + app. 4 p. Research Report; VTT-R-04632-13.
[5] EN 14135. Coverings - Determination of fire protection ability. CEN 2004.
[6] EN 13501-2. Fire classification of construction products and building elements – Part 2: Classification using test data from resistance fire tests, excluding ventilation services.
[7] EN 1995-1-2. Eurocode 5. Design of timber structures. Part 1-2: General. Structural fire design.

MATEC Web of Conferences **46**, 05003 (2016)
DOI: 10.1051/matecconf/20164605003


# Fire protection concepts for Timber-Glass Composite façades

Andreas Schleicher

Vienna University of Technology Vienna, Austria

**Abstract.** The issues of this article are subject of the research project "Timber-Glass Composite Façades | Behavior in Case of Fire | Fire Protection Concepts" which is performed by Brandrat ZT GmbH in cooperation with the Department of Architectural Sciences and Timber Engineering, Vienna University of Technology.

It is part of the research program "Urban Solutions 2014" and is funded by the Vienna Business Agency. Start of the project took place in October 2014 and it should be completed in March 2017. The project management is carried out by the author of this paper.

**The main objectives can be summarized as follows:** Growth and densification in urban areas require the development of intelligent and resource-efficient building systems for "Smart Cities" of the future. By using timber-glass composites (TGC) the primary energy demand of buildings can be reduced substantially. This research project examines the feasibility of applications of this new technology in multi-story and high-rise buildings.

Critical aspects concerning fire protection such as flammability of timber elements, fire spread and failure of façade elements with bracing capacity will be analyzed. Different strategies will be developed in case studies and validated by structural analysis. Large scale mock ups of TGC façade elements will be checked on their suitability in fire tests.

The findings of this research will lead to innovative fire safety concepts for building systems with TGC façades. Compliance with the high safety standards for multi-story buildings in urban areas like Vienna is one of the main objectives of this work. The adaptation of these fire safety concepts to the national standards of the neighboring countries will be continued subsequently. The gained knowledge should lead to further cooperation with companies for serial productions with TGC technology.

## 1. INTRODUCTION

### 1.1 Basic principles

The specific design feature of TGC elements is a continuous adhesive bonding between the glass pane and an adapter frame made of timber. Approved serial applications are using bond lines made of silicone adhesives – however, current research projects are also investigating alternative adhesives with increased stiffness (e.g. polyurethane, acrylate, epoxy adhesives). The connection between the glazing and the adapter frame enables TGC façades to serve as a structural element especially for horizontal forces like wind or earthquake loads. Thereby other bracing elements like walls or crossings in the façade level can

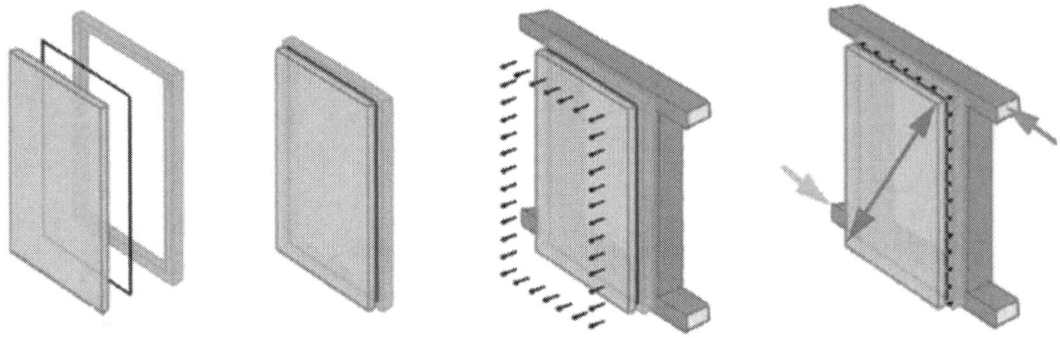

**Figure 1.** Schematic illustration of the basic principles of TGC façade elements. Picture credits: Winter, W., Tragende Holz-Glas-Klebeverbindungen, Grundlagen und Anwendungsmöglichkeiten, 5. Europäischer Kongress für effizientes Bauen mit Holz, 2012.

**Figure 2.** TGC system with flat adapter frame (shear). Picture credits: Uniglass Handbuch für die Planung und Erstellung von Uniglass | Facade Holz-Glas-Verbundelementen, 2014.

be reduced, depending on the main structure of the building. The basic principles of TGC elements can be described as follows and are depicted in Fig. 1.

- adhesive bonding of glass pane and adapter frame
- mounting of the composite element to a substructure
- force-locked connection of façade with floor slabs
- horizontal load transfer by TGC element.

Higher load capacities can be achieved by load transmission by shear (adhesive bond line) in combination with contact by blocking the glass pane. Additional blocking forces are concentrated in the corners between the adapter frame and the glazing and lead to higher stiffness with less deformation. Modeling and calculation of these structural elements with both shear forces in the bond line and a compression diagonal in the glass pane were subject of research projects and dissertations by the Department of Architectural Sciences and Timber Engineering, Vienna University of Technology [1, 2]. In figure 2 and 3 the different details of a TGC system with a flat adapter frame (shear) and a TGC system with a u-shaped adapter frame (shear and contact) are illustrated.

## 1.2 Motivation for TGC-applications in multi-story buildings

Small scale applications of TGC for windows and conservatories are already used by the building industry. Pilot projects of two-story family houses represent the current state of the art. Hence TGC

**Figure 3.** TGC system with U-shaped frame (shear and blocking contact). Picture credits: author (Glasstec 2014 – Düsseldorf).

façades should also be spread to applications in larger buildings due to the positive aspects of this technology listed below:

- transparent architecture with façade systems without visible bracing elements
- double use of glazing as a climate shell and as a structural element
- solar energy gains through transparency without thermal bridges
- sustainability, reduced primary energy consumption and recyclability.

### 1.3 Specific risks concerning fire protection issues and TGC façades

In order to implement TGC applications in multi-story and high-rise buildings, both European and national laws, standards and guidelines must be considered as an assessment basis. The European Regulation No. 305/2011 [3] gives the following essential requirements for the limitation of fire risks:

*"The construction works must be designed and built in such a way, that in the event of an outbreak of fire*

- *the load bearing resistance of the construction can be assumed for a specified period of time,*
- *the generation and spread of fire and smoke within the works are limited,*
- *the spread of fire to neighboring construction works is limited,*
- *the occupants can leave the works or can be rescued by other means,*
- *the safety of rescue teams is taken into consideration".*

On the national level the "Richtlinien des Österreichischen Instituts für Bautechnik (OIB Richtlinien)" [4] provide the current state of the art concerning building technology and fire safety in Austria. Fire protection requirements concerning façades of multi-story buildings can be identified as follows:

- prevention of fire spread across the façade to the second floor level, located above the fire
- no danger to people and rescue teams by the falling of large façade parts
- prevention of spread of fire through gaps or joints and voids in the range of floor slabs.

Due to the specific design of TGC façades the following risks have to be considered and, where appropriate, additional measures have to be taken into account:

- flammability of individual components of the façade system (timber substructure and adapter frame, adhesives)
- maintaining the same level of protection for high-rise buildings, while using combustible materials in the façade area
- failure of TGC façade elements with bracing capacity in case of fire.

## 2. FIRE PROTECTION CONCEPTS FOR TGC-APPLICATIONS

Fire safety regulations by the OIB Richtlinien [4] can be complied by TGC façades for objects up to building classification 5 (flight level up to 22 m). Due to the increased level of security for buildings with a flight level of more than 22 m (high-rise buildings) the requirement, that only non-combustible material must be used, cannot be fulfilled with TGC façades, mainly because of components such as timber and adhesives. Any deviation from these requirements is only acceptable if the same level of protection can be demonstrated in conjunction with corresponding compensation measures (e.g. specific sprinkler systems). Depending on structure, usage and building classification one of the following three strategies might be used to develop object-specific fire protection concepts for buildings using TGC technology.

### 2.1 Protection of TGC façades in case of fire

In order to fulfil the requirement of preventing fire spread across the façade, the implementation of various fire protection measures into TGC façade systems is being investigated in this research. Basically fire resistant glazing could be used for TGC façades, however, due to economic considerations this is not a realistic solution. Besides the issue of how to protect the bonding in case of fire remains unsolved.

The design of TGC façades with fire resistant parapets seems to be more appropriate and can be solved either with façade-integrated fire-resistant panels or with parapets being part of the main structure of the building. Depending on the details of these measures it is possible to realize fire resistance between 30 and 120 minutes. With a height of at least 120 cm, these elements meet the requirement of preventing vertical fire spread in accordance with the current OIB Richtlinien [4]. However, this measure is not designed to protect TGC façades against reduction or loss of bracing capacity in case of fire.

Another option to prevent horizontal and vertical fire spread across the façade is to use automatic sprinkler systems, especially when they are designed with the objective to prevent fire spread in the façade area. In this case no additional structural measures of the façade have to be taken into account. For high-rise buildings with a flight level of more than 22 m automatic sprinkler systems might also be considered as a compensatory measure for the use of combustible materials in TGC façades in some cases, depending on the type of façade and the sprinkler system. Besides in fire tests of façades in conjunction with sprinkler systems temperatures in the range of 200 °C in the fire room were observed – TGC façades with silicone adhesives will still serve as a bracing element (with reduced capacity) in this temperature range.

### 2.2 Partial failure of TGC façades in case of fire

When using TGC façades as structural bracing elements (e.g. in building structures without a core element for primary bracing) the bracing capacity of the façade has to be ensured in case of fire. The division into fire compartments or small cells by vertical components is more effective to retain the bracing capacity of the building structure (at least with reduced capacity) in case of fire. On the other hand building structures with fire compartments divided by floor slabs might not retain any bracing capacity due to the fact that in case of fire TGC elements will fail along all sides of the building in a worst case scenario. This would lead to a total failure of the bracing capacity of the building, even if

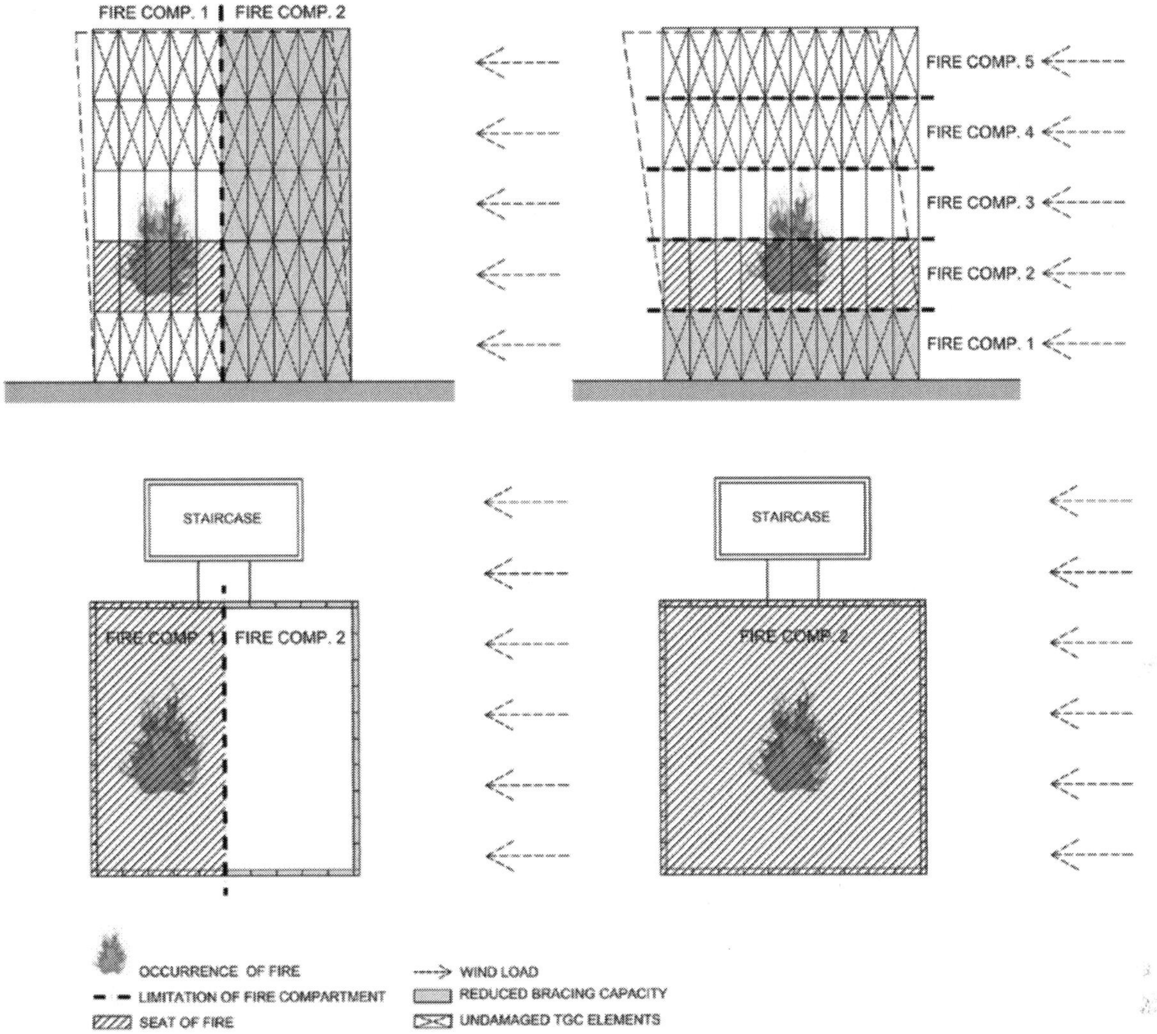


**Figure 4.** Schematic illustration of the relationship between the layout of fire compartments and the bracing capacity of the TGC façade. Picture credits: author.

the floor levels above the seat of fire would remain undamaged. The relation between the layout of fire compartments and the bracing function by TGC façade is depicted in Fig. 4.

## 2.3 Total failure of TGC façades in case of fire

TGC elements might be appropriate for building constructions with interior bracing core elements to absorb high horizontal forces of wind and earthquake loads and thereby relieve the core. Particularly in multi-story buildings with timber structures, which usually show higher elasticity and larger deformation compared to reinforced concrete structures, TGC façades can be used economically [1].

In the current research project various building structures consisting of a primary bracing by an internal core component with different stiffness (wood and wood-composite structures, reinforced concrete structures) and a secondary bracing by TGC façade elements are being investigated in case studies. The aim of this ongoing work is to evaluate the relationship between primary and secondary bracing and resulting optimizations.

In case of fire bracing capacity of TGC façades could fail if the primary bracing by the inner core is designed with fire resistance corresponding to the building classification. The secondary bracing does not necessarily require a defined fire resistance due to the fact that fire is considered an accidental situation. Only those actions, whose occurrence in case of fire is likely to happen, have to be taken into account – a simultaneous occurrence of other accidental actions doesn´t have to be considered [5].

In a structural analysis based on the partial factor method according to European standards Eurocode EN 1990 [6] leading and accompanying actions have to be considered for accidental design situations like fire. In this case the design value $E_d$ of the relevant effects of actions will be determined according to the following Eqs. (1) and (2).

$$E_d = E\,\{G_{k,j}\,; P; A_d\,; (\psi_{1,1} \text{ or } \psi_{2,1})\, Q_{k,1}\,; \psi_{2,i} Q_{k,i}\}\, j \geq 1;\ i > 1. \tag{1}$$

The combination of actions in brackets { }can be expressed as:

$$\sum_{j\geq 1} G_{k,j} \text{ “ + ” } P \text{ “ + ” } A_d \text{ “ + ” } (\psi_{1,1} \text{ or } \psi_{2,1}) Q_{k,1} \text{ “ + ” } \sum_{i\geq 1} \psi_{2,i}\, Q_{k,i} \tag{2}$$

$E_d$ *Design value of effect of actions according to EN 1990*
$G_{k,j}$ *Characteristic value of permanent action j*
$P$ *Relevant representative value of a prestressing action*
$A_d$ *Design value of an accidental action*
$Q_{k,1}$ *Characteristic value of the leading variable action 1*
$Q_{k,i}$ *Characteristic value of the accompanying variable action i*
$\psi_1$ *Factor for frequent value of a variable action*
$\psi_2$ *Factor for quasi-permanent value of a variable action.*

In a structural analysis regarding accidental action of fire the variable actions (e.g. wind load) have to be considered either with the factor for quasi-permanent value or, if it is a leading variable action, with the factor for frequent value. These factors are specified by the national annexes to EN 1990 – in Austria these values are defined by $\psi_2 = 0$ (factor for quasi-permanent value) and $\psi_1 = 0{,}2$ (factor for frequent value) according to ÖNORM B 1990–1 [7]. Hence TGC elements do not need to fulfil any requirements to fire resistance, if these parameters are considered in the design and calculation of the building construction.

## 3. TGC EXPERIMENTS UNDER TEMPERATURE STRESS

To examine the behaviour of TGC elements when being exposed to different temperature loads a series of small scale timber-glass samples with different adhesives were tested at the beginning of the research project. The experiments were conducted in the laboratory of the Department for Building Construction and Technology of the Vienna University of Technology.

After reaching the targeted temperature load, the specimens were taken from the oven and put into a universal testing machine (Zwick, Type Z250) to perform an individual pressure test.

The selected types of adhesive for the tests are being used for serial TGC façades on the one hand and on the other hand examined in current research projects. Hence they constitute a representative choice in accordance with the current state of the art and science. Three series of specimens were produced with the following adhesives:

- Nolax C44.8505: two-component adhesive on EP basis
  Component A: Silane Polymer formulated
  Component B: Epoxy Resin formulated

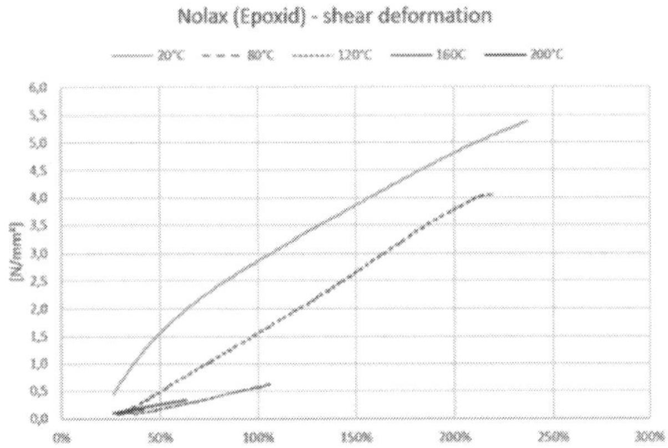


**Figure 5.** Graph of shear deformation of TGC specimens with Epoxy bond line. Picture credits: author.

- SikaFast-5221 NT: two-component adhesive based on acrylate
  Component A: SikaFast-5221 NT
  Component B: SikaFast-5200
- Ottocoll S 660: two-component silicone adhesive on Alcoxy base.

The experiments were carried out under continuous loading rate at different temperature levels (20 °C, 80 °C, 120 °C, 160 °C, 200 °C) until failure of the specimens. During the tests the following failure modes occurred:

- fracture in the bond line (cohesive failure)
- fracture between bond line and timber element (adhesive failure)
- fracture between bond line and glass element (adhesive failure)
- mixed fracture (cohesive and adhesive failure)
- fracture in the timber element.

The evaluation of the measured data shows, amongst others, significant changes in shear deformation under temperature stress. The reduction in shear strength substantially depends on the type of adhesive as shown in the following graphs. Specimens with epoxy bond line rapidly change their initial properties (Fig. 5) whereas samples with silicone bonding show a more constant behaviour but with significantly lower shear strength at the initial temperature of 20 °C (Fig. 6).

## 4. OUTLOOK TO FIRETESTS WITH TGC FAÇADE MOCK UPS

In the current work package of the project a typology of different TGC façade systems will be designed and evaluated. Focus of the assessment is on one hand the compliance of fire safety regulations, on the other hand to increase the bracing capacity with respect to the intended applications in multi-story buildings. The following drawings offer an insight of the current status of the design process by means of a single and a double façade with TGC elements.

In the further progress of this research the following measures to increase the bracing capacity of the TGC elements will be evaluated:

- blocking of the glazing in the inner façade level
- timber-glass composites bonded with modified epoxy adhesives in the inner façade level
- timber-glass composites bonded with silicone adhesives in the outer façade level

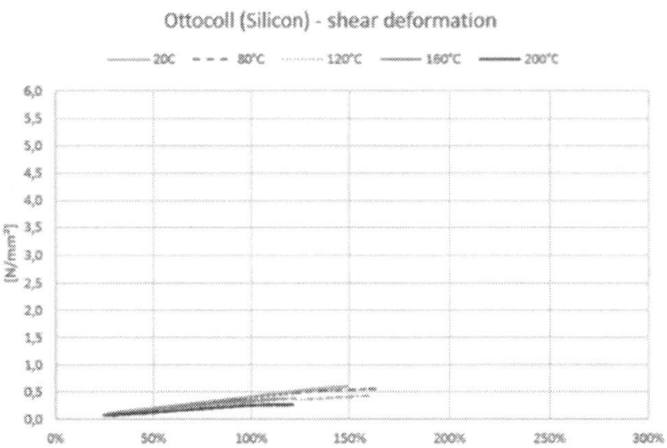


**Figure 6.** Graph of shear deformation of TGC specimens with Silicone bond line. Picture credits: author.

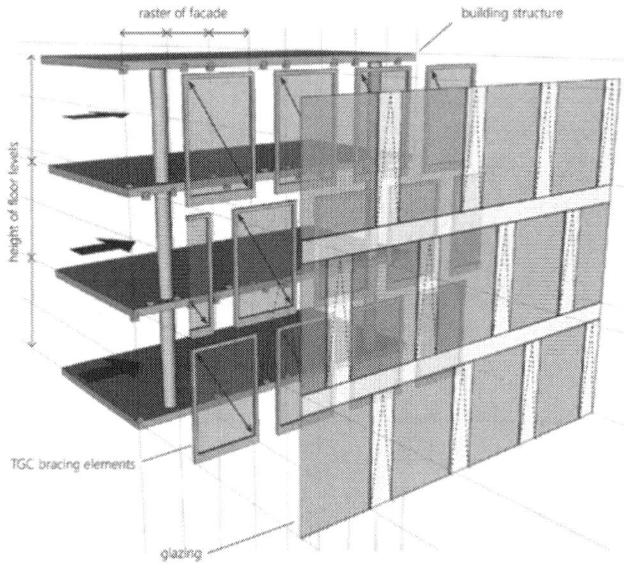


**Figure 7.** Schematic drawing of a single façade with TGC elements. Picture credits: author.

- framework made of veneer laminated timber products
- coupling of TGC elements to increase bracing capacity.

Verification of the fire safety requirements concerning TGC applications in multi-story buildings up to classification 5 and for buildings with a flight level above 22 m will be done on the basis of mock ups in fire tests. The following tests will be provided:

- testing according to ÖNORM B 3800-5 [8] or -6 [9] with respect to the criteria of fire spread across the façade and falling of large façade parts
- testing according to ÖNORM EN 13501-2 [10] to prove the criteria EW 30-ef.

## 5. DISCUSSION

The previously completed work packages of the research project "Timber-Glass Composite Façades | Behaviour in Case of Fire | Fire Protection Concepts" and the experimental series "Bonded

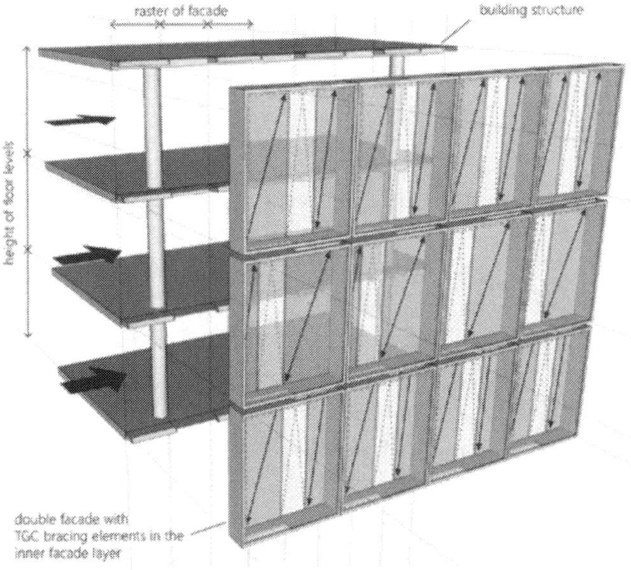


**Figure 8.** Schematic drawing of a double façade with TGC elements. Picture credits: author.

Timber-Glass Composites under Thermal Stress" lead to the following conclusions:

- TGC bond lines with the investigated epoxy or acrylic adhesives can absorb significantly higher shear stresses than silicon bond lines at room temperature.
- As temperature loads above 80 °C are likely in the façade area even without exposure to fire, TGC applications with the investigated acrylic adhesive are not appropriate and will not be investigated furthermore in the course of this project.
- TGC applications using the investigated epoxy adhesives cannot record any significant shear strength under a temperature load that occur in case of fire and hence cannot serve for bracing in this case
- TGC applications with the examined silicone adhesive provide a very constant behavior up to a temperature load of 200 °C with little reduction of shear strength. Further research is needed to examine the behavior of TGC with silicone adhesives at higher temperature loads which occur on the façade area in case of fire.
- The strategy of "Protection of TGC façades in case of fire" seems to be appropriate only in conjunction with an automatic sprinkler system due to the significant reduction of shear strength of epoxy and acrylate adhesives at a temperature load of 200 °C.

The following interim results from the currently running work packages "Fire Scenarios and TGC Structures" and "TGC Façade Prototypes" are important for the further progress of the project:

- Structural analysis from current case studies of multi-story buildings with TGC façades show that a combination of reinforced concrete structures with an internal core element (primary bracing) and a TGC façade for secondary bracing offers very little reduction of the primary bracing due to the high stiffness of the reinforced concrete structure. A combination of structures made of wood or wood composite with TGC façades appear to be more advantageous due to the lower stiffness of these structures and will be investigated in the ongoing project.
- Double façade system using TGC elements have positive aspects in terms of their intended usage in multi-story buildings. The outer façade level might be used to protect the inner façade level both from climate conditions (e.g. rain, wind, sunlight) and from thermal effects by fire, when the design is appropriate.

- Higher bracing capacity can be achieved by coupling of TGC façade elements into a multi-story TGC façade structure.

## References


[1] Winter, W., "Application of Load Bearing Glass Timber Composites", 4th Forum Wood Building Nordic, Växjö, 2015.

[2] Hochhauser, W., "Ein Beitrag zur Bemessung und Berechnung von geklebten und geklotzten Holz-Glas-Verbundscheiben", Dissertation, 2011.

[3] Regulation (EU) No 305/2011 of the European Parliament and of the Council of 9th March 2011 laying down harmonised conditions for the marketing of construction products and repealing Council Directive 89/106/EEC.

[4] Österreichisches Institut für Bautechnik (OIB), "Richtlinien des Österreichischen Instituts für Bautechnik", `http://www.oib.or.at/de/oib-richtlinen`, 2015.

[5] Hosser, D., *Brandschutz in Europa – Bemessung nach Eurocodes, Erläuterungen und Anwendungen zu den Brandschutzteilen der Eurocodes 1 bis 5*, Beuth Verlag GmbH, 2012.

[6] ÖNORM EN 1990:2003-03-01, "Eurocode – Grundlagen der Tragwerksplanung".

[7] ÖNORM B 1990-1:2013-01-01, "Eurocode – Grundlagen der Tragwerksplanung – Teil 1: Hochbau, Nationale Festlegungen zu ÖNORM EN 1990 und nationale Ergänzungen".

[8] ÖNORM B 3800-5:2013-04-15, "Brandverhalten von Baustoffen und Bauteilen, Teil 5: Brandverhalten von Fassaden, Anforderungen, Prüfungen, Beurteilungen".

[9] ÖNORM B 3800-6:2013-04-15, "Brandverhalten von Baustoffen und Bauteilen, Teil 6: Brandverhalten von Kasten-Doppelfassaden (2-schalige Fassade), Anforderungen, Prüfungen, Beurteilungen".

[10] ÖNORM EN 13501-2:2010-02-15, "Klassifizierung von Bauprodukten und Bauarten zu ihrem Brandverhalten, Teil 2: Klassifizierung mit den Ergebnissen aus den Feuerwiderstandsprüfungen, mit Ausnahme von Lüftungsanlagen".

MATEC Web of Conferences **46**, 05004 (2016)
DOI: 10.1051/matecconf/20164605004


# Large glazing in curtain walls – study on impact of fixing methods on fire resistance

Jacek Kinowski, Bartłomiej Sędłak and Paweł Sulik

Fire Research Department of Building Research Institute, Ksawerów 21, 02-656 Warsaw, Poland

**Abstract.** Paper presents curtain walls fire resistance tests results comparison. Considered tests specimens were of similar system solution regarding profiles material, shape, dimension and insulation, glass pane family group and dimensions and glazing arrangement, fixing of framing system (anchoring) but differ in glazing fixing method. Collected study material allowed reliable comparison of glazing fixing method impact on whole construction fire resistance. Comparison was made for curtain walls glazed with glass panes fixed in standard way (by means of pressure plates with covering profiles) and for curtain walls with structural glazing (fixed by means of steel clamps and structural silicone).

## 1. INTRODUCTION

Glazed curtain walls play essential role in modern building industry. Ease of installation and aesthetic effect of the final provides great popularity of this type of solutions, especially in a tall [25] and high-rise [24] buildings. This external cover of the building, which consists of vertical and horizontal structural members, connected together and fixed to the floor supporting structure has to fulfil the basic aspects like climatic and environmental influence as well as those connected with protection against burglary and fire. Facing up these problems is a huge challenge for both designers and contractors, especially in case of the increasingly popular buildings with great floor to floor distance [19] and facades with large glass panes [14, 15].

According to the regulations of many countries curtain walls, as a non-loadbearing part of the building, should be designed and constructed in a way that in case of fire it will limit the spread of fire inside the building, ensure the safety of rescue team and allow the evacuation of users. Facades should for a specific period of time prevent development of fire between the adjacent floors of the building, between the adjacent objects, and in specific cases also between the adjacent rooms in the building (connection of the curtain wall with inner wall of the building). Moreover, its fixing shall be made in a way that prevents falling of the parts of element during the rescue team activities and evacuation of the users [6, 9, 21, 23].

Requirements listed above quite often results in necessity of use the curtain walls with specific fire resistance class. In order to ensure the insulation and reduce the adverse impact of thermal effects, those constructions are equipped with special accessories such as intumescent gaskets [12], profile insulation inserts [8] and obviously fire resistant glazing [16, 18]. However, even a vast knowledge

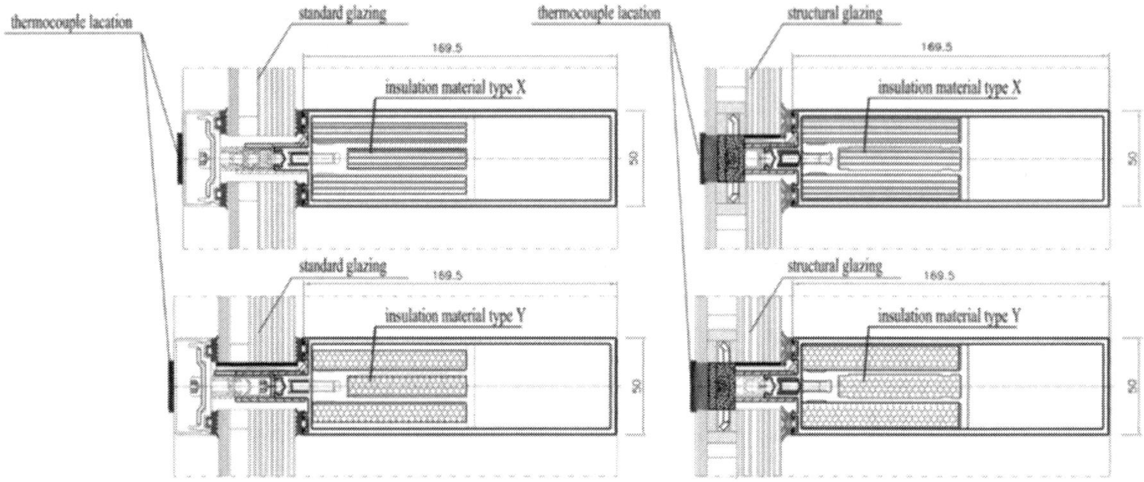


**Figure 1.** Transoms cross – section, construction details and thermocouples location.

of systems, used solutions and the behaviour of glazed curtain walls in fire conditions isn't enough to calculate or asses its fire resistance class, therefore the only way to determine this characteristic is to perform the fire resistance test. The European methodology of fire testing, as well as detailed information regarding the selection of appropriate test specimens, has been widely discussed in the existing papers [1, 10, 11, 13, 22]. Curtain walls as an external cover of the building can be exposed for the fire inside as well as outside the object and that's why two different methods of heating and test specimen configuration depending on the side of fire exposure (external or internal) are given in the testing procedure. The impact of the side of exposure on fire resistance of glazed curtain walls is presented in papers [17, 20].

Fire resistance of glazed curtain walls depends on many factors of the system solution to name only: profile material and shape, type of insulation inserts used, type of glazing and glazing fixing method, (especially important in case of large glass panes). Study on impact of latter factor mentioned on fire resistance is the main topic of this paper.

## 2. FIRE RESISTANCE TESTS

Four test specimens of glazed aluminium curtain walls in full configuration, were tested in the Fire Testing Laboratory of Building Research Institute (ITB) in Pionki. All tests were performed on straight curtain wall type B (in accordance with EN 1364-3 [3]) in full configuration, with internal fire exposure heated under the standard fire curve. All four curtain walls had the same framing system fixing (anchoring system). Two test specimens were tested with glass panes fixed standard way (by means of pressure plates with covering profiles), and two with structural glazing (fixed by means of steel clamps and structural silicone). All tested specimens were of the same transom – mullion structure with aluminium profiles sections of:

- $165 \times 50$ mm (mullions, Fig. 2),
- $169,5 \times 50$ mm (transoms, Fig 1).

Dimension of exposed surface of tested specimens was $3350 \times 4500$ mm (width × height). General view of tested specimens is presented in Fig. 3 (drawing), 4, 5 (photos). In performed fire resistant tests two types of insulations inserts (X, Y) were used in profiles and two types of glass panes (in terms of fire resistant glass pane) in glazing units, each type for pair of tests. Test specimens details specification

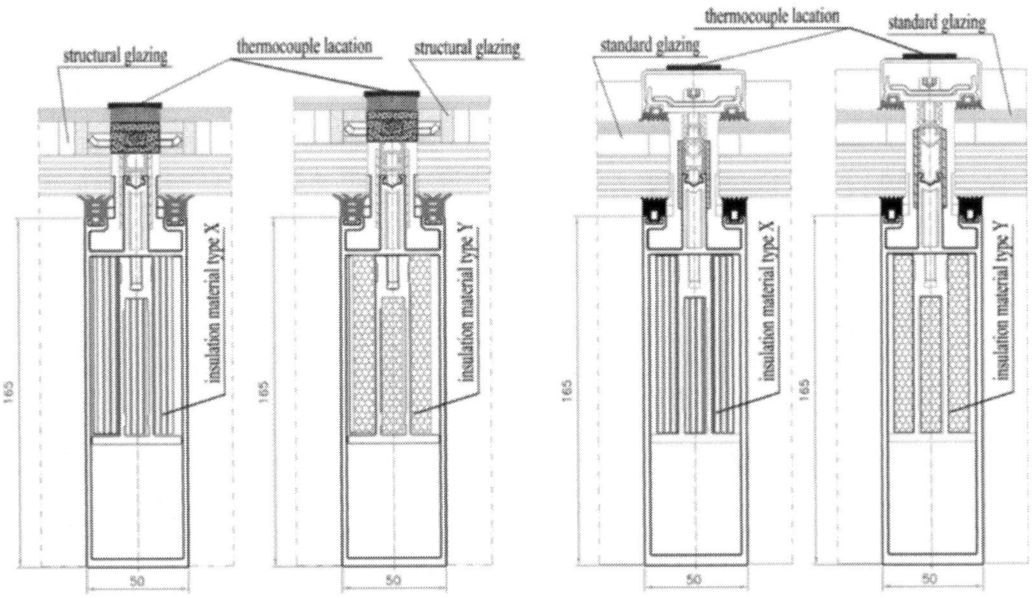


**Figure 2.** Mullions cross – section, construction details and thermocouples location.

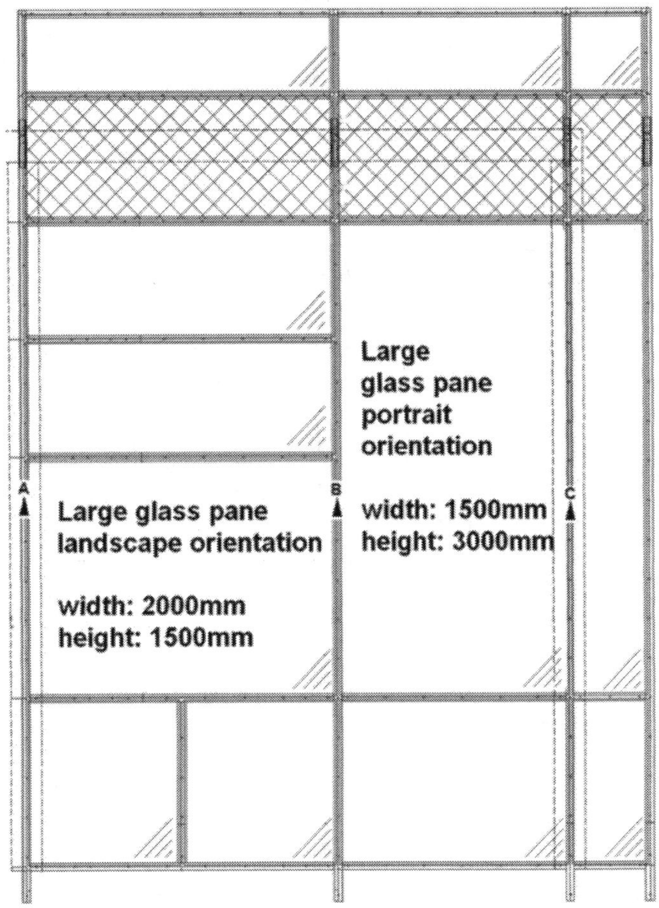


**Figure 3.** General view of tested specimens (with location of points of deformation measurements).

**Table 1.** Test specimens.

| Test specimen | Glass pane fixing method | Type of fire resistant glass pane | Type/ application of glazing unit | Type of insulation inserts |
|---|---|---|---|---|
| **1** | Standard | 1 | Standard glazing (C) | X |
| **2** | Standard | 2 | Standard glazing (C) | Y |
| **3** | Structural | 1 | Structural glazing (S) | X |
| **4** | Structural | 2 | Structural glazing (S) | Y |

**Figure 4.** The view of the unexposed surface of the test specimen during the test (structural glass pane fixing).

is presented in Table 1. Dimensions of the glass panes were (width x height):

- 1500 × 3000 mm – for portrait orientation,
- 2000 × 1500 mm – for landscape orientation.

Glazing units manufactured using fire resistant glass pane of type 1, specification and composition:

- thickness of 34 mm [20 mm fire glass, 8 mm steel frame and 6 mm toughened glass]; (C),
- thickness of 42 mm [20 mm fire glass, 16 mm steel frame and 6 mm toughened glass]; (S).

Glazing units manufactured using fire resistant glass pane of type 2, specification and composition:

- thickness of 39 mm [25 mm fire glass, 8 mm steel frame and 6 mm toughened glass], (C),
- thickness of 47 mm [25 mm fire glass, 16 mm steel frame and 6 mm toughened glass] (S).

The insulation inserts type 1 were made of gypsum plasterboards and insulation inserts type 2 were made of special cooling plates.

**Figure 5.** The view of the unexposed surface of the test specimen during the test (standard glass pane fixing).

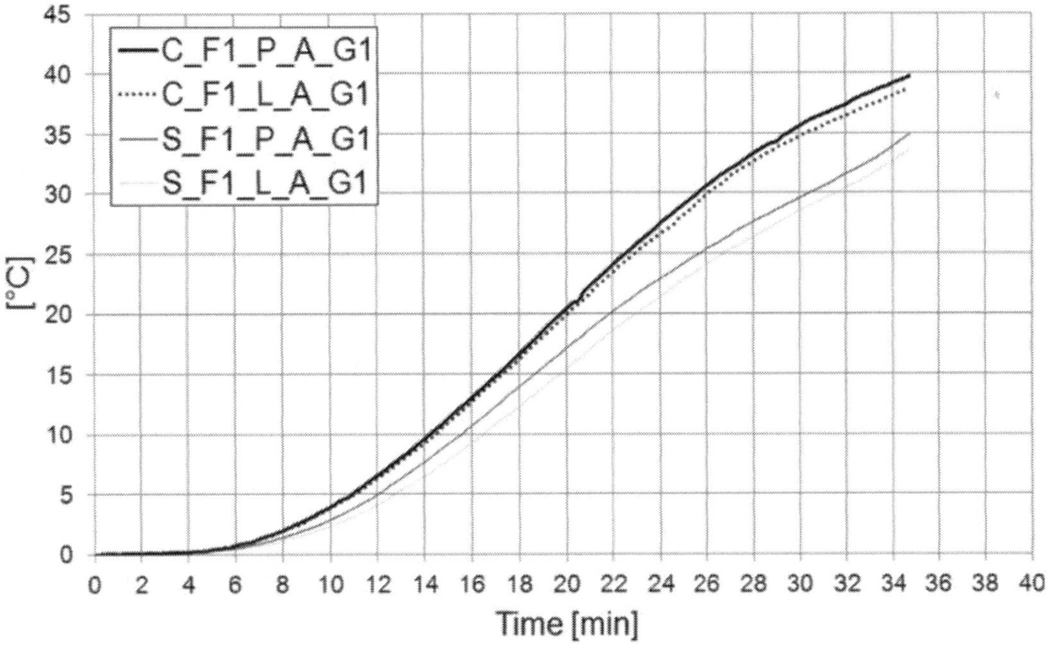


**Figure 6.** Comparison of average temperature rise on the glass panes (C_F1_P_A_ G1 – glass pane type 1, standard fixing method, portrait orientation; C_ F1_ L_ A_ G1 – glass pane type 1 standard fixing method, landscape orientation; S_ F1_ P_ A_ G1 – glass pane type 1, structural fixing method, portrait orientation; S_ F1_ L_ A_ G1 – glass pane type 1 structural fixing method, landscape orientation).

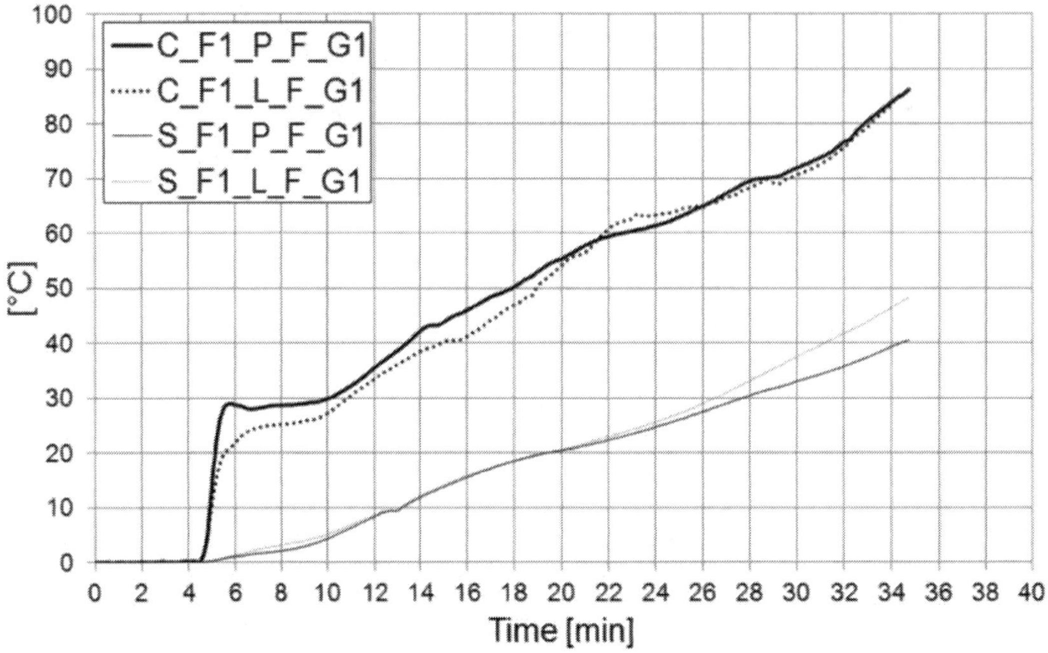


**Figure 7.** Comparison of average temperature rise on the framing (C_ F1_ P_ F_ G1 – glass pane type 1, standard fixing method, portrait orientation; C_ F1_ L_ F_ G1 – glass pane type 1 standard fixing method, landscape orientation; S_ F1_ P_ F_ G1 – glass pane type 1, structural fixing method, portrait orientation; S_ F1_ L_ F_ G1 – glass pane type 1 structural fixing method, landscape orientation).

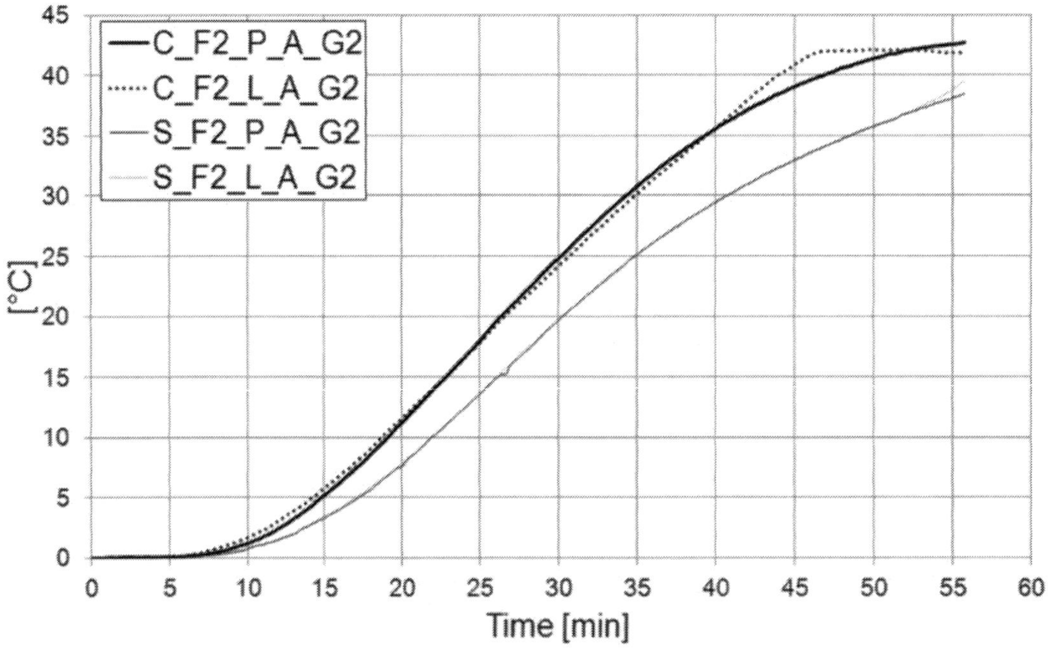


**Figure 8.** Comparison of average temperature rise on the glass panes (C_ F2_ P_ A_ G2 – glass pane type 2, standard fixing method, portrait orientation; C_ F2_ L_ A_ G2 – glass pane type 2 standard fixing method, landscape orientation; S_ F2_ P_ A_ G2 – glass pane type 2, structural fixing method, portrait orientation; S_ F2_ L_ A_ G2 – glass pane type 2 structural fixing method, landscape orientation).

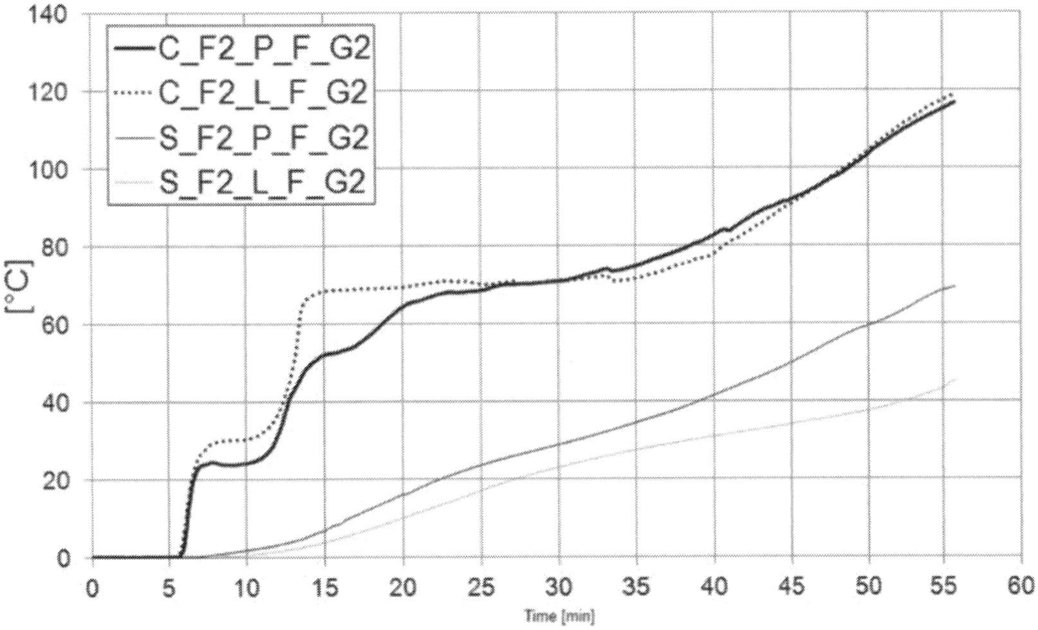


**Figure 9.** Comparison of average temperature rise on the framing (C_ F2_ P_ F_ G2 – glass pane type 2, standard fixing method, portrait orientation; C_ F2_ L_ F_ G2– glass pane type 2 standard fixing method, landscape orientation; S_ F2_ P_ F_ G2 – glass pane type 2, structural fixing method, portrait orientation; S_ F2_ L_ F_ G2 – glass pane type 2 structural fixing method, landscape orientation).

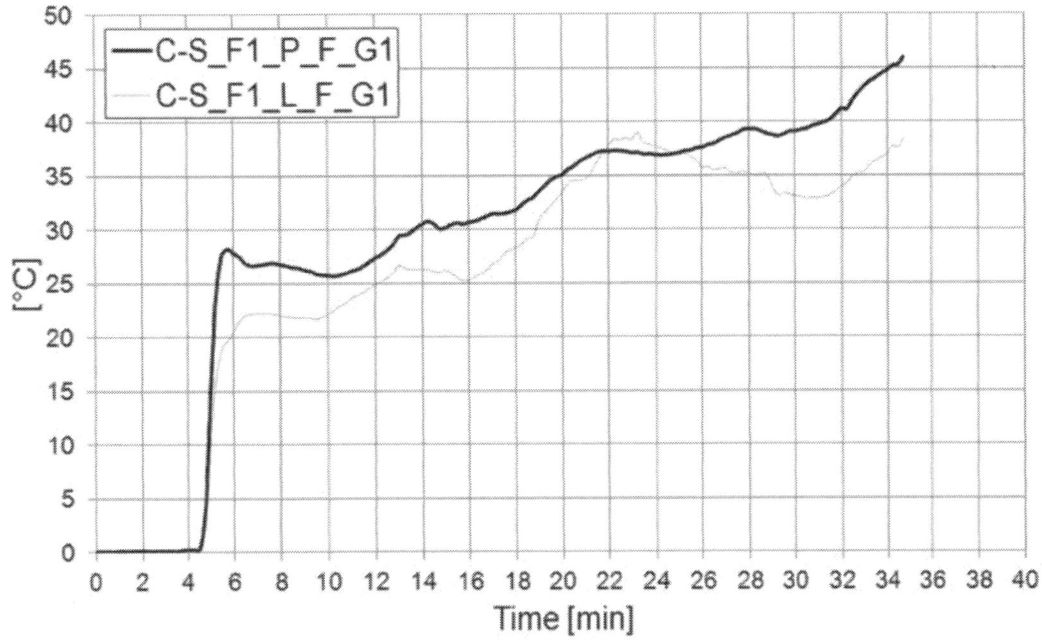


**Figure 10.** Differences between the average temperature rises on the unexposed surface of glass pane framing elements (C-S_ F1_ P_ F_ G1 – glass pane type 1, portrait orientation; C-S_ F1_ L_ F_ G1 – glass pane type 1, landscape orientation).

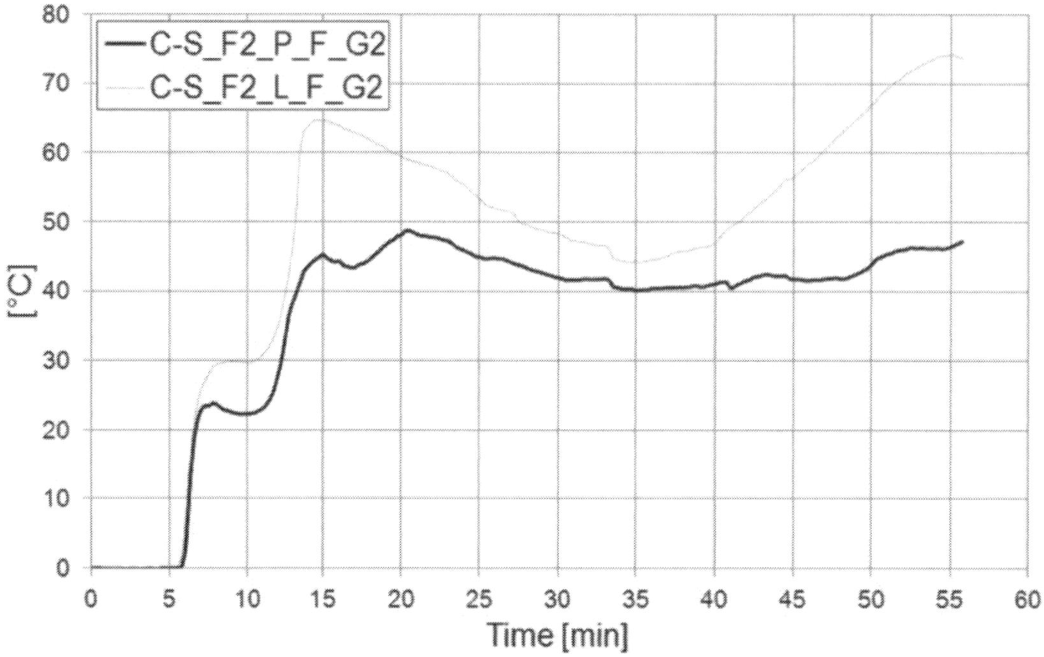


**Figure 11.** Differences between the average temperature rises on the unexposed surface of glass pane framing elements (C-S_ F2_ P_ F_ G2 – glass pane type 2, portrait orientation; C-S_ F2_ L_ F_ G2 – glass pane type 2, landscape orientation).

## 3. INSTRUMENTATION AND TESTS RESULTS

The comparison was made for average temperature rises on unexposed surface of largest glass panes (in portrait and landscape orientation) and its framing – aluminium covering profiles in case of standard glass pane fixing and fireproof silicone in case of structural glass pane fixing. The temperature rise on unexposed surface of the test specimens was measured with surface thermocouples type K specified in EN 1363-1:2012 [1]. Places of temperature rises measurements were arrange in accordance with EN 1364-3:2014 [2].

Comparison of average temperature rise on the glass panes is presented in Fig. 6 (glazing type 1) and Fig. 8 (glazing type 2). Comparison of average temperature rise on the framing is presented in Fig. 7 (glazing type 1) and Fig. 9 (glazing type 2). Figures 10 and 11 presents the differences between the average temperature rises on the unexposed surface of glass pane framing elements.

## 4. DISCUSSION

Glazing units for curtain walls with structural glazing have improved insulation compared to the solution with standard glazing presumably due to their construction – wider frame provides larger gap of air, thus improves insulation in case of internal heating, Figs. 6, 8. Also framing system designed for curtain walls with structural glazing have improved insulation compared to the solution with standard glazing. Appropriately applied silicone fugue is a perfect sealing for all framing joints, but entails greater risk of installation errors which are much easier to avoid in case of using standard fixing system with pressure plates and covering profiles, Figs. 7, 9. Behaviour of both types of curtain walls (in terms of glazing fixing method) during the fire resistant tests were similar – there are no significant differences in test elements (deflection difference of max. about 25 mm for the elements height of 4500 mm cannot be

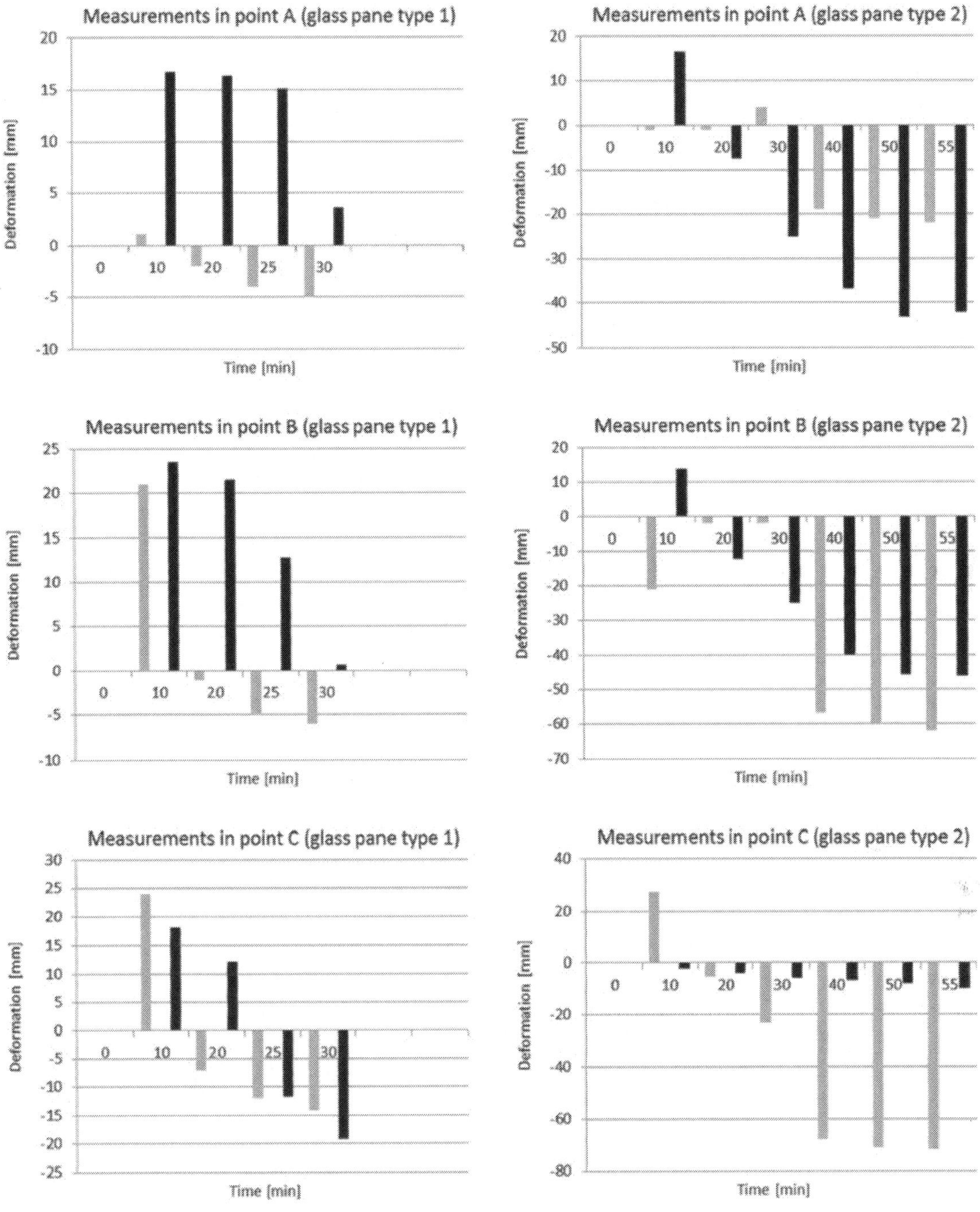


**Figure 12.** Deformation comparison (grey colour – curatin wall with standard glazing; black colour – curatin wall with structural glazing).

considered as large), Fig. 12. Perhaps fixing of framing system (anchoring) method could cause greater impact on the deflections [26]. According to EN 1364-3:2014 [3] test results of specified pressure plate fixing system (standard fixing system) are applicable to other solutions only with minor changes [5, 7]. Glazing unit can be exchanged for one with wider steel frame, but without additional fire test it is not possible to expand field of direct application. A relatively new method of structural glazing fixing is not jet analysed in Standard [3].

## References


[1] Borowy A., "Fire resistance testing of glazed building elements", *Požární ochrana 2014: Sborník přednášek XXIII. ročníku mezinárodní konference, Ostrava, 03-04.09.2014*, 15–17.

[2] EN 1363-1:2012 Fire resistance tests – Part 1 General requirements.

[3] EN 1364-3:2014 Fire resistance tests for non-loadbearing elements – Part. 3: Curtain walling – Full configuration (complete assembly).

[4] EN 1363-2:1999 Fire resistance tests – Part 2 Alternative and additional procedures.

[5] Gravit M., Gumenyuk V. Nedryshkin O. "Fire resistance parameters for glazed non load-bearing curtain walling structures. Extended application", *Procedia Engineering*, **117**, 114–118, 2015.

[6] Kinowski J., "Bezpieczeństwo pożarowe przeszklonych ścian osłonowych (kurtynowych)", *Świat Szkła*, **20**(5), 8–14, 2015 *(in polish)*.

[7] Kinowski J., Sędłak B., Sulik P., Izydorczyk D.,"Fire resistance glazed constructions classification, changes in field of application", *Applications of Structural Fire Engineering, 15–16 October 2015, Dubrovnik, Croatia*, 460–465, 2015. DOI: dx.doi.org/10.14311/asfe.2015.073.

[8] Kinowski J., Sędłak B., Sulik P., "Izolacyjność ogniowa aluminiowo – szklanych ścian osłonowych w zależności od sposobu wypełnienia profili szkieletu konstrukcyjnego", *Izolacje*, **20**(2), 48–53, 2015 *(in polish)*.

[9] Kinowski J., Sulik P., "Bezpieczeństwo użytkowania elewacji", *Materiały Budowlane*, **9**, 38–39, 2014 *(in polish)*.

[10] Sędłak B., "Badania odporności ogniowej przeszklonych ścian osłonowych – Część 1", *Świat Szkła*, **17**(9), 52–54, 201 *(in polish)*.

[11] Sędłak B., "Badania odporności ogniowej przeszklonych ścian osłonowych – Część 2", *Świat Szkła*, **17**(10), 53–58,60, 2012 *(in polish)*.

[12] Sędłak B., "Systemy przegród aluminiowo szklanych o określonej klasie odporności ogniowej", *Świat Szkła*, **18**(10), 30–33,41, 2013 *(in polish)*.

[13] Sędłak B., "Badania odporności ogniowej przeszklonych ścian osłonowych wg nowego wydania normy PN-EN 1364-3", *Świat Szkła*, **19**(7-8), 49–53, 2014 *(in polish)*.

[14] Sędłak B., "Odporność ogniowa ścian osłonowych z dużymi przeszkleniami – Część 1", *Świat Szkła*, **19**(3), 16–19,25, 2014 *(in polish)*.

[15] Sędłak B., "Odporność ogniowa ścian osłonowych z dużymi przeszkleniami – Część 2", *Świat Szkła*, **19**(5), 28–31, 2014 *(in polish)*.

[16] Sędłak B., Kinowski J., "Badania odporności ogniowej ścian osłonowych – przyrosty temperatury na szybach", *Świat Szkła*, **18**(11), 20–25, 2013 *(in polish)*.

[17] Sędłak B., Kinowski J., Borowy A., "Fire resistance tests of large glazed aluminium curtain wall test specimens–results comparison", *MATEC Web of Conferences*, 9, p. 02009, EDP Sciences, 2013, DOI: 10.1051/matecconf/20130902009.

[18] Sędłak B., Sulik P., "Odporność ogniowa pionowych elementów przeszklonych", *Szkło i Ceramika*, R.**66** (nr. 5), 8–10, 2015 *(in polish)*.

[19] Sędłak B., Sulik P., "Odporność ogniowa wielkogabarytowych pionowych elementów przeszklonych", *Materiały Budowlane*, (nr. 7), 26–28, 2015. DOI: 10.15199/33.2015.07.06 *(in polish)*.

[20] Sulik P., Kinowski J., Sędłak B., "Fire resistance of aluminium glazed curtain walls, test results comparison depending on the side of fire exposure", *Applications of Structural Fire Engineering, 15–16 October 2015, Dubrovnik, Croatia*, 478–483, 2015. DOI: dx.doi.org/10.14311/asfe.2015.076.

[21] Sulik P., Sędłak B., "Bezpieczeństwo pożarowe przeszklonych elewacji", *Materiały Budowlane*, (nr. 9), 18–20, 2015. DOI: 10.15199/33.2015.09.04 *(in polish)*.

[22] Sulik P., Sędłak B., Kinowski J., "Bezpieczeństwo pożarowe ścian zewnętrznych – Część 1 – Elewacje szklane: wymagania, badania, przykłady", *Ochrona przeciwpożarowa*, **4**(50), 10–16, 2014 *(in polish)*.

[23] Sulik P., Sędłak B., Kinowski J., "Bezpieczeństwo pożarowe ścian zewnętrznych – Część 2 – Mocowanie okładzin elewacyjnych", *Ochrona przeciwpożarowa*, **1**(51), 9–12, 2015 *(in polish)*.

[24] Sulik P., Sędłak B., Turkowski P., Węgrzyński W., "Bezpieczeństwo pożarowe budynków wysokich i wysokościowych", *Budownictwo na obszarach zurbanizowanych – nauka, praktyka, perspektywy*, 14–19 September 2014, Krynica, Poland, 105–120, 2014 *(in polish)*.

[25] Umarani G., Premalatha E, Aruna Malini. T.P., "Facades of Tall Buildings – State of the art", *Modern Applied Science*, **4**, No. 12, 116–125, 2010.

[26] Wang Y., Wu Y., Wang Q., Liew K.M., Chen H., Sun J., He L., "Numerical study on fire response of glass facades in different installation forms", *Construction and Building Materials*, **61**, 172–180, 2014.

MATEC Web of Conferences **46**, 05005 (2016)
DOI: 10.1051/matecconf/20164605005


# Durability of the reaction to fire performance for fire retardant treated (FRT) wood products in exterior applications – a ten years report

Birgit Östman and Lazaros Tsantaridis

SP Wood Building Technology Stockholm, Sweden

**Abstract.** Several long term experimental studies on the maintained reaction to fire performance of fire retardant treated (FRT) wood products over time are presented. They are performed according to a European system based on earlier Nordic and North American systems and include accelerated ageing according to different procedures and natural weathering up to ten years. Main conclusions are:

- The hygroscopic properties are unchanged compared to untreated wood for most FRT wood products used commercially.
- The reaction to fire properties of FRT wood may be maintained after accelerated and natural ageing if the retention levels are high enough, but several FRT wood products loose most of their improved reaction to fire properties during weathering.
- Paint systems contribute considerably to maintain of the fire performance at exterior application and are usually needed to maintain the fire performance after weathering.

## 1. INTRODUCTION

### 1.1 Durability classes for fire performance – Principles and methods

Fire retardants may considerably improve the reaction to fire properties of wood products, but the durability in interior and exterior applications needs to be addressed. Requirements on durability of the fire performance are not mentioned in most building codes. This is probably partly caused by unawareness of the problem, but may also be due to the lack of procedures.

The problems with maintained reaction to fire performance over time have been known for a long time in the US and the UK, but are not so well known in the rest of Europe. A US study on exterior exposure of North American products during ten years [1] and a literature review [2] have been published.

Two cases of durability of the fire retardant treatment of wood products can be identified. One is the risk for high moisture content and migration of the fire retardant chemicals within the wood product and salt crystallisation on the product surface. These hygroscopic properties of the treated wood-based product can be evaluated by exposure to high relative humidity.

**Table 1.** Requirements for DRF classes of FRT wood products according to CEN/TS 15912 [3].

| DRF class | Intended use | Fire class Initial | Performance requirements for different end uses | |
|---|---|---|---|---|
| | | | **Hygroscopic properties** | **Fire performance after weather exposure** |
| ST | Short term | Relevant fire class | - | - |
|  | Interior, dry applications | - " - | Limited moisture content Minimum visible salt | - |
|  | Interior, humid applications | - " - | Limited moisture content Minimum visible salt | - |
|  | Exterior applications | - " - | - " - | Maintained fire performance |

Further details are given in CEN/TS 15912.

The other case is the risk for decreased fire performance due to loss of the fire retardant chemicals by leaching or other mechanisms. This case is mainly for exterior applications, e.g. as façade claddings. Maintained fire performance over time has to be verified.

A European system with Durability of Reaction to Fire performance (DRF) classes has been developed in order to guide the potential users to find suitable FRT wood products [3], see Table 1. The system is based on a North American system [4, 5] and a previous Nordic system [6–8]. It consists of a classification system for the properties over time of FRT wood and suitable test procedures.

The technical specification CEN/TS 15912 is currently being transformed to a full European standard. This paper is an extended version of an earlier paper [9] and includes more products and test data.

The relevant fire class shall be verified according to EN or IMO systems [10, 11]. Maintained fire performance after weather exposure shall be verified according to ISO 5660 [12] or the European system [10].

## 2. EXPERIMENTAL STUDY

### 2.1 Hygroscopic properties

The hygroscopic properties of wood products have been determined according to [7] or the equivalent [3]. The method includes calculation of equilibrium moisture content at two climates, 50% RH at 23 °C and 90% RH at 27 °C. The requirement for DRF class INT is moisture content below 30% in the humid climate. Untreated wood has moisture content below 20% in that climate. In addition, possible salt crystallisation at the wood surface and exudation of liquid in the wet climate is observed.

### 2.2 Accelerated and natural weathering

For DRF class EXT, the durability of the fire performance at exterior applications has been studied by accelerated ageing according to [8] or the equivalent [3]. Both Method A, i e exposure to 12 one-week cycles of simulated rain and drying and Method B, i e exposure to 6 one-week cycles of which also includes UV radiation have been used.

**Figure 1.** Natural weathering field in Stockholm.

Natural field exposure has been performed with wood panels facing south, both at vertical (90 °) and at 45° slope. In both cases the rear sides of the panels were open. The test field is in the Stockholm area, Sweden, see Figure 1. Results are presented for 1, 2, 3, 5 and 10 years.

### 2.3 Fire performance

The reaction to fire performance has been determined according to ISO 5660 [12] at 50 kW/m$^2$. Duplicate tests have been run in most cases and the repeatability was very good. The time to flashover in the room corner test has been predicted [13].

### 2.4 Mass loss during accelerated and natural weathering

The weight of the FRT wood panels has been measured after conditioning at 50% RH at 23 °C, both initially and after the weathering procedures and is expressed in percentage of the total panel weight.

## 3. WOOD PRODUCTS TESTED

The wood products tested are all wood panelling products, mainly spruce, vacuum pressure impregnated with different fire retardant chemicals. Untreated wood panelling has been used as reference. The initial reaction to fire performance for the FRT wood products has been determined or estimated to

**Table 2.** Commercial FRT wood products.

| *FR ID* | *Mean FR kg/m³* | *Thick-ness mm* | *Width mm* | *Approximate initial density kg/m³* | *Type of panel* |
|---|---|---|---|---|---|
| 0 | – | | | | Tongue and groove with notched rear side |
| X | 25–60 | 20 | 135 | 540 | -"- |
| Y | 45–110 | 20 | 135 | 510 | -"- |
| ZA | 95–190 | 22 | 130 | 690 | Tongue and groove |
| ZG | 170 | 22 | 130 | 660 | Tongue and groove, factory primed |
| VF | 70–140 | 20 | 135 | 520 | Tongue and groove with notched rear side |
| CI | na | 21 | 145 | 410 | Panel with straight edges |
| CM | na | 22 | 150 | 470 | -"- |

**Table 3.** Development FRT wood products.

| *FR ID* | *FR kg/m³ ** | *Thickness mm* | *Width mm* | *Approximate initial density kg/m³* | *Type of panel* |
|---|---|---|---|---|---|
| – | – | 20 | 100 | 540 | Panel with straight edges |
| BS | 107–347 | 20 | 100 | 640–830 | -"- |
| FP | 71–228 | 20 | 100 | 570–670 | -"- |
| DQ | 89–290 | 20 | 100 | 570–875 | -"- |
| BH | 87–236 | 20 | 100 | 635–790 | -"- |
| MA | 216 | 20 | 100 | 585 | -"- |
| SI | 84–368 | 20 | 100 | 625–850 | -"- |
| FF | 92–169 | 20 | 100 | 647–657 | -"- |
| PhF | 413–667 | 20 | 100 | 826–1007 | -"- |
| BZ-15 | 78– 282 | 20 | 100 | 470–520 | -"- |
| BZ-40 | 67–191 | 20 | 100 | 469–558 | -"- |
| BZ-30 | 75–229 | 20 | 100 | 531–544 | -"- |
| NF | 115–206 | 20 | 100 | 541–711 | -"- |
| AF | 120–251 | 20 | 100 | 560–677 | -"- |
| DF | 125–556 | 20 | 100 | 629–1152 | -"- |
| LG | 114–536 | 20 | 100 | 614–673 | -"- |
| NS | 139–619 | 20 | 100 | 549–680 | -"- |

* Three retention levels per FR chemical, the range is given in the table.

be equivalent to class B or at least class C according to the Euroclass system [10]. Untreated wood is Euroclass D. These data have been used as reference for the fire performance after weathering.

Two main sets of wood products have been studied, one set of commercial FRT wood products X, Y, Z, CI and CM intended for exterior application, a former commercial product VF, and another set of development products [14]. All are wood panels impregnated with fire retardants to different levels of retention, see Tables 2 and 3.

The size of the specimen for ageing may have some influence on the effects of ageing due to edge effects. 800 mm long specimen were found to be superior to shorter specimen in a pre-study [15] and have then been used together with edge seals both at accelerated and natural exposure. After the exposures the panels have been cut in a systematic way (with test pieces at the same position from the panels to secure consistency) and used for fire and other tests.

## 3.1 Paint systems

Paint systems are usually needed to maintain the fire performance properties of FRT wood products for exterior applications. Four different paint systems have been used, see Table 4. The panels for weathering

**Table 4.** Paint systems used.

| *Paint number* | *Paint type* | *Number of coats* | *Total amount g/m²* | *Comments* |
|---|---|---|---|---|
| 1 | Alkyd | 4 * | 530–650 ** | 1 coat primer oil, 1 coat alkyd primer, 2 alkyd top coats (all products solvent borne) |
| 3 | Red paint | 1 | 200 | Swedish red paint (water based) |
| 4 | Linseed oil | 3 | 400–550 ** | 2 coats with diluted paint, 30 and 15% resp, 1 top coat of undiluted paint |
| 6 | Wood oil | 2 | 180–200 ** | Pigmented oil |

* Only 2 top coats on factory primed panel, total amount 250 g/m²;

** The higher amount on panel type Z

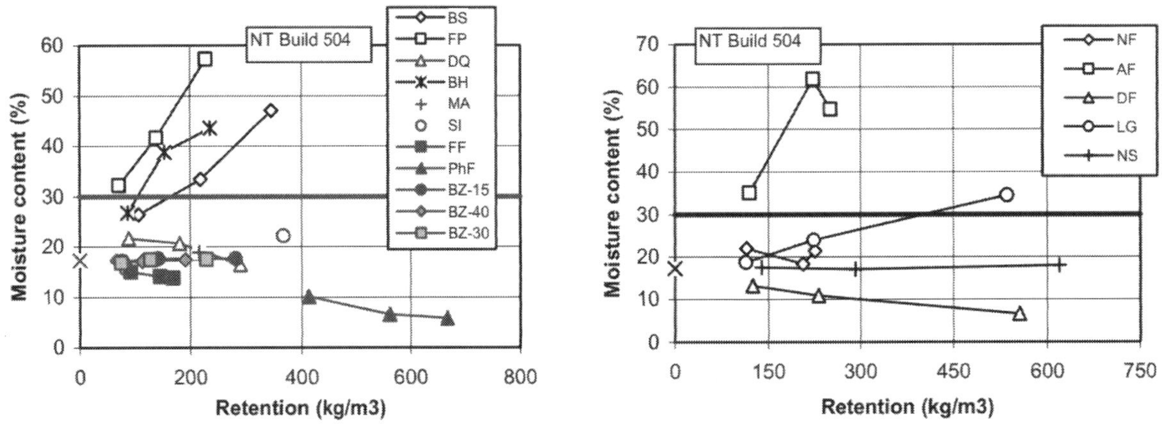


**Figure 2.** Moisture content at high RH vs retention for FR chemicals and for untreated pine sapwood.

were 800 mm long and edge sealed, first with an alkyd solvent borne primer and then with a thick coat of a silicone sealer.

## 4. HYGROSCOPIC PROPERTIES

The hygroscopic properties expressed as moisture content at higher relative humidity according to [3] of FR treatments as a function of the retention level are illustrated in Fig. 2. The moisture content is unchanged compared to untreated wood for several FRT wood products.

It is evident from the test data that the moisture content may increase with increased amount of fire retardant chemical added. It is thus important to optimise the FR content not only from an economical point of view, but also to reach the intended fire performance with a safety margin to maintain the fire performance during service life of the product and not jeopardizing the moisture resistance.

## 5. COMPARISONS OF REACTION TO FIRE PERFORMANCE

The reaction to fire performance has been determined in the cone calorimeter ISO 5660 [12] at 50 kW/m². All results for the initial reaction to fire performance and after accelerated ageing and natural weathering are summarised and compared in Figs. 3–6. The comparison is based on predicted time to flashover [13]. Several products exhibit high initial reaction to fire performance, but it may be reduced over time during both accelerated and natural weathering.

Comparison of natural weathering exposures at vertical (90°) and at 45° slope is presented in Fig. 4. No major difference can be observed.

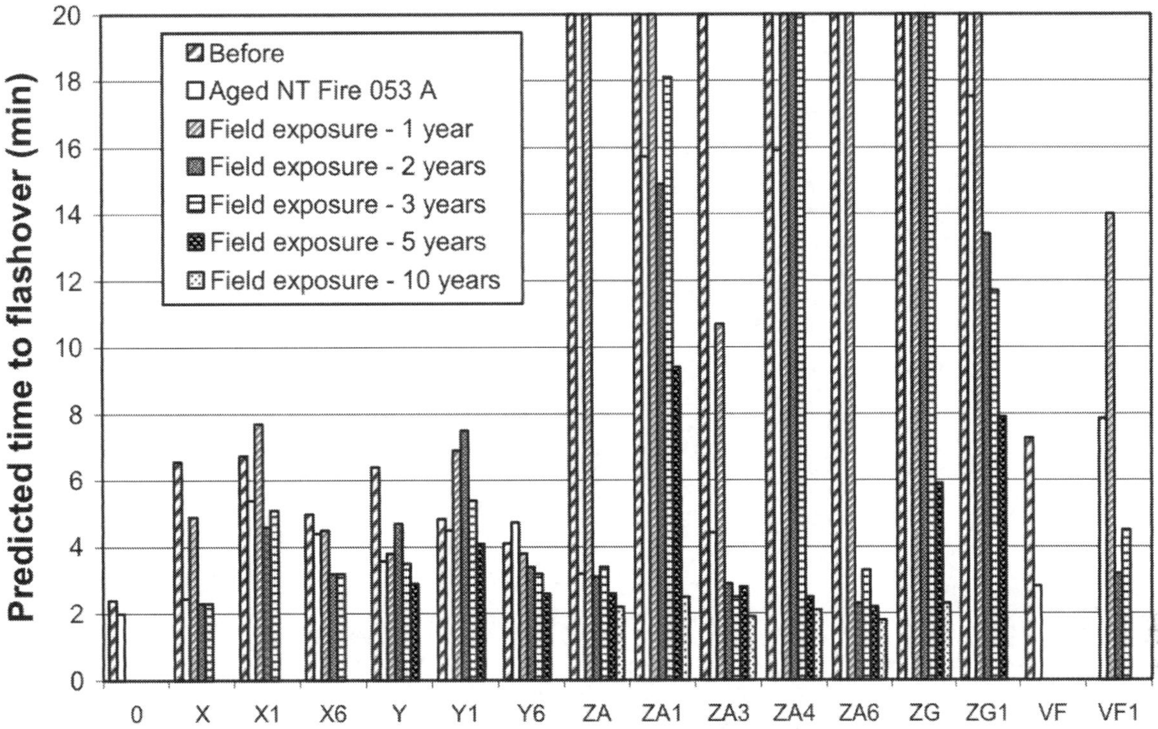


**Figure 3.** Reaction to fire performance (as predicted time to flashover) before and after accelerated ageing according to NT FIRE 053 Method A, and after natural weathering at 45° slope during up to 10 years. Untreated spruce (0) and treated (X, Y, ZA, ZG and VF) spruce. Surface coatings with paints number 1, 3, 4 and 6 are included.

Comparison of accelerated weathering exposures according to Method A and Method B is presented in Figs. 5–6. No major difference can be observed.

## 6. DURABILITY OF THE REACTION TO FIRE PERFORMANCE

The reaction to fire performance is reduced both after accelerated ageing and natural field exposure for most of the FRT products. Only a few FRT products maintain a high fire performance after accelerated and natural exposures. The best performance is found at high retention levels and for FRT products with paint as a protective surface coat. Among the paints, the linseed oil paint (number 4) and the factory primed product ZG were found to exhibit a full maintenance of the reaction to fire performance up to three years. The other FRT products were more or less degraded during the weathering exposure, regardless of a protective coat or not. For products with low retention of FR chemicals and low initial fire class, the maintenance of the fire performance could not be evaluated.

The accelerated ageing thus seems to be equivalent to maximum five years of natural field exposure. However, it should be noticed that the field exposure includes also a certain degree of acceleration. The 45° exposure was intended to include some acceleration, but no major difference to the vertical (90°) orientations was found, see Fig. 4. Some indications of a more severe degradation at the vertical orientation were found in a previous study [9]. This may be explained by the lack of protection on the rear side of the vertical panels, which were open to the weather exposure. On the other hand, the panels at 45° slope, were at least partly protected on the rear side from direct influence of rainfall and snow. In a real end use, e.g. as a facade cladding, the rear side is totally protected. Such conditions have to be studied further before a more clear guidance on the accelerating factors can be established.

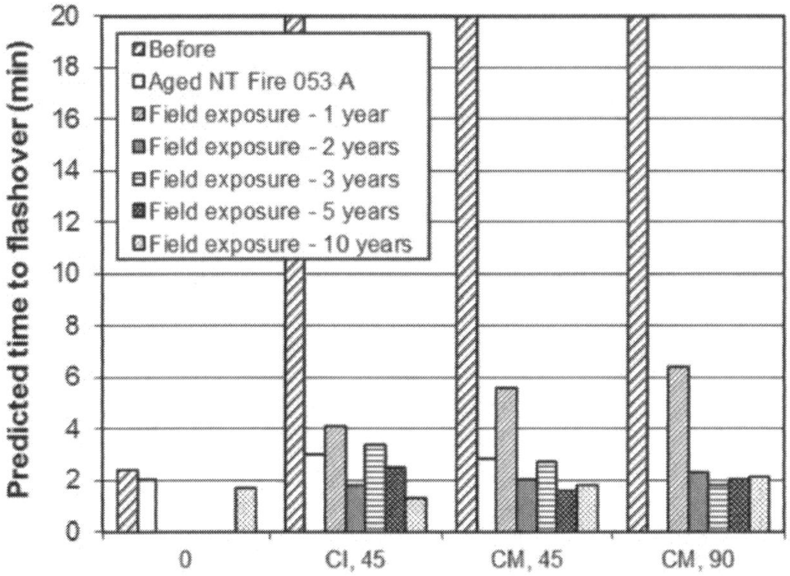


**Figure 4.** Reaction to fire performance (as predicted time to flashover) before and after accelerated ageing according to NT FIRE 053 Method A, and after natural weathering with wood panels vertical (90°) and at 45° slope during up to 10 years. Untreated spruce (0) and treated (CI and CM) spruce.

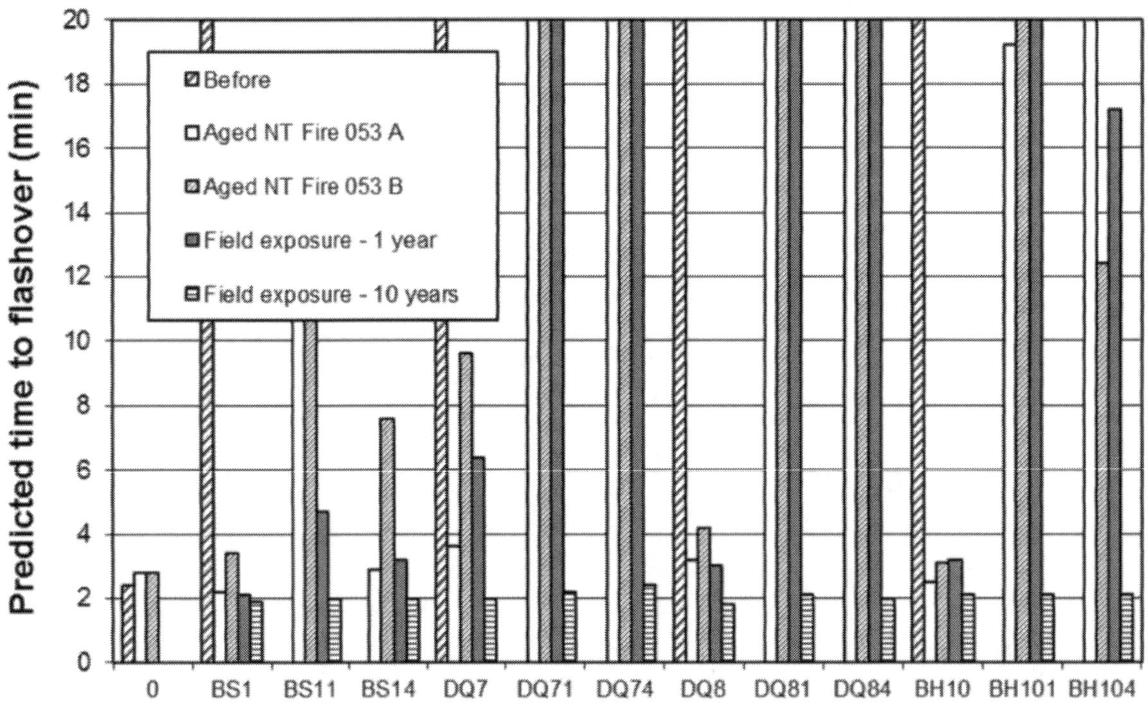


**Figure 5.** Reaction to fire performance (as predicted time to flashover) before and after accelerated ageing according to NT FIRE 053 Method A and B, and after natural weathering at 45° slope during up to 10 years. Untreated spruce (0) and FR treated (BS, DQ and BH) spruce. Surface coatings with paints number 1 and 4 are included.

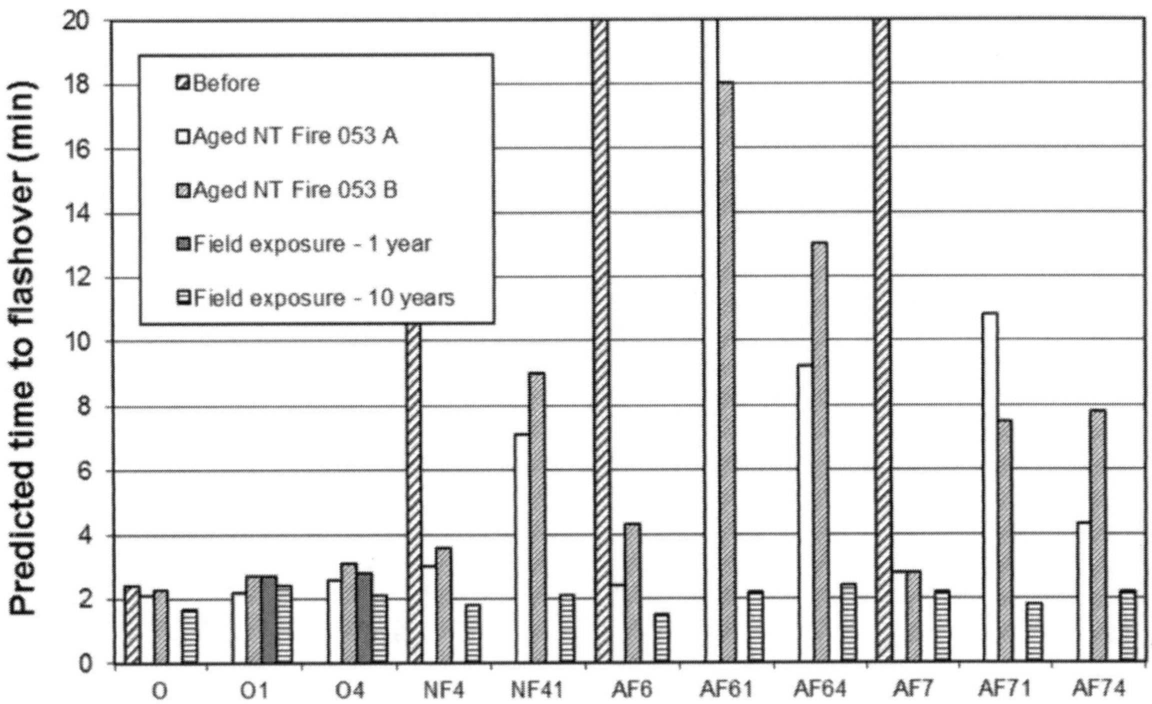


**Figure 6.** Reaction to fire performance (as predicted time to flashover) before and after accelerated ageing according to NT FIRE 053 Method A and B, and after natural weathering at 45° slope during up to 10 years. Untreated spruce (0) and FR treated (NF and AF) spruce. Surface coatings with paints number 1 and 4 are included.

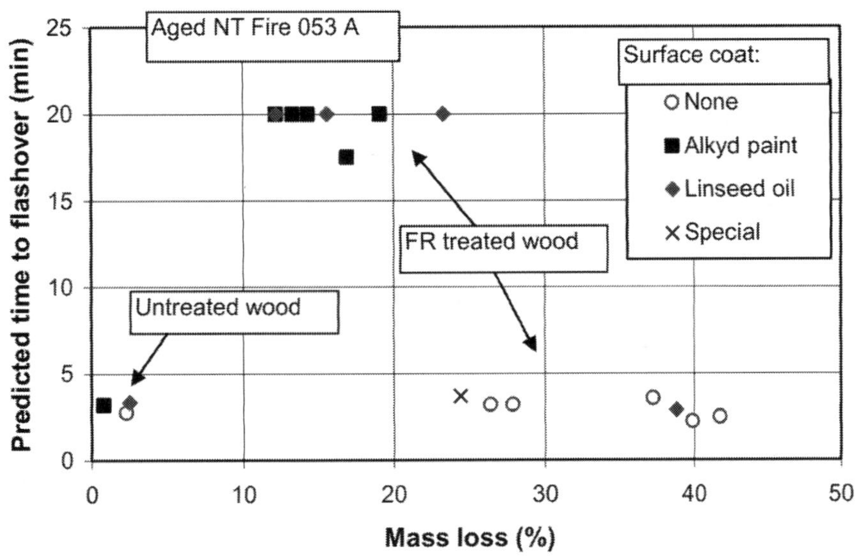


**Figure 7.** Reaction to fire performance (as predicted time to flashover) vs mass loss during accelerated ageing of FRT and untreated wood according to NT Fire 053 Method A [8].

## 7. MASS LOSS DURING WEATHERING

The mass loss during accelerated ageing and natural weathering may be used as an indicator of the maintained reaction to fire performance over time. Some data are presented in Figs. 7–8.

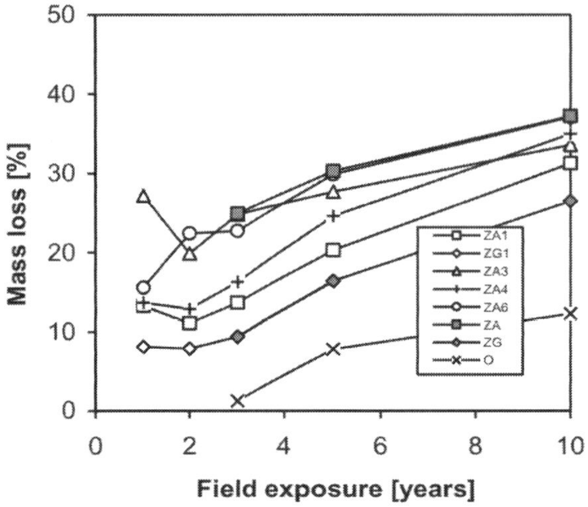


**Figure 8.** Mass loss during natural weathering of FRT and untreated wood up to ten years.

# 8. CONCLUSIONS AND SUGGESTIONS FOR FURTHER WORK

Main conclusions are:

- A system with Durability of Reaction to Fire performance (DRF) classes to evaluate the fire performance of FRT wood over time at humid and exterior conditions has been developed. It provides a very useful supplement to requirements on the fire performance in national building codes and enables to guide potential users to find suitable and reliable FRT wood products.
- The hygroscopicity of sufficiently durable FRT wood is about the same as for untreated wood, but much higher for simple inorganic salts.
- The fire properties of FRT wood may be maintained after accelerated ageing and natural weathering if the retention levels are high enough, but several FRT wood products loose most of their improved reaction to fire properties during weathering.
- Paint systems contribute considerably to maintain of the fire performance at exterior applications and are usually needed to maintain the fire performance after weathering.
- The mass loss during accelerated aging and natural weathering may be used as an indicator of the maintained reaction to fire performance over time.
- More experience with correlation of natural field testing and accelerated ageing methods is needed.

## References

[1] LeVan S, Holmes C A. Effectiveness of fire-retardant treatments for shingles after 10 years of outdoor weathering. Research Paper FPL 474, 1986.

[2] Östman B, Voss A, Hughes A, Hovde P J, Grexa O. Durability of fire retardant treated wood at humid and exterior conditions. Review of literature. *Fire and Materials*, **25**, 95–104, 2001.

[3] CEN/TS 15912. Durability of reaction to fire performance of fire-retardant treated wood-based products in interior and exterior end-use applications, European Technical Specification, 2012.

[4] ASTM D 2898, Standard Methods for Accelerated weathering of fire-retardant-treated wood for fire testing. American Society for Testing and Materials.

[5] ASTM D 3201, Standard Test Method for Hygroscopic properties of fire-retardant treated wood and wood-based products. American Society for Testing and Materials.

[6] Nordtest method NT FIRE 054. Durability of Reaction to Fire – Performance classes of fire-retardant treated wood-based products in interior and exterior end use applications, 2006.

[7] Nordtest method NT BUILD 504. Hygroscopic properties of fire-retardant treated wood and wood-based products, 2003.
[8] Nordtest method NT FIRE 053. Accelerated weathering of fire-retardant treated wood for fire testing, 2003.
[9] Östman B, Tsantaridis L. Durability of the reaction to fire performance for FRT wood products in different end use applications – a ten years report, Proc. 12th International Conference on Durability of Building Materials and Components, XII DBMC, Porto, Portugal, 2011.
[10] EN 13501-1. Fire classification of construction products and building elements – Part 1: Classification using test data from reaction to fire tests.
[11] IMO FTP Code (MSC 61/67), International Code for Application of Fire Test Procedures. International Maritime Organization.
[12] ISO 5660-1. Fire tests – Reaction to fire – Part 1: Rate of heat release from building products.
[13] Östman B A-L, Tsantaridis L D. Correlation between cone calorimeter and time to flashover in the room fire test, *Fire and Materials*, **18**, 205–209, 1994.
[14] Östman, Tsantaridis, Mikkola, Hakkarainen, Belloni, Brumer and Piispanen. Innovative eco-efficient high fire performance wood products for demanding applications. Final report for Vinnova-Tekes project InnoFireWood. SP Technical Research Institute of Sweden, SP Report 2006:30, 2006.
[15] Östman B, Tsantarids L, Mikkola E, Hakkarainen T, Nilsen T-N, Evans F, Grexa O. Durability of fire retardant wood – New test methods and round robin. Nordtest project 1527-01. Trätek Swedish Institute for Wood Technology Research, Report P 0211040, 2002.

MATEC Web of Conferences **46**, 05006 (2016)
DOI: 10.1051/matecconf/20164605006


# Influence of fire barriers on fire performance of facades with combustible insulation

Dubravka Bjegovic[1], Ivana Banjad Pecur[1], Birgitte Messerschmidt[2], Bojan Milovanovic[1] and Marina Alagusic[1]

[1] University of Zagreb, Faculty of Civil Engineering, Croatia
[2] Fire Safe Europe, Belgium

**Abstract.** In today's modern society sustainability has become a key driver for innovation. Sustainability in buildings is, or at least it should be, strongly interconnected with energy efficiency and fire safety. Unfortunately, cases of fires in buildings all over the world are a wakeup call to start thinking of fire safety as an inevitable part of energy efficiency in buildings. Thermally enhanced building envelopes, due to materials used, can represent significantly increased fire load on façades. To determine their fire performance, it is necessary to move on from material level to the whole building level, i.e. full scale tests. This paper presents findings from comprehensive research on influence of fire barrier on fire performance of External Thermal Insulation Composite System – ETICS with combustible thermal insulation. To confirm findings, tests on identical ETICS systems were conducted in March 2014 and May 2014.

## 1. INTRODUCTION

Having sustainable and energy efficient buildings are exciting new trends which are driving the innovation of new construction products and systems. Unfortunately what is often forgotten is that if a building is to be truly sustainable it should not be built to burn. Examples of fires in different building types all over the world are proof of an urgent need to start considering fire safety as an integral part of sustainable building design. Sustainable buildings cannot be achieved without a holistic approach, which implies that sustainable buildings are at the same time energy efficient and fire safe. It can be said that buildings are sustainable only when they perform better in terms of energy consumption, are resilient and provide safety for various users and use types of the building, while at the same time they minimize the need to replace material resources over the building's lifetime.

Considering that in Europe buildings use 41% of all energy, and 2/3 of this energy consumption is due to heating and cooling, buildings are an obvious and necessary target for energy saving measures. One of the obvious ways of reducing energy consumption in buildings is to add thermal insulation to the envelope of the building. Often this leads to the use of highly insulating façade systems. To secure the fire safety of buildings it is therefore necessary to be able to test how highly insulated façade systems perform in fire. Due to the complex nature of these systems it is not sufficient to only consider the performance of individual materials on their own. The interaction of different materials within the

**Table 1.** Description of test specimens.

| TEST SPECIMEN | Thermal insulation material and thickness | Render | Fixing method | Reaction to fire classification |
|---|---|---|---|---|
| TS_ 1 | Expanded polystyrene (EPS) – 150 mm | Basic render reinforced with glass fibre mesh and final organic (acrylic) render – 5 mm | Adhesive and anchoring | B-s2,d0 |
| TS_2 | Expanded polystyrene (EPS) 150 mm + fire barrier 200 mm high; directly above combustion chamber | | | B-s2,d0 (A2-s1,d0 barrier) |

system on building level are key to how it will perform in fire and it is therefore necessary to test façade systems in a test that replicates a realistic fire scenario.

The main aim of this paper is twofold: first, to provide deeper understanding of the fire performance of *External Thermal Insulation Composite Systems* (ETICS) with combustible thermal insulation, which are nowadays one of the most commonly used technologies to thermally enhance building envelopes in Europe [1] and second, to investigate whether the fire barrier constructed above openings influence fire performance of ETICS system with combustible thermal insulation.

## 2. EXPERIMENTAL PART

### 2.1 Experimental methodology and set-up

In the absence of a harmonised EU standard for large scale façade fire testing, fire performance test of ETICS system in these experiments was performed according to BS 8414-1 test method [2]. This standard describes a method of assessing the behaviour of façade systems applied to the face of a building with a masonry wall and exposed to external fire under controlled conditions. The fire exposure is representative of an external fire source or a fully-developed (post flashover) fire in a room, venting through an opening such as a window aperture that exposes the cladding to the effects of external flames [2].

Test results presented and analysed in this paper, were obtained during research conducted in March 2014 and May 2014, i.e. Test 1 and Test 2, respectively. Both in Test 1 and Test 2, respectively, identical systems were constructed and tested (Table 1). By conducting measurements on identical full scale test specimens during two tests, the repeatability of testing ensured comparability of obtained results and therefore confirmation of fire performance of tested systems, i.e. influence of fire barrier on fire performance of ETICS system with combustible thermal insulation.

Both test specimens were L-shaped, 8 m high, with one leg 2.6 m long forming the main test wall (main face) and the other leg 1.5 m long forming the return wall (wing) as shown in Fig. 1. The L-shape of the specimen represents an internal corner of a building. The only difference between test specimens was the fire barrier made of non-combustible stone wool (SW) thermal insulation that was constructed above the opening of one test specimen. Composition of test specimens and their classification according to the reaction to fire classification as declared by the manufacturer [3, 4] is shown in Table 1.

Fire barrier in the test specimen TS_2 was positioned directly above the combustion chamber which simulates an opening in a building, Fig. 2a. Lintel protection above each opening prevents fire penetration in the thermal insulation layer. A schematic drawing of the lintel protection above the opening on test specimen TS_2 is shown in Figure 2b. It was constructed with respect to the best rules of practice. According to literature [5], this is one of the possible ways of constructive "fire protection", i.e. passive fire protection in ETICS systems with combustible thermal insulation.

The heat source placed in the combustion chamber was in accordance with BS 8414-1. This heat source releases a nominal total heat output of 4500 MJ over 30 min at a peak rate of $(3 \pm 0.5)$ MW. It needs to be pointed out that BS 8414-1 standard was used only as a "guide" for preparing and

**Figure 1.** Test specimens before the test start.

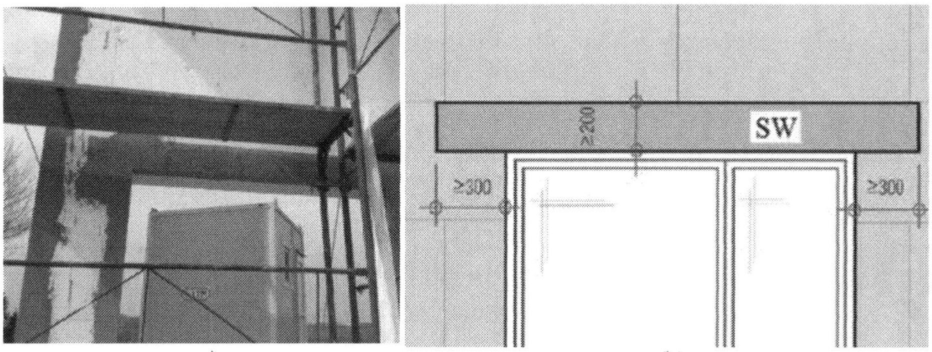


*a)* *b)*

**Figure 2.** Lintel protection: a) Applied on test specimen TS_ 2, b) General scheme [5].

**Table 2.** Description of additional measurements performed.

| Measuring instrument | Purpose | Number and position | *Note |
|---|---|---|---|
| Additional internal thermocouples (In TCadd) 1.5 mm | Measuring temperature distribution in thermal insulation layer | 10 thermocouples per test specimen (7 on main face and 3 on wing) | Used only in Test 2 |
| Load cell under the support platform legs (LC) | Measuring weight loss rate of the wood crib during the test; i.e. burning rate of fire source | Placed below each leg of platform onto which the wood crib was placed. | Used in Test 1 and Test 2 |

conducting this test, since the main aim was to gain new knowledge about complex fire performance of ETICS system with combustible thermal insulation on a building level and influence of fire barrier on its fire performance. The deviations from BS 8414-1 were that the tests were performed outside and in addition to the standard temperature measurements defined by BS 8414-1, additional measurements were conducted, as described further in this chapter.

External thermocouples were positioned on the main face of the façade and on the wing, both at Level 1 and Level 2 (Fig. 3a) in accordance with BS 8414-1. Internal thermocouples required by BS 8414-1 were positioned at Level 2, on the main face of the façade and on the wing.

Table 2 shows the additional measurements done during the tests with Fig. 3b showing the position of the additional internal thermocouples and load cells. The results of these are part of the analysis in this paper.

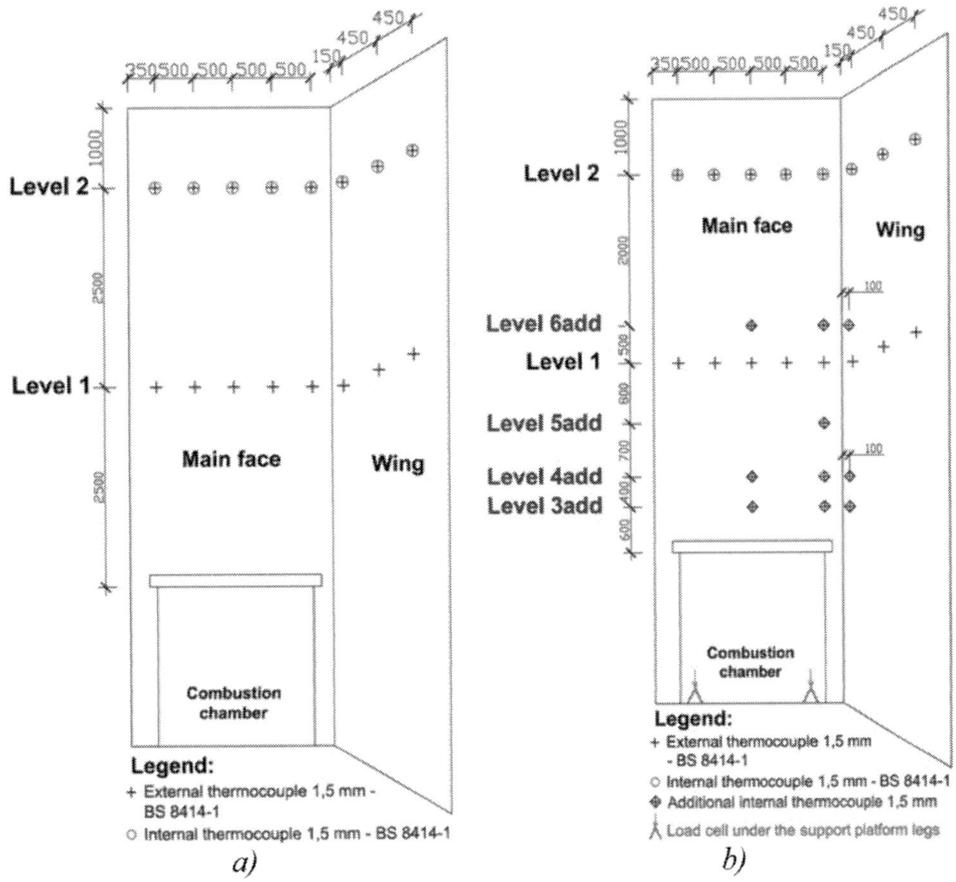


**Figure 3.** Positioning scheme of: a) thermocouples required by BS 8414-1 [2]; b) load cells and additional internal thermocouples.

For both test specimens, i.e. TS_1 and TS_2, the wood crib has been ignited at the same time with the aim that the fire development and fire spread on both test specimens could be visually comparable.

Since the tests were performed outside, the weather conditions, i.e. wind speed, wind direction and air temperature, were monitored in order to be able to take into account their possible influence on testing and obtained results. During the Test 1, the air temperature at the start of the test was within the range 11.2–12.3°C. The air velocity was 1.8 m/s at the start of the tests, while during the test air velocity was measured up to 5.4 m/s in N–E direction. Air temperature during the Test 2 was within the range of 20.5–22.2°C and the air velocity was within the range of 2.2–4.5 m/s in N, N–W direction. Direction of north is shown in Fig. 1.

## 2.2 Test results

Test results indicating fire performance of ETICS systems with combustible thermal insulation, with and without SW fire barrier, will be presented in terms of developed temperature profiles, mass loss of fire source, i.e. wood crib, and visual appearances on test specimens. Behaviour of tested ETICS systems in this type of fire scenario (fire occurs inside a building and spreads to the façade due to venting through an opening) will be compared, thus the influence of SW fire barrier on the fire performance of ETCIS system with combustible thermal insulation will be evaluated. Furthermore, test results obtained in Test 1 and Test 2 will be shown parallel, so the temperature evolution behaviour of test specimens could be easily noticed.

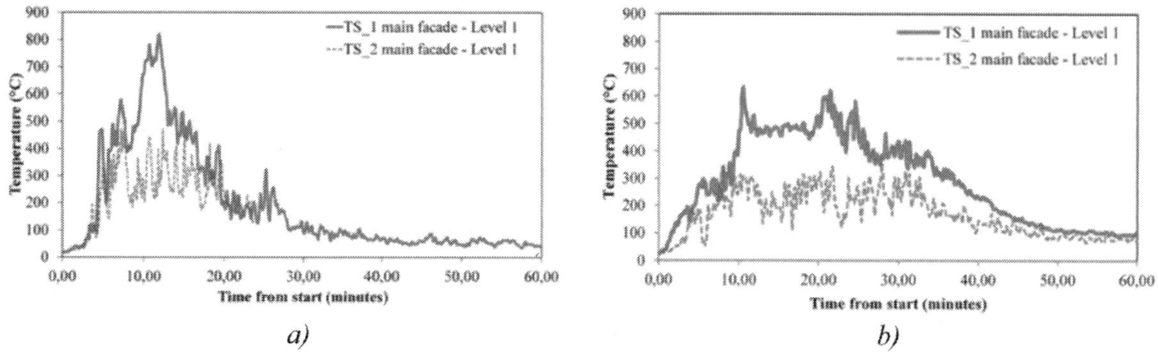


**Figure 4.** Average surface temperature at Level 1 on main facade: a) Test 1, b) Test 2.

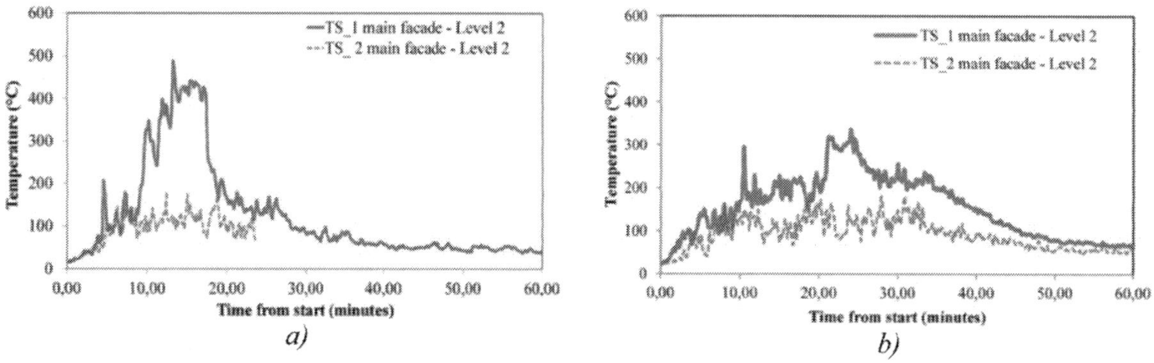


**Figure 5.** Average surface temperature at Level 2 on main facade: a) Test 1, b) Test 2.

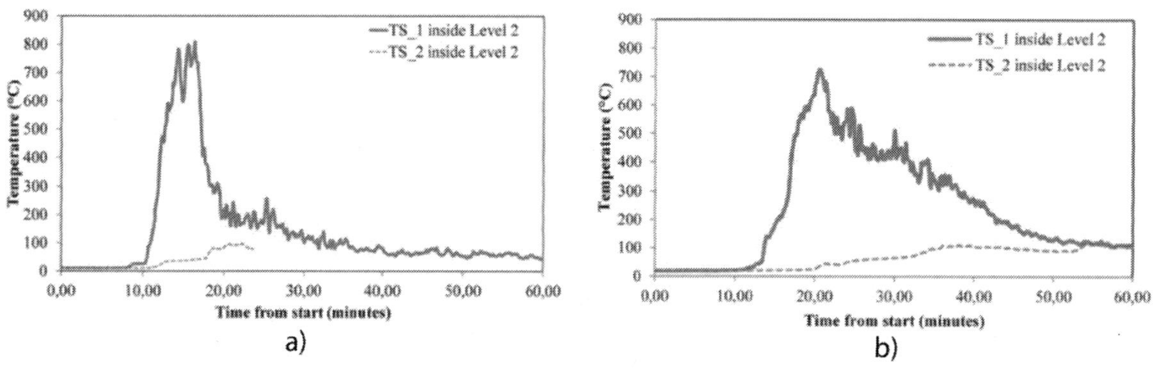


**Figure 6.** Average temperature within thermal insulation layer at Level 2 on main facade: a) Test 1, b) Test 2.

Figure 4 and Fig. 5 are presenting temperature profiles based on average temperatures measured according to the standard [2] with external thermocouples on test specimens at Level 1 and Level 2, respectively. Figure 6 presents the average temperature measured within the thermal insulation layer at Level 2, according to the standard. Average temperatures imply an average of several thermocouples at the same level.

Development of surface temperatures on test specimens TS_1 and TS_2 in Test 1 is comparable and has practically the same increase rate and values up to 8 min from the test start (Fig. 4a and Fig. 5a). After those 8 min average surface temperatures on test specimen TS_1 are considerably higher compared to the test specimen TS_2. In case of Test 2, such a correspondence of temperatures in first

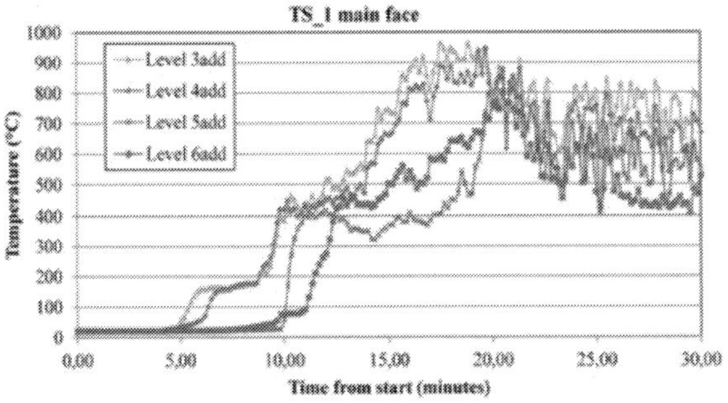


*a)*

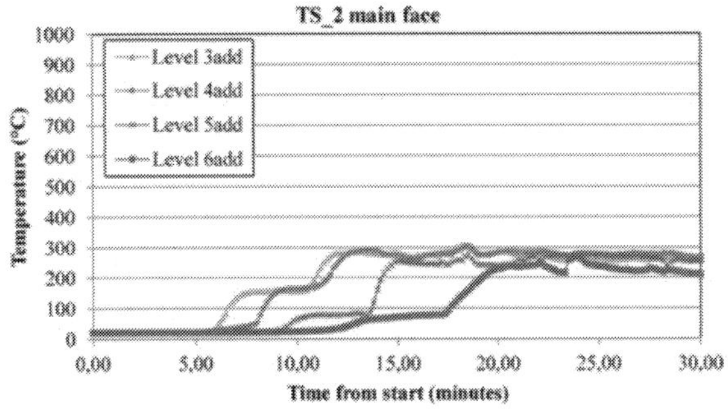


*b)*

**Figure 7.** Average temperature within thermal insulation layer at additional levels on main façade from Test 2: a) specimen TS_1, b) specimen TS_2.

few minutes from test start is not present (Fig. 4b and Fig. 5b). Although, surface temperatures on test specimen TS_1 during the Test 2 were higher compared to test specimen TS_2, thus confirming the behavior from Test 1. As it can be observed in Fig. 4 and Fig. 5, temperature profiles of test specimen TS_1 are characterized with stronger peaks, while the temperature profiles of test specimen TS_2 are characterized with a multitude of oscillations and significant peaks cannot be recognized. Temperature profiles of both test specimens, i.e. TS_1 and TS_2, follow form of the fire development curve, but the fire development curve of TS_2 has quite lower amplitude, i.e. the stages corresponding to the flashover and fully developed stage are mitigated due to presence of SW fire barrier.

When results from Test 1 and Test 2 are being compared, the differences in developed temperatures in each test can be noticed on both test specimens. Differences in surface temperatures on test specimen TS_1 reach up to 200°C at Level 1 (Fig. 4a and Fig. 4b) and up to 150°C at Level 2 (Fig. 5a and Fig. 5b). On test specimen TS_2, when Test 1 and Test 2 are being compared, differences in surface temperatures reach up to 150°C at Level 1 (Fig. 4a and Fig. 4b), while at Level 2 temperatures remain practically the same (Fig. 5a and Fig. 5b). It can be assumed that the differences are caused by changes in weather conditions and different burning rate of fire sources, which will be further discussed in this paper.

Analysis of average temperatures within the thermal insulation layer shows that test specimen TS_2 has considerably lower temperatures compared to test specimen TS_1. Obviously, fire barrier has limited the fire development within the insulation in test specimen TS_2 and temperatures remained below or around 100°C at Level 2 (both Test 1 and Test 2). The adverse effect occurred on TS_1, i.e. the thermal insulation in test specimen TS_1 caught fire and peak temperatures reached around 800°C in Test 1 and around 720°C in Test 2 at Level 2 (Fig. 6a and Fig. 6b). The first peak in test specimen TS_1 appeared around 14 min from the start in Test 1 (Fig. 6a), while in Test 2 it appeared around 20 min from the start (Fig. 6b). Possible explanation for this occurrence of peak's time shift could be change in weather conditions and different burning rate of fire source.

Analogously to the above presented tendencies of average surface temperatures, the average temperatures within the thermal insulation layer are constant and have the same value in both test specimens up to the first 8 min from the start in Test 1 (Fig. 6a) and up to the first 10 min from the start in Test 2 (Fig. 6b).

Results of additional measurement are shown in Fig. 7 and Fig. 8.

Additional measurements of temperature within thermal insulation layer were conducted only during Test 2 (Fig. 7). Additional levels marked in Fig. 3b were defined to gain an insight into the vertical fire spread within combustible insulation, with and without fire barrier. The results show that the fire barrier in specimen TS_2, caused the temperatures within thermal insulation layer up to Level 6 add (6.0 m above the combustion chamber) to stay below 300°C. However, this indicate that EPS in specimen TS_2 has melted away at least 6 m above the combustion chamber, as EPS starts to melt at temperatures

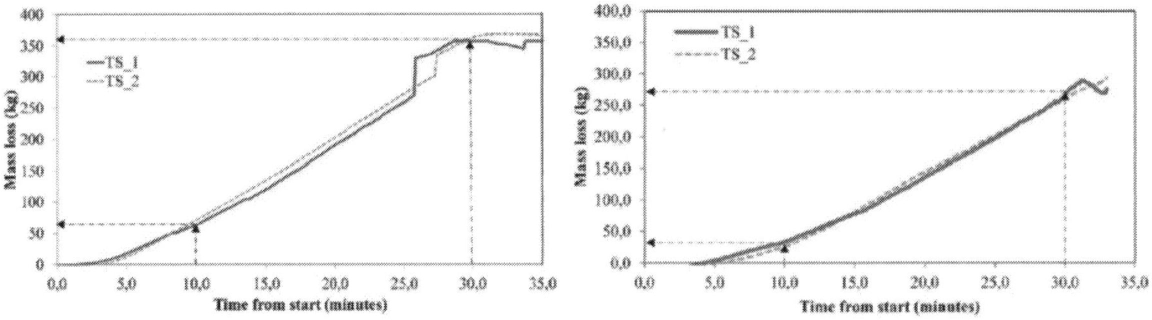


*a)* *b)*

**Figure 8.** Mass loss of wood crib: a) Test 1, b) Test 2.

above 80°C [6–8]. EPS melting vertically underneath uncracked render is confirmation of accelerating internal fire spread because of the occurring stack effect.

As expected the fire barrier has no influence on burning rate of the fire source, i.e. the mass loss of wood crib was practically the same for test specimens TS_1 and TS_2. That finding from Test 1 was confirmed in Test 2 (Fig. 8a and Fig. 8b). On the other hand, there is a noticeable difference in mass loss measured in Test 1 and the one measured in Test 2. In Test 2 the burning rate is decreased (Fig. 8b) compared to that in Test 1 (Fig. 8a). It should be noted that the standard [2] only defines to "nominal" volume of wood crib used as a fire source. For the same geometry of wood crib and the same type of softwood used (fir), in Test 2 the initial mass of wood crib was approximately 100 kg heavier for TS_1 and approximately 80 kg heavier for TS_2, compared with the mass of wood cribs in Test 1. Even for the same type of softwood, its density can vary and thus affect the mass of wood crib. Also, different moisture content in the wood will affect the mass of the crib. Variation in density of softwood and its moisture content, together with changes in weather conditions, resulted with wood cribs in Test 2 that were burning more slowly, i.e. the mass loss was lower compared to the mass loss in Test 1 (Fig. 8a and Fig. 8b). Consequently, lower temperatures were developed on the surface of the test specimens and within the thermal insulation layers in Test 2 compared with temperatures developed in Test 1 (Figs. 4–6).

Since, at the moment, there is a strong effort to establish a harmonised European full scale test method for determining fire performance of building facades, it has to be highlighted that the fire source in future EN full scale test method needs to be carefully defined, whereas it can greatly affect the outcome of testing.

Figures 9–12 shows the development and fire spread on both specimens during Test 1. By visual observation during the test, it has been noted that the fastest fire spread, both vertically and horizontally, was present on the test specimen TS_1 accompanied with significantly more smoke production compared to specimen TS_2. On test specimen TS_2, fire spread was mitigated due to the fire barrier above the combustion chamber, which is also reflected in the temperature profiles (Figs. 4–7). The fire barrier however, did not prevent the fire from consuming the test specimen. The insulation above the fire barrier melted as the temperature passed the melting point and pooled on top of the fire barrier. Here it collected until the render at the lintel level broke and allowed melted material to drip out as burning droplets as well as burning on top of the fire barrier creating a delayed smoke production. The main risk with burning droplets is that they can spread the fire downwards and can also endanger firefighters in real fire cases. Smouldering was present on specimen TS_2 after the wood crib had burned out. This is assumed to be the residual melted EPS insulation slowly combusting within the fire barrier.

In both tests, after the fire sources have self-extinguished, firefighters hosed down the wood cribs and test specimens for safety reasons. For test specimen TS_1, only the glass fibre mesh and the finishing

**Figure 9.** Visual observations of test specimens during the first 10 min of testing.

**Figure 10.** Visual observations of test specimens from 13 min to 28 min from the test start.

render has left, while all the EPS thermal insulation had burned out, Fig. 12. On test specimen TS_2 the EPS thermal insulation had melted only partially, Fig. 12.

All specific appearances observed during Test 1 (Figs. 9–12) have also occurred during Test 2, but at slightly later times. Nevertheless, a pattern of behaviour in this type of fire scenario can be established for tested ETICS systems as follows:

1) ETICS with EPS (TS_1): As the fire extends from the combustion chamber the organic render ignites and starts to crack. This is followed by the render opening up allowing the EPS underneath to ignite. As the EPS starts to burn the smoke emissions increase until a thick black smoke is observed, and it is followed by occurrence of pool fire and burning droplets. All the EPS in the specimen is consumed by the fire before the wood crib is burned out.
2) ETICS with EPS and fire barrier (TS_2): As with TS_1 the fire extends from the combustion chamber and ignites the organic render. The fire is "localized" on the façade above the opening. The EPS above the fire barrier starts to melt. The fire spreads up the façade in part due to melted EPS igniting and starting to burn at the level of the fire barrier. This causes a delayed smoke production as well as significant amount of burning droplets as the fire barrier can no longer contain the melted EPS. The EPS is melted only partially along the specimen height.

Time shift in specific appearances during Test 1 and Test 2 can possibly be caused by changes in weather conditions and different burning rates of fire source. Different weather conditions may cause different

**Figure 11.** Visual observations of test specimens 30 min from the test start.

**Figure 12.** Test specimens TS_1 and TS_2 at the end of testing.

test outcome for the same systems tested, i.e. may cause different evaluation of fire performance. The same applies for the differences in wood cribs.

## 3. DISCUSSION

The results presented in this paper show what kind of influence a non-combustible stone wool fire barrier positioned above the window opening can have on the fire performance of an ETICS system with combustible insulation, i.e. EPS. The stone wool fire barrier has two major functions during the test:

1. It keeps the render and mesh in position longer delaying cracking of the render.
2. It absorbs melted EPS delaying the spread of fire and burning droplets.

However, the results also shows as whereas a stone wool fire barrier can slow down and decrease the fire spread in EPS ETICS it cannot fully prevent the EPS from participating in the fire. When the fire barrier gets saturated with melted EPS it will allow some of this to spill out through the cracks that has developed in the render resulting in a flow of burning droplets. Depending on the thickness of EPS involved it could lead to a pool fire in front of the façade as seen in this test. As more oxygen is allowed in through the cracks the melted EPS in the fire barrier will start to burn causing significant smoke production.

Consequently EPS ETICS with stone wool fire barriers cannot be considered as fire safe as systems with non-combustible insulation and hence are not suitable for all building types. When deciding on the type of ETIC system to use on a building it is important to know how the system will perform in case of fire by having it tested in an appropriate large scale test. Only then can a decision be made if the system will lead to safety in case of fire for the building and occupants in question. This emphasizes the importance of having a harmonised large scale fire test for facades in EU.

## 4. CONCLUSION

Non-combustible stone wool fire barriers if positioned above the window opening can delay the fire spread and smoke production of EPS ETICS in case of a fully developed fire impinging on the façade. Specifically it was seen that

- *EPS burned out completely if no fire barrier was present, while it was melted under the render and pool fire occurred on the ground near specimen when a stone wool fire barrier was in place over the window.*
- *The fire barrier delayed the smoke production and burning droplets compared to the sample with no fire barrier.*
- *Temperature profiles show considerably higher surface temperature and temperature within thermal insulation layer on specimen when no fire barrier is used.*

Therefore, in systems like the ones tested a non-combustible stone wool fire barrier can possibly be used as passive fire protection. It has to be highlighted that further research is needed to generalise this hypothesis to other systems as well.

The fact that the system with the fire barrier allowed considerable fire spread and smoke production, although later in the test, shows that the safety margin of these types of systems are limited. They should therefore not be used on buildings such as high rise buildings and buildings where occupants need additional time to escape such as hospitals, schools and nursing homes.

The testing highlights that appropriate large-scale testing of façade systems is needed in order to fully understand their behaviour in fire. It also shows that in order to sustain repeatability of the testing procedure, it is necessary that environmental factors such as wind and ambient temperature are controlled and within certain limits.

Based on the experiences form these tests the authors considers the test BS8414 as an appropriate large scale façade fire test to be used for the basis of harmonisation in EU.

### References

[1] Pasker, R., "WDVS in Europa", European Association for External Thermal Insulation Composite Systems EAE, EUMEPS Master Class ETICS, Vienna, Austria, 2013.

[2] BS 8414-1:2002 "Fire performance of external cladding systems. Test methods for non – loadbearing external cladding systems applied to the face of building".

[3] Šilhard Mihaljević, S., Weigand I., Jelčić M., "Potvrđivanje sukladnosti povezanih sustava za vanjsku toplinsku izolaciju (ETICS) za primjenu u zgradarstvu" Hrvatska normizacija i srodne djelatnosti – tehničko usklađivanje na putu prema Europskoj uniji, Plitvička jezera, Croatia, 2005.

[4] HRN EN 13501-1:2010 "Fire classification of construction products and building elements – Part 1: Classification using data from reaction to fire tests" (EN 13501-1:2007+A1:2009).

[5] Brillux, "Technische Info 5b04, WDVS Brandschutz Grundlagen und Begriffe, brandschutzrelevante Verarbeitungsdetails", Münster, Germany, 2011.

[6] Hens, H., *Performance Based Building Design 1, From Below Grade Construction to Cavity Walls,* Ernst & Sohn A Wiley Company, 2012.

[7] Banjad Pečur, I., Bjegović, D., Boström, L., Milovanović, B., Hajduković, M., "ETICS Fire Performance test", Proceedings of the Fifth International Workshop on Performance, Protection & Strengthening of Structures under Extreme Loading, 964-971, Michigan, USA, 28-30 June, 2015.

[8] Bjegović, D., Banjad Pečur, I., Milovanović, B., Jelčić Rukavina, M., Alagušić, M., "Comparison of fire performance of ETICS systems in full scale test", *Građevinar*, 2016. – *accepted for publishing*.

MATEC Web of Conferences **46**, 05007 (2016)
DOI: 10.1051/matecconf/20164605007


# Fire safety of ETICS with wood fibreboards for multi-storey buildings – first research and development results

Judith Küppers[1], Jochen Zehfuß[1], Felix Steeger[2] and Björn Kampmeier[2]

[1] Technische Universität Braunschweig, Institute of Building Materials, Concrete Construction and Fire Protection (iBMB), Braunschweig, Germany
[2] Hochschule Magdeburg-Stendal (HSMD), Division of Fire Safety and Structural Design, Magdeburg, Germany

**Abstract.** Wood fibreboards can serve as alternative sustainable insulation material for external thermal insulation component systems (ETICS). In Germany, the application of ETICS with wood fibreboards is restricted to low buildings. The restrictions are mainly caused by the smouldering tendency of the wood fibreboards. Thus, the development of an ETICS with wood fibreboards for multi-storey buildings complying with the requirements would provide a new scope of application for this sustainable insulation material.

This paper presents first results of standardised and innovative investigation methods as basis for the development. These investigations and analyses concern the fire behaviour, especially smouldering processes, the plaster system, other constructive protection measures as well as risk and safety analyses.

## 1. INTRODUCTION

External thermal insulation component systems (ETICS) serve as outside insulation of exterior walls. For many constructions, they are indispensable under the applicable increased requirements for thermal protection and energy efficiency of buildings in Germany. This applies not only to new buildings, but in particular to existing buildings.

In addition to energy efficiency, there are fire protection requirements valid for the construction of facades. The model building code of Germany (representative of the state building codes) generally divides buildings by their height and utilization units' area into five building-classes (Table 1). For building-class 1 to 3, e.g. single family houses, building materials with normal flammability are allowed. In building-class 4 and 5, e.g. multi-storey apartment buildings, the requirements for combustible insulation materials of ETICS are higher (low flammability). Within the European classification system the C-class represents the minimum requirements valid for low flammability according to the German model building code and the Building Rules List. It has also to be considered that for sustainable materials like wood fibreboards, their smouldering tendency is the main reason why they do not comply with the requirements of the application as ETCIS in higher building-classes.

**Table 1.** Building-classes and fire protection requirements for ETICS according to the model building code of Germany.

| 1 | 2 | 3 | 4 | 5 |
|---|---|---|---|---|
| freestanding building height of floor $\leq 7$m; less than 2 units | height of floor $\leq 7$m | | height of floor between 7 and 13m, size of unit $< 400\text{m}^2$ | height of floor $> 13$m (except special constructions and utilisations) |
| | less than 2 units | more than 3 units | | |
| 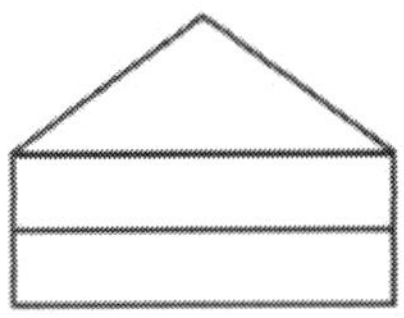 | 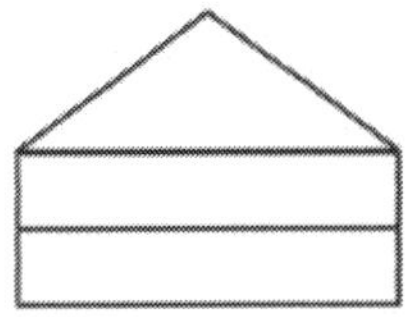 | | 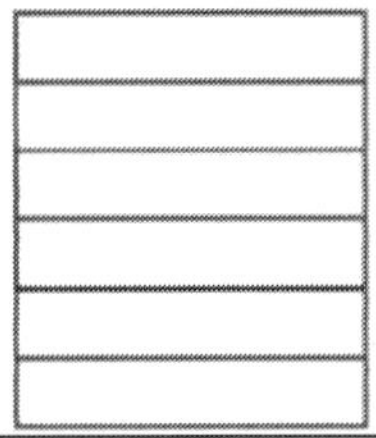 | 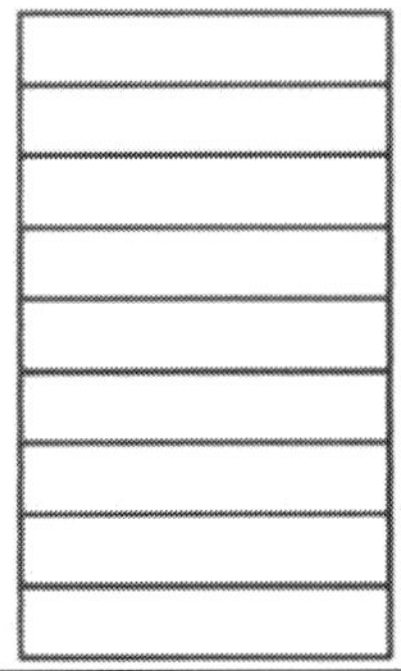 |
| normal flammability | | | low flammability | |

The objective of the presented research project is to improve the wood fibreboards for reducing smouldering or stopping it with barriers at an exactable level, thus developing an ETICS for multi-storey buildings complying with the requirements. The German Federal Ministry of Economics and Technology funds this research within the Central Innovation Program for Small- and Medium-sized Enterprises (ZIM). The operating institutions in this research project are the iBMB of the Technische Universität Braunschweig, the University of Applied Sciences in Magdeburg (HSMD), the Fraunhofer Institute for Wood Research (WKI) and industrial partners.

In general, two ways are pursued developing the ETICS: On the one hand, a "thick layer system" is examined, where the plaster systems mainly should protect the combustible materials. On the other hand, the whole system and mainly the wood fibreboards shall be enhanced to reduce fire behaviour and smouldering processes to an acceptable level.

## 2. RESEARCH AND DEVELOPMENT RESULTS PROVIDED BY INDUSTRY

The involved wood fibreboard producers provided information about several fire tests. Small- and medium-size tests like the single-flame source test (ISO 11925-2), "Brandschachttest" (DIN 4102-15/16) and SBI test (EN 13823) yield information about the fire behavior. They are standardised and provide comparable results. These tests of different (equipped) wood fibreboards are taken into consideration for planning and analysing the tests within this project. In addition, results of three facade tests were provided, which did not yet lead to a system that complies with all requirements, but yield useful results for further development.

## 3. FIRE BEHAVIOUR AND SMOULDERING PROCESSES

As first step of the developing process, the fire behaviour and smouldering processes of the wood fibreboards are studied. In addition to the aforementioned results, literature about smouldering processes, test methods and results, e.g. [1–7], was studied and test series were planned.

The most influencing parameters have to be studied for analysing the fire behaviour and especially smouldering for getting a better understanding of this complex mechanism. Flame-retardants are an

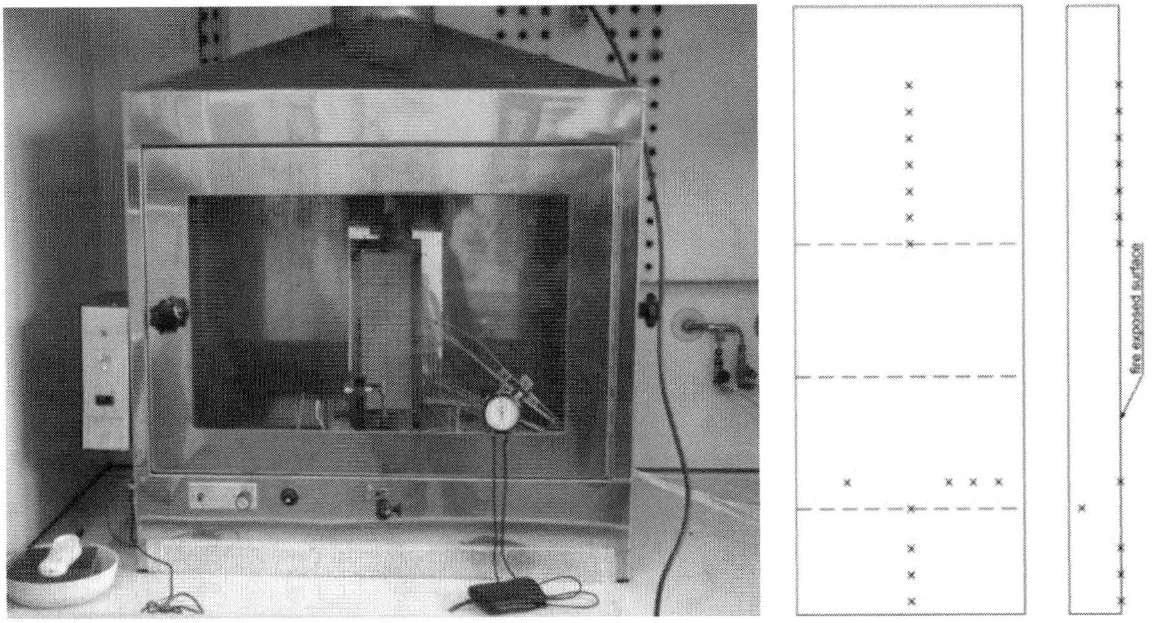


**Figure 1.** Modified single-flame source test and example of thermocouple placement [iBMB].

important influencing parameter and well known for enhancing wood fibreboards and other sustainable building materials. Another main parameter influencing the smouldering processes is the density of the wooden boards.

In order to investigate the aforementioned and further influencing parameters, different test series are in process. For example, in the cone-calorimeter according to ISO 5660 the ignition temperature, mass loss rate and heat release rate (by oxygen calorimetry) are measured comparing different equipped boards. Samples for cone-calorimeter-tests are small ($10 \times 10\,cm^2$) and the smoke gases are analysed. Due to this, these tests are effective for test series with a large amount of samples. On the other hand, the horizontal placement and the full-faced fire exposure of the samples does not correspond with the application and expected fire scenario. Due to this, additional tests are in process.

In addition to the single-flame source tests provided by the project partners, some more tests were performed. The test-procedure was modified for orienting tests with a longer fire exposure and samples equipped with thermocouples (Fig. 1). In these tests, the thermocouples were not only placed upwards. To receive more detailed information, e.g. about the smouldering velocity in different directions (in or against the upward current and lateral propagation) additional, thermocouples were placed. These tests were performed for orientation and specification of further tests.

In European standard fire tests for building materials, performance criteria for smouldering are missing. Thus, in Germany the smouldering of materials is investigated by performing the 'Brandschachttest' according to DIN 4102-15/16. At the European level, with the draft standard prEN 16733 (Fig. 2) smouldering shall be assessed. Both test methods require an equipment of the samples with thermocouples for assessing the measured temperature.

Different equipped boards of $300 \times 800\,mm^2$ are investigated in this test device and analysed. Typical temperature-time curves of the thermocouples placed in the samples are shown in Fig. 3. The results of the modified single-flame source tests (equipped with thermocouples) lie in the same range like the tests according to prEN 16733. It is characteristical that the smouldering process starts with a rapid temperature increase in or close to the initial area (TC 1 and 2), followed by a plateau at around 80 to 110°C and an additional, but faster, temperature increase. While smouldering, the first increase is

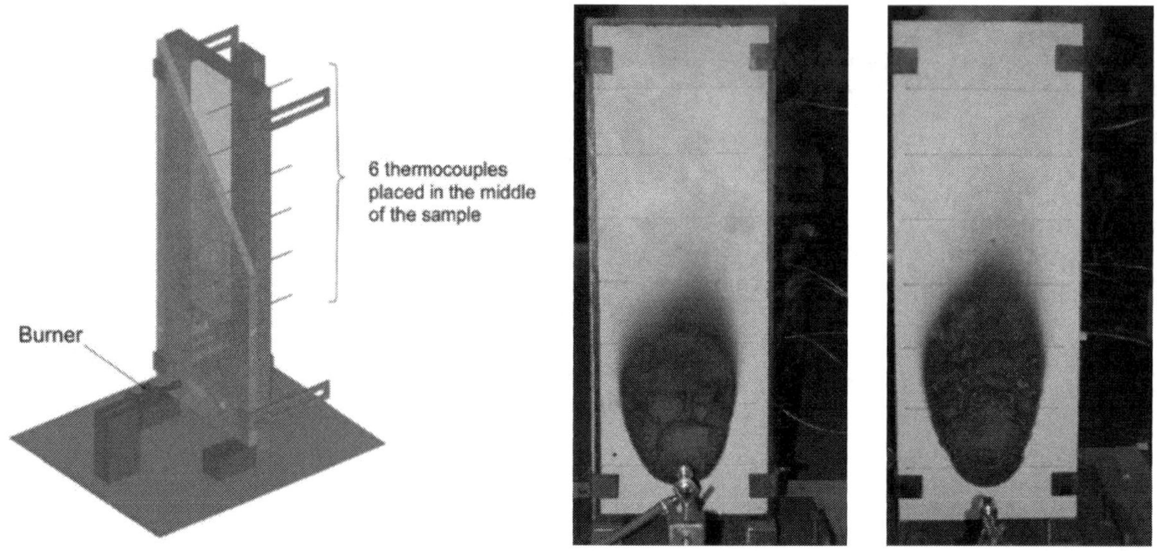


**Figure 2.** Test device (left) and wood fibreboards with (middle) and without flame-retardants (right) shortly after fire exposure [drawing:iBMB, pictures:HSMD].

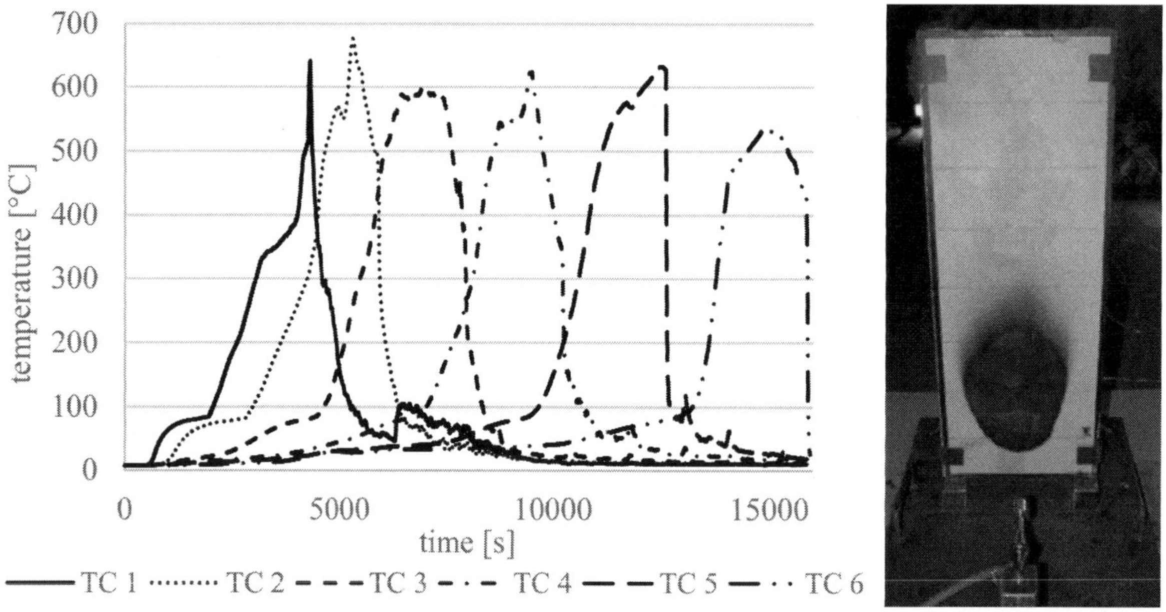


**Figure 3.** Temperature-time curves of a wood fibreboard with flame-retardants and a density of ca. 180 kg/m$^3$ tested according to prEN 16733 and picture shorty after fire exposure [HSMD].

slower than in (or close to) the initial area and the inflexion point at around 80 to 110°C is followed by a temperature increase nearly as fast as in the initial area.

These test results can be used for determining the smoldering velocity. Therefore only temperatures higher than 110°C should be considered. Although, right now only some tests were carried out, first results thick boards are showing the expected influence of the density: The smouldering velocity decreases, when the density increases. In detail, the smoldering velocity was determined, for example, as 0.11 m/h up to 0.18 m/h for densities of around 130 kg/m$^3$ up to 190 kg/m$^3$ by estimating the time difference for temperature steps of 200°C to 400°C at the two highest thermocouples (TC 5 and 6).

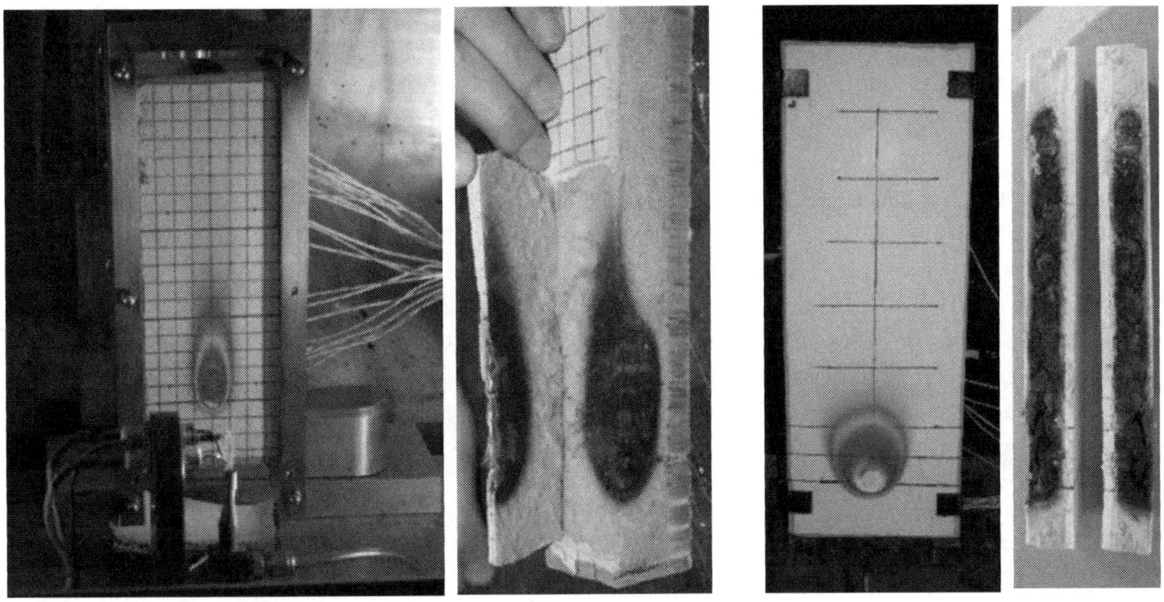

**Figure 4.** Plastered wood fibreboard performing the modified single-flame source test (left) [iBMB] and the smouldering test (right) [HSMD].

Currently only two samples with flame-retardants and different density are tested and analysed, so the influence cannot be estimated in detail yet. The measured temperatures indicate no decrease of the smouldering velocity. These results do not imply that the flame-retardants are not effective as they are designed for decreasing the heat release rate as well as flame spread. This complies with the observed smaller initial areas (black/decomposed area) of equipped boards (Fig. 2) and the delayed temperature increase measured at the lowest thermocouples (TC 1 and 2). Another general observation was that the equipped boards left more residues than the boards without flame-retardants.

First results of the analyses of board with different thickness are showing the influence of this parameter. Observing the smouldering front, after the nearly circular smouldering at the beginning until reaching the edges of the specimen, a parabolic course was determined. It is supposed, that this course, which was observed at all surfaces (front, back and sides), is caused by cooling effects at the edges. A further insight is the influence of the surrounding temperature during the test, which affects obviously the smouldering velocity.

In general, further tests are needed to provide a greater data base for a more detailed analysis. These tests shall underline the mentioned theses. Furthermore, they are planed with additional thermocouples to get more detailed information about the smouldering velocity in different directions for comparison with the abovementioned modified single-flame source tests.

One of the most important parameters for the system development is the minimum temperature of starting decomposition processes of the wood fibreboards (protected by the plaster system) in the ETICS. A first impression of this characteristic is given by the test method described in [8]: In the cone-calorimeter the temperature is measured between a gypsum board (representing a protective layer) and the samples of wood fibreboards. The method was developed for estimating protection criteria for (sustainable) combustible materials for their application as insulation in construction elements, e.g. timber panel wall systems for building-class 4, where any decomposition should be prevented. First occurrence of discoloration of the insulation material is the criteria of the method according to [8]. Because of the different application (as ETICS), it was decided for this project to accept little discoloration, but to receive mass loss and oxygen calorimetry as indication for pyrolysis. The tests were performed with temperature steps of 25°C and led to a critical temperature of 350°C for all examined wood fibreboards (with and without flame-retardants), except a board with high density (of ca. 190 kg/m$^3$ and a critical temperature of 375°C). This will be verified by further tests (of high density boards for example). The first test results assume, that smouldering processes start, if this temperature is reached behind the plaster while a fire exposure to an ETICS. Therefore, this temperature can be defined as initial temperature for smouldering processes. Currently, test with plastered samples are in process to find out, if and how long respectively how far the initialized smouldering is going on in this enclosed system (see below).

DIN 4102-15/16 and prEN 16733 define the end of a smouldering process, when no temperature above 50°C is measured. According to the temperature-time curves shown in Fig. 3 with the inflexion point at around 80 to 110°C, temperatures below 50°C seem to be a good criterion for no occurrence of smouldering. Related to the aforementioned cone-calorimeter results of 'protected' wood fibreboards, 50°C should be only a criterion for progression but not for initialising.

In addition to the smouldering tests analysing the wood fibreboards themselves respectively comparing them, plastered boards were investigated. For these tests, the wood fibreboards were affixed to a gypsum board and plastered with a standard plaster system at the front and a plaster at the sides to close the system. To ensure a started smouldering process, a thermocouple was installed between the plaster and the wood fibreboard at the same height where the burner is placed. The burner flame is applied until a temperature of 350°C is measured here. During the following observation time, further thermocouples measure the smouldering process.

Figure 4 is showing wood fibreboards, plastered with a thin layer system of ca. 6 mm, performing the smouldering test according prEN 16733 after the fire exposure of ca. 12 min until 350°C were reached behind the plaster. During the modified single-flame source test, a fire exposure of almost 25 min was necessary to reach the "initial temperature" of 350°C. No visible cracks arose and the smouldering process seemed to be stopped ca. 60 min after the fire exposure. The plaster system could not prevent a decomposition of the wood fibreboard in the initial area, but it seemed to stop the smouldering process, probably by reducing the oxygen entry into the system sample. This result underlines the relevance of the plaster system as a protection measure. In contrast, the fire exposure in the smouldering test device arises cracks in the plaster in the initial area, a "real" smouldering process started and lasted around 55 h. As next step, a smouldering test with a thick layer system (15 mm) is planned to find out, if this plaster system can prevent crack development for the mentioned test conditions and may stop the decomposition process earlier.

These test results are needed to analyse the complex fire behaviour of an ETICS with wood fibreboards and dimensioning protection measures. This means the improvement of wood fibreboards and the dimensioning of further protection measures e.g. the plaster system or the smouldering barriers.

## 4. PLASTER SYSTEM

Plaster and claddings are commonly used as fire protection for steel and timber constructions and there are existing classifying methods, e.g. EN 13501-2. Due to these classification tests and research results carried out at iBMB [9–11], the protection effects are known. Compared to fire protection claddings for construction elements, the plastering of ETICS is much thinner and not as effective. On the other hand, the test conditions for ETICS are different compared to construction elements. However, in this project, the maximum applicable thickness of the ETICS plaster is investigated, as well as fire protection additives for the plaster. This leads to various plaster compositions.

As first step of this part of the research project, the heat transfer through the plaster is measured in the cone-calorimeter. These tests are for orientation and comparison of samples with one another, so that a reasonable choice can be made for more expensive standard fire tests. Although the test area of cone-calorimeter samples being $10 \times 10\,cm^2$ a size of $16 \times 16\,cm^2$ was chosen to reduce the lateral heat transport (outside of the test area). Mineral wool was used as insulation material below the sample.

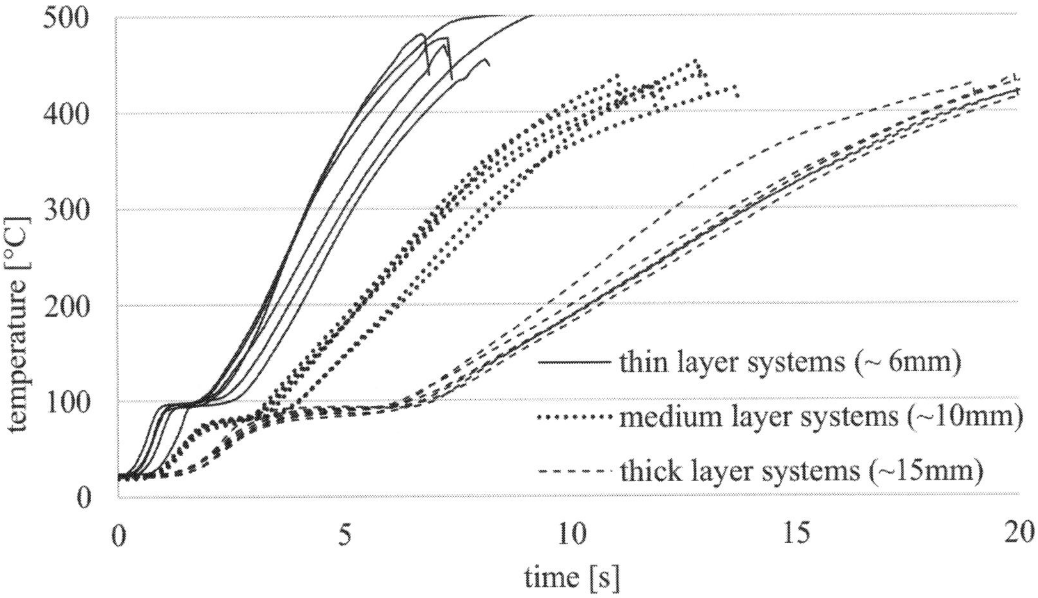


**Figure 5.** Heat transfer through different plaster systems measured in the cone-calorimeter [iBMB].

This noncombustible insulation was chosen instead of a combustible one because of the test objective (assessment of the heat transfer through the plaster and not the combustion of the insulation material). The heater of the cone-calorimeter produces a nearly continuous heat flux. For these investigations, the heat flux was set to 40 kW/m$^2$; this is nearly equal to the heat flux of the standard fire according to E DIN 4102-20 in a height of 1 m above the opening [12]. These test settings also allow a comparison with the former measurements [9–11]. In each test, five thermocouples placed between the plaster sample and the insulation measured the heat transfer. The five resulting temperature-time curves were close to each other, so that the average was chosen for comparison.

For the first test series, the project partner Endress produced plaster samples of ca. 6, 10 and 15 mm thickness with different composition of additives with emphasis on volumetric comparability. The measurements (Fig. 5) are underlining the results of the former project [9–11], that mostly the thickness influences the protection effects. Considering these results and having a look at the little differences, Endress produced plaster samples for the second test series, which will be tested in the cone-calorimeter again under the same conditions.

Furthermore, optimisation and application tests are in process to enlarge the maximum applicable thickness in a cooperation between Endress as plaster producer and Homatherm as wood fibreboard producer. Because of the limited plaster thickness, it is possible that the described measures will not prevent or stop the smouldering processes as required. Therefore smouldering protection barriers shall be developed.

## 5. FURTHER PROTECTION MEASURES

As further measures, smouldering protection barriers are conceivable. These barriers shall stop the smouldering process at an acceptable level, thus they have to be noncombustible. To define the placement of barriers, the fire behaviour and the smouldering velocity of the wood fibreboard have to be known as well as the expected fire exposure and the safety requirements resulting from risk analyses (see below).

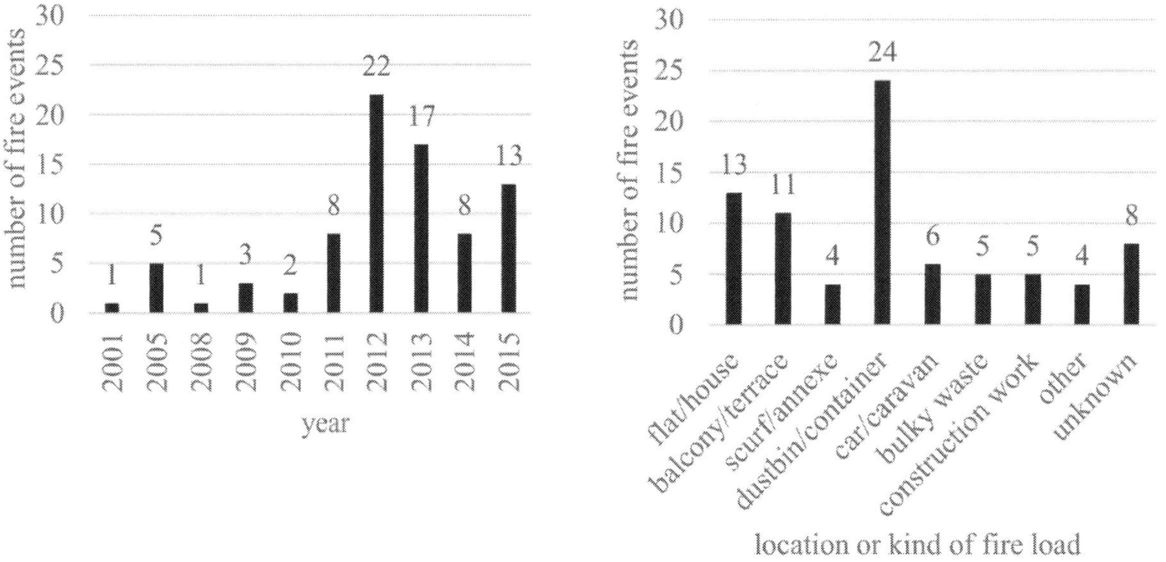


**Figure 6.** Occurrence and location or kind of fire load of fire events.

In addition, the requirements for thermal protection have to be considered. Currently, the material properties of different noncombustible materials are analysed with regard to their suitability as smouldering protection barrier.

## 6. SAFETY AND RISK ANALYSES

The mentioned requirements of the German model building code represent the fire safety level of exterior wall cladding, which increases with the building-class. Insulation materials of ETICS for building-class 4 and 5 have to exhibit their low flammability. One of the wood fibreboards of the project partner Homatherm is classified as euroclass C, but does not comply with the German smouldering criteria. Due to this, the use of ETICS with wood fibreboards is prohibited in the building-classes 4 and 5.

In Germany for application of an ETICS for building-class 4 and 5 the insulation materials do not only have to pass the standard fire test (reaction for fire), the whole system has to perform an additional (nearly) real-scale test (according to E DIN 4102-20). This is reasoned by the practical application as an ETICS consisting of different components.

Currently, several facade fire events arose a controversial discussion about fire safety of facades. One reaction was a research project commissioned by the building ministers. Within this project some tests with a fire load (wood crib) of ca. 2 MW, placed at the ground in front of an ETICS with EPS (socle fire scenario), were carried out [13, 14]. Before this, the window fire scenario of E DIN 4102-20 was expected as covering the socle fire scenario [15, 16]. The mentioned fire tests resulted in an instructtion paper [17] requiring more fire protection barriers in EPS systems.

The fire brigade of Frankfurt (Main) collects ETICS fire events with EPS in Germany [18]. Due to a standardised questionnaire on their homepage, the fire events can be reported. The current list (status august 2015) was analysed in dependence to [13].

Analysing this data it has to be considered that the reports are carried out on a voluntary basis. There is no commitment of reporting facade fire events and no standard fire statistic in Germany. Due to this, it is intended to generate an additional database in this research project. The fire brigade of Braunschweig tested first versions of a modified questionnaire developed by the iBMB. Currently, other fire brigades shall be convinced for supporting this investigation by participation. The cooperation with different fire brigades shall be the basis for a nationwide extrapolation and the estimation of the relevant risk parameters, e.g. criteria for an acceptable fire behaviour.

In addition to this kind of damage and risk analyse, wood fibreboard unions and organisations as well as a central insurance organisation [19, 20] were questioned about fire events with wood fibreboard ETICS. The only reply was a fire damage during the building process of a wood fibreboard ETICS in 2010. This complies with the mentioned fire events with wood fibreboards in [20], where no facade fire is listed. The author approved that the damage report still represents the current situation [19].

As main reason for the little number of registered fire events with wood fibreboard ETICS, the little distribution of such ETICS is assumed. That is why the damage analyses in this project cannot focus on fire damages at wood fibreboard ETICS. In addition to the mentioned damage analyses and statistic, theoretical and numerical analyses are in process, e.g. [13].

The results of the safety and risk analyses as well as the research and development results will be discussed with representatives of the building authority, insurance organisations, fire brigades and industry. A successive integration of these partners in the developing process shall lead to an objective- and performance-based design.

## 7. CONCLUSION AND OUTLOOK

This paper presents first results of standardised and innovative investigation methods as basis for the development of an ETICS with wood fibreboards for multi-storey buildings. The current research work focuses on the smouldering behaviour of wood fibreboards as the main reason why they do not comply with the requirements for application in building-class 4 and 5. The results imply that a value of 350°C as initialising temperature of smouldering processes is valid for most of the investigated wood fibreboards and temperatures lower than 50°C could be confirmed as criteria for a stopped or decreasing smouldering process. Although the first test results could not identify a decrease of the smouldering velocity when flame-retardants are used, a smaller initial area and delayed temperature increase in or close to the initial area was measured. These and other influencing parameters are studied in ongoing tests in more detail. In addition to these investigations, safety and risk analyses are needed for dimensioning protection measures.

The plaster system, for example, serves as protection measure. Plaster samples with different additive compositions are studied in the cone-calorimeter. The first results underline the results of the former project [9–11]: The plaster thickness influences mainly the performance criteria for fire protection. Additional compositions will be examined in the cone-calorimeter and the maximum applicable plaster thickness is investigated in a cooperation between the involved plaster and wood fibreboard producers. Other constructive measures, e.g. smouldering protection barriers, are also considered and will be investigated.

In addition to the mentioned investigations of smouldering, risk and safety analyses are needed for defining the performance criteria of these measures and dimensioning the system for application. All components have to be arranged according to their performance. At first, small single specimen will be investigated and then, the assembled components will be tested under more realistically conditions. At the end of the project, the newly developed system has to perform a large-scale fire test.

The research project was funded within the Central Innovation Program for Small- and Medium-sized Enterprises (ZIM) of the German Federal Ministry of Economics and Technology, for which the authors are grateful. Furthermore, thanks to the involved partners: Fraunhofer Institute for Wood Research (WKI), fire brigade of Braunschweig, Wolfgang Endress Kalk- und Schotterwerk GmbH & Co. KG, Holzbau Weizenegger GmbH, HOMATHERM GmbH and the further industrial partners GUTEX Holzfaserplattenwerk H. Henselmann GmbH & CO. KG, STEICO SE and UNGER-DIFFUTHERM GmbH.

## References


[1] Ohlmiller, T. J., "Smouldering Combustion". In *SFPE Handbook of Fire Protection Engineering* (P. J. DiNenno, Ed.), Springer, Westford, 2002.

[2] Rein, G., "Smouldering Combustion". In *SFPE Handbook of Fire Protection Engineering* (M. J. Hurley, Ed.) Springer, New York, 2016.

[3] Hölemann, H., "Schwel- und Glimmversuche an Baustoffen", *wksb*, **18/1984**, 1–9, 1984.

[4] Hölemann, H., "Entwicklung eines Prüfverfahrens für die Bewertung von schwel- und glimmfähigen Baustoffen", *vfdb-Zeitschrift*, **1/97**, 11–19, 1997.

[5] Hölemann, H., "Schwelbrände – Abgrenzung, Entstehung, Ausbreitung, Brandbekämpfung", vfdb-Zeitschrift, **3/2009**, 124–132, 2009.

[6] Gertlová, Z., Bergmeister, K., Marková, I., Cunderlík, I., Wegener, G., and Windeisen, E., "Schwel- und Glimmverhalten von Dämmstoffen", Institut für Konstruktiven Ingenieurbau, Universität für Bodenkultur Wien, Veröffentlichung des Institutes, Heft Nr. 49, Wien, Österreich, 2001, ISSN 1028–5334, ISBN 3-900962-27-8.

[7] Adamietz, P., Ehrlenspiel, R., Richter, K., "Experimental Study on Smoldering Combustion of Construction Products", Proceedings of Fire and Materials, 297–308, San Francisco, USA 28–30 January, 2013.

[8] Kampmeier, B., "Risikogerechte Brandschutzlösungen für den mehrgeschossigen Holzbau", Dissertation, Braunschweig, Germany, 2008.

[9] Wachtling, J., Hosser, D., Zehfuß, J., "Fire protection of multi-storey straw bale buildings", Proceedings of Research and applications in Structural Engineering, Mechanics and Computation, (A. Zingoni, Ed.), 707–708, Cape Town, 2–4 September, 2013, ISBN 978-1-138-00061-2.

[10] Hosser, D., Zehfuß, J., Wachtling, J., "Direktverputzte Strohballenbauteile für die Gebäudeklasse 4 - Teilprojekt Brandschutztechnische Untersuchungen", Abschlussbericht ZIM-Förderkennzeichen: KF 2178804 KI0, 2014.

[11] Küppers, J., Hosser, D., Zehfuß, J., "Neue Forschungsergebnisse zum Brandschutz für mehrgeschossige Strohballenbauten", *vfdb-Zeitschrift*, **1/2015**, 20–27, 2015.

[12] White, N., Delichatsios, M., "Fire Hazards of Exterior Wall Assemblies Containing Combustible Components", Final Report, Fire Protection Research Foundation, Quincy, Massachusetts, USA, 2014.

[13] Kaudelka, S., Hofmann-Böllinghaus, A., Hauswaldt, S., Krause, U., "Auswirkungen von Zündquellen und Systembeschaffenheit auf das Brandverhalten EPS-basierter Wärmedämm-Verbundsysteme", *Bauphysik*, **37**, 205-212, 2015, (`http://onlinelibrary.wiley.com/doi/10.1002/bapi.201510023/epdf`).

[14] Zehfuß, J., Riese, O., Northe, C., Küppers, J., "Experimentelle und numerische Erkenntnisse zum Brandverhalten von WDVS-Fassaden auf Polystyrol-Basis", *Bauingenieur*, **12-2015**, 567–574, 2015.

[15] Kotthoff, I., "Wie brennt was? – Brandszenarien am Beispiel der Gebäudeaußenwand", *Fachzeitschrift zuschnitt*, **14**, 2004, (`http://www.proholz.at/zuschnitt/14/wie-brennt-was/`).

[16] Hofmann-Böllinghaus, A., Kaudelka, S, "Wärmedämmverbundsysteme – Neue Erkenntnisse aus Versuchen und Analysen aktueller Brandereignisse", Proceedings of 62. Jahresfachtagung der Vereinigung zur Förderung des Deutschen Brandschutzes e. V., 291–304, Dortmund, Germany, 16–18 Juni, 2014.

[17] Hinweis des Deutschen Instituts für Bautechnik (DIBt): WDVS mit EPS-Dämmstoff, Konstruktive Ausbildungen zur Verbesserung des Brandverhaltens von als "schwerentflammbar" einzustufenden Wärmedämmverbundsystemen mit EPS, Stand 16. Dezember 2014. `https://www.dibt.de/de/Fachbereiche/data/Merkblatt_Sicherstellung_der_Schutzwirkung_von_EPS-WDVS_18062015.pdf` Fankfurther Liste

[18] Homepage of the fire brigade of Frankfurt (Main) and the project reporting fires with EPS-ECTIS (`http://www.feuerwehr-frankfurt.de/index.php/projekte/wdvs`)

[19] Engfeld, A., "Baustoff Holzweichfaserplatten – Ein ökologischer, aber brandgefährlicher Baustoff", *Schadenprisma*, **1/2009**, 4–8, 2009.

[20] Engfeld, A., Institut für Schadenverhütung und Schadenforschung der öffentlichen Versicherer, Interview transcript, 2 September, 2015.

MATEC Web of Conferences **46**, 05008 (2016)
DOI: 10.1051/matecconf/20164605008


# Reaction to fire of ETICS applied on wood particle board

Antonio Bonati and Vittorio Galimberti

Istituto per le Tecnologie della Costruzione – Italian National Council of Research San Giuliano Milanese (Milan), Italy

**Abstract.** As well known the ETICS are diffusely used both for energy saving and thermal insulation reasons. They have been applied recently in wood buildings and in regions of southern Europe too due to green building and sustainability reasons.

ITC-CNR has tested a lot of building materials and developed good knowledge about reaction to fire since the 1980 and currently, ETICS fixed directly to particle wood panels have been investigated with several SBI tests.

In the case study are presented the main factors that can influence the fire reaction results when applied on wood structure are highlighted: the thickness of the insulating material, the presence of accidental damage, the flame attack from the inside.

From the results obtained by tests on samples prepared with simulated accidental damages and fire from inside, some considerations are made about the hazard due to this specific construction technology and others on limits of the type of actually used standards product classification.

## 1. MATERIALS AND SYSTEMS USED IN THE STUDY

The study was carried out at ITC Fire Laboratories in the 2015 on some ETICS of current production, usually sold on the market in kits, to be applied on masonry or wood based supports. For samples assembling were used the following components:

- Adhesive layer (5 mm) and base coat (4 mm) cement-based with calcium carbonate fillers, density 1750 kg/m$^3$.
- EPS insulating panels, pentane expanded, Fire Reaction Class E in accordance with EN 13501-1:2007+A1:2009 [1], density 18 kg/m$^3$.
- Reinforcement by glass fibber net with mesh 4, 15 × 3, 80 mm, mass per unit area 155 g/m$^2$.
- Key coat and red finishing coat made of synthetic resin binder with 2 mm fillers added.
- Plastic anchors.

## 2. PERFOMED TESTS

The fire behaviour analysis of the different samples for comparison has been based on results from tests that have been done in accordance with EN 13823:2010+A1:2014 [2].

**Table 1.** Samples used in the SBI tests.

| SAMPLE | DESCRIPTION |
|---|---|
| A | ETICS sample with thick EPS 200 mm applied on calcium silicate boards |
| B | ETICS sample with thick EPS 200 mm applied on wood particle boards |
| C | ETICS sample with thick EPS 40 mm applied on wood particle boards |
| D | Sample with wood particle boards only |
| E | ETICS with thick EPS 40 mm applied on wood particle boards, with artificial damage (little shallow cavity in the long wing surface of rendering placed at 15 cm from the corner and 15 high from the burner) 20 × 50 × 15 mm |
| F | ETICS sample with thick EPS 40 mm applied behind on wood particle boards: sample reverse assembled (flame against particle board) |
| G | ETICS sample with thick EPS 200 mm applied behind on wood particle boards: sample reverse assembled (flame against particle board) |

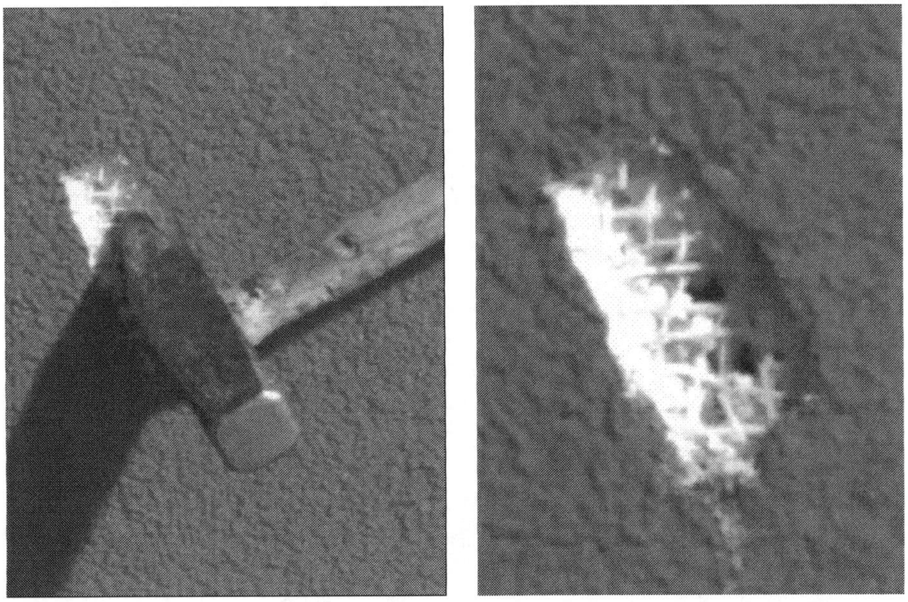

**Figure 1.** Damage manually done by hammer on Sample E.

The substrates that have been used in accordance with EN 13238:2010 [3] were the following:

- Calcium silicate boards, thickness 12 mm, density 875 kg/m$^3$.
- Particle wood boards, thickness 12 mm, density 650 kg/m$^3$.

All the samples have been assembled in accordance with annex D, ETAG 004 [4] in the laboratory and have been left there for a curing period of 28 days. Then samples have been placed for 8 weeks more in a conditioning room (23 ± 2°C; 50 ± 5% RH).

On the basis of the experimental data coming from the SBI tests done, some significant parameters have been compared. The data reported in this paper are not the result of an average data processing but the results of the tests that have been done on specific samples as listed in Table 1, samples selected, from a large number of sample tested, as to be representative of the fire behaviour of one samples typology.

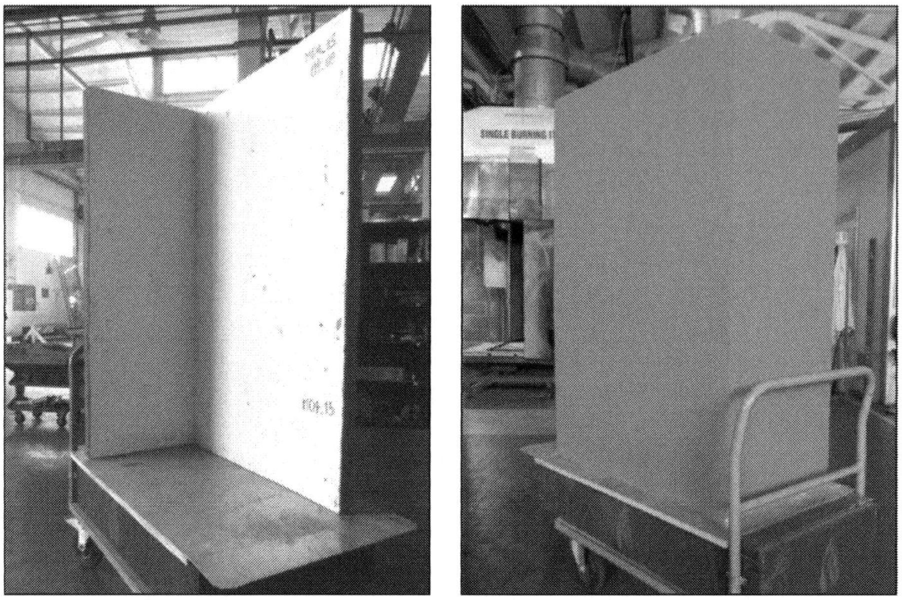

**Figure 2.** Inside and outside view of sample F reverse assembled.

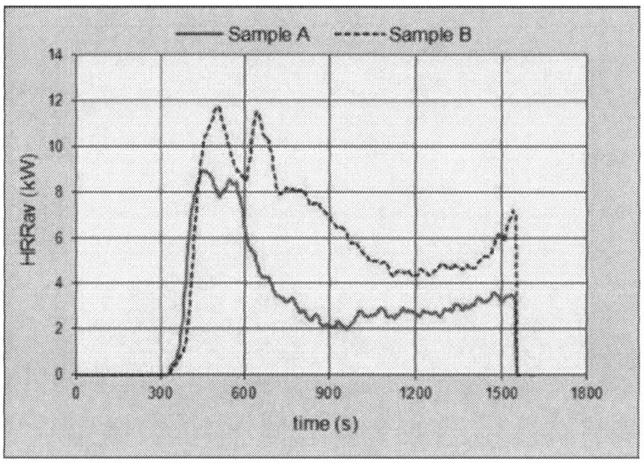


**Figure 3.** $HRR_{av}$ of samples A and B.

**Table 2.** Main data from SBI tests of samples A and B.

| Sample | Peak HRRav (kW) | $t-t_0$ (s) | $THR_{ta}$ (MJ) | Figra (W/s) |
|---|---|---|---|---|
| A | 8.9 | 153 | 2.8 | 64.60 |
| B | 11.8 | 207 | 4.5 | 65.67 |

## 3. RESULTS: SUBSTRATE IN COMPARISON

The following results have been processed from tests of samples "A" and "B" and aimed to point out the different fire behaviour of the same ETICS when assembled on calcium silicate boards (incombustible material) or on wood particle boards (fire reaction class D-s1, d0).

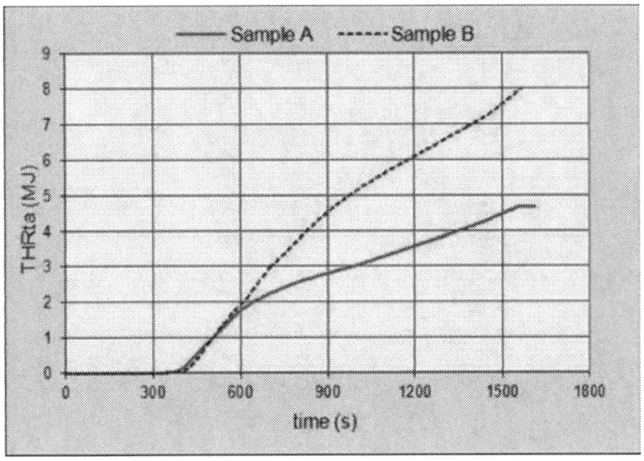


**Figure 4.** THR of samples A and B.

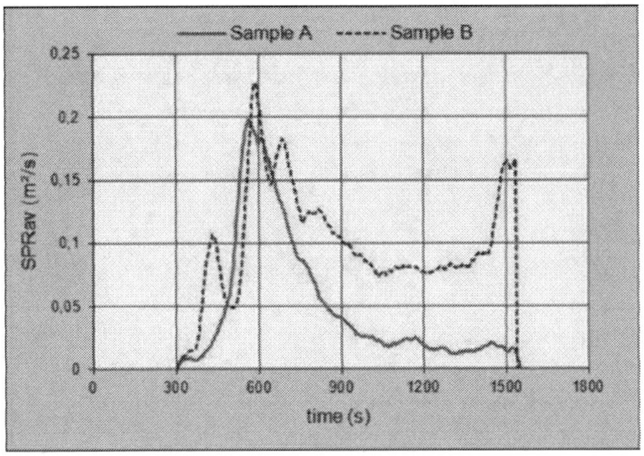


**Figure 5.** $SPR_{av}$ of samples A and B.

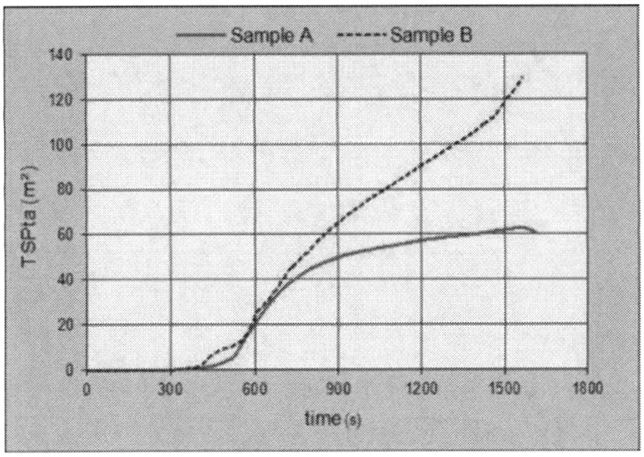


**Figure 6.** TSP of samples A and B.

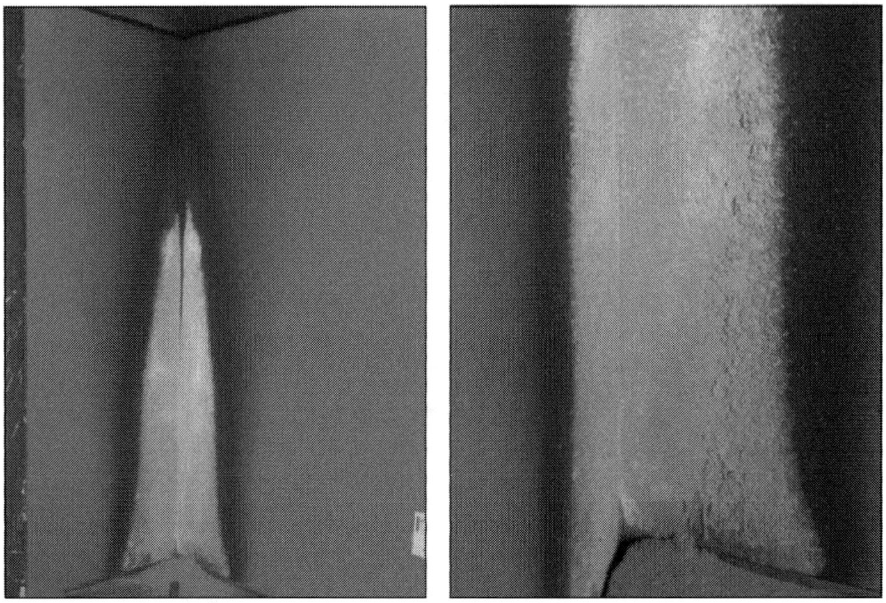

**Figure 7.** Sample A: typical burned surface after the test where no holes occurred. Damage from burning, soot deposition and discoloration.

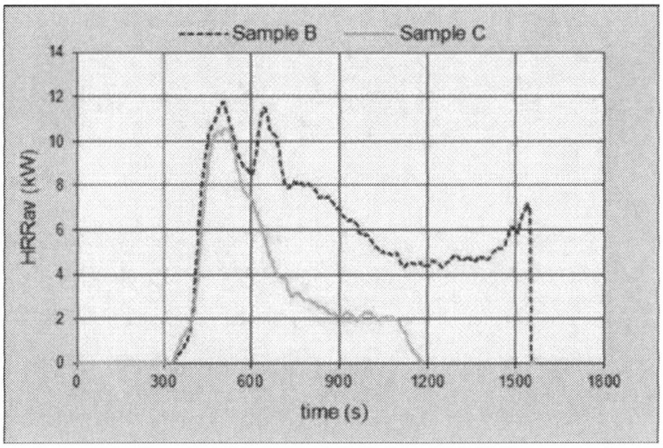


**Figure 8.** $HRR_{av}$ of samples B and C.

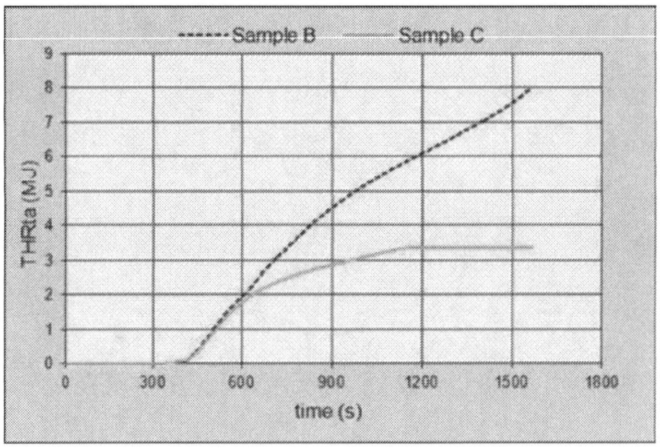


**Figure 9.** THR of samples B and C.

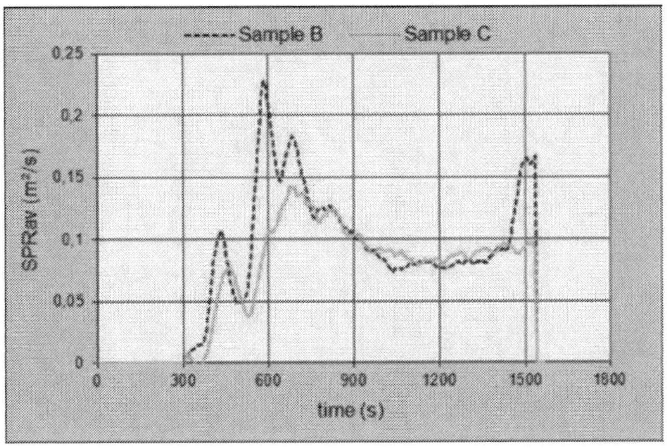


**Figure 10.** $SPR_{av}$ of samples B and C.

**Table 3.** Main data from SBI tests of samples B and C.

| Sample | Peak HRRav (kW) | $t$-$t_0$ (s) | $THR_{ta}$ (MJ) | Figra (W/s) |
|---|---|---|---|---|
| B | 11.8 | 207 | 4.5 | 65.67 |
| C | 10.6 | 225 | 2.9 | 58.92 |

### 3.1 Considerations

$HRR_{av}$ samples trend is quite similar: initial peak is a not of a high level, due to organic component of the rendering layer, followed with strong decreasing.

The maintenance of a certain value of HRR which tends to increase slightly on continued testing it is due to an increase in surface temperature (modest flame spread on the surface) that makes liquefy modestly but continues the insulating material: melting, then decomposition followed by evaporation of low molecular weigh product [5].

The rendering layer, when, as in this case, does not tears and maintains its consistency (falls apart only in the cooling phase, at test end, under the action of the jet of water required to extinguish the burning sample) and protect the insulation material from the direct attack of flame (only burning at the foot of the melting party, the area just wide of the burner). This behaviour does not allow to call into question the type of substrate as trends resulting are very similar.

Same considerations can be made about the production of smoke.

## 4. RESULTS: INSULATION LAYER THICKNESS IN COMPARISON

The following results have been processed from tests of samples "B" and "C" and aimed to point out the different fire behaviour of the same ETICS but with different thickness (40 and 200 mm) of the insulation layer when assembled on wood particle boards (fire reaction class D-s1, d0).

### 4.1 Considerations

As has been observed in the tests being compared before (Test A and B with different substrates) the integrity of the rendering layer ensures a similar behavior, even in the case of different thickness of insulation (otherwise it might happen at the same thickness but with different density [6]). Only little difference is seen in the maintenance phase of the HRR, following the initial peak, due to the different amount of insulating material which liquefies at the foot of the sample and participates in the combustion

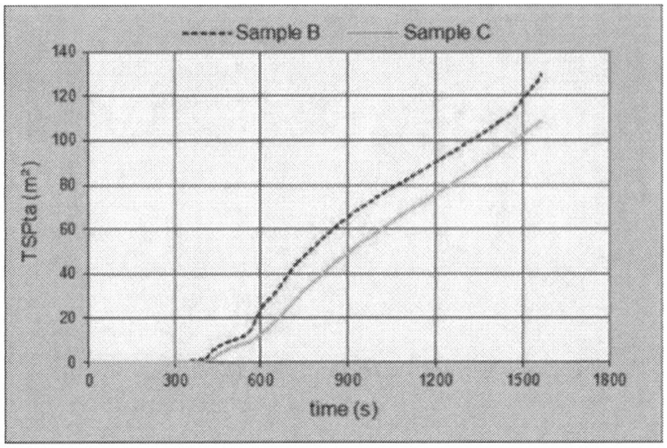


**Figure 11.** TSP of samples B and C in comparison.

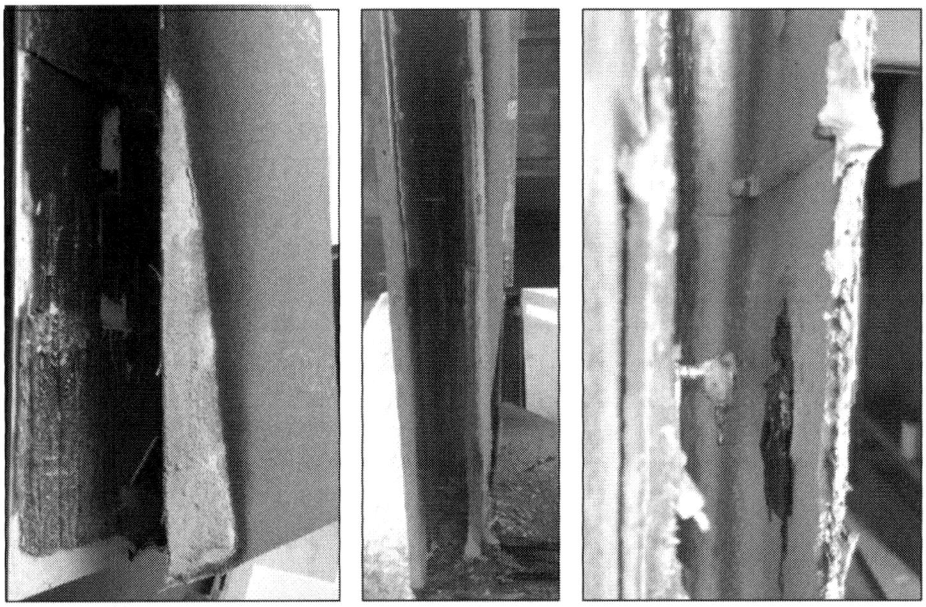

**Figure 12.** Samples B and C after removing short wings: gaps resulting after insulation melting.

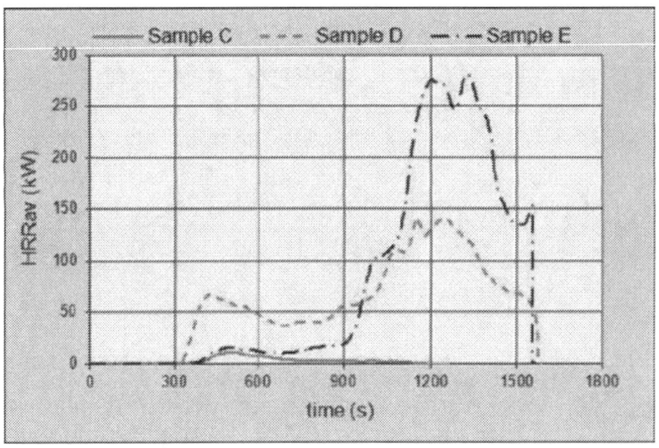


**Figure 13.** $HRR_{av}$ of samples D, C and E.

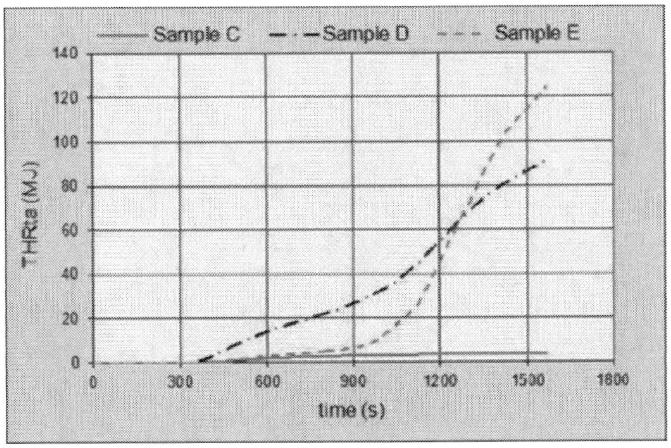


**Figure 14.** THR of samples D, C and E.

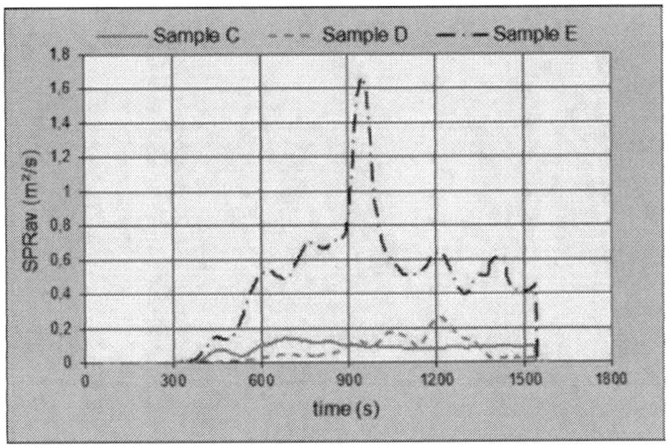


**Figure 15.** $SPR_{av}$ of samples D, C and E.

**Table 4.** Main data from SBI tests of samples D, C and E.

| Sample | Peak HRRav (kW) | $t\text{-}t_0$ (s) | $THR_{ta}$ (MJ) | Figra (W/s) |
|---|---|---|---|---|
| D | 141.7 | 849 | 26.8 | 598.93 |
| C | 10.6 | 225 | 2.9 | 58.92 |
| E | 281.1 | 1026 | 6.6 | 306.24 |

(at the end of the tests it has been noted that the amplitude of the affected part to liquefaction remains almost identical with changes only in the thickness).

## 5. RESULTS: EFFECTS OF ACCIDENTAL DAMAGE

The following results have been processed from tests of samples "D", "C" and "E" and aimed to point out the effects on fire behaviour of the same ETICS (insulation thickness = 40 mm) with or without an artificial damage on the surface when assembled on wood particle boards (fire reaction class D-s1, d0).

### 5.1 Considerations

The presence of a little damage in the rendering layer totally changes the fire behavior of the insulation system. After an initial phase ($t\text{–}t_0$ 300 s) in which the samples behave essentially in the same way

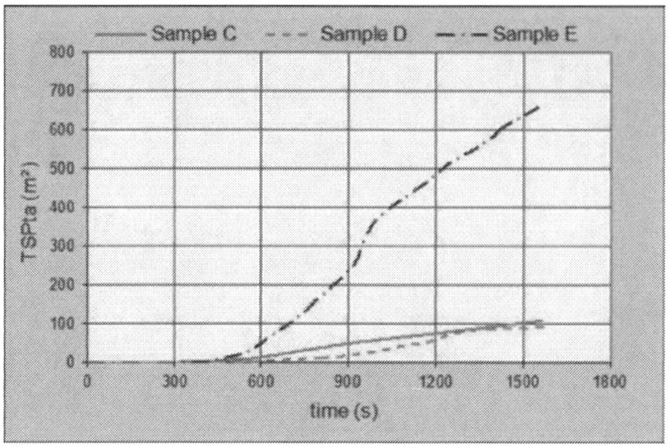


**Figure 16.** TSP of samples D, C and E.

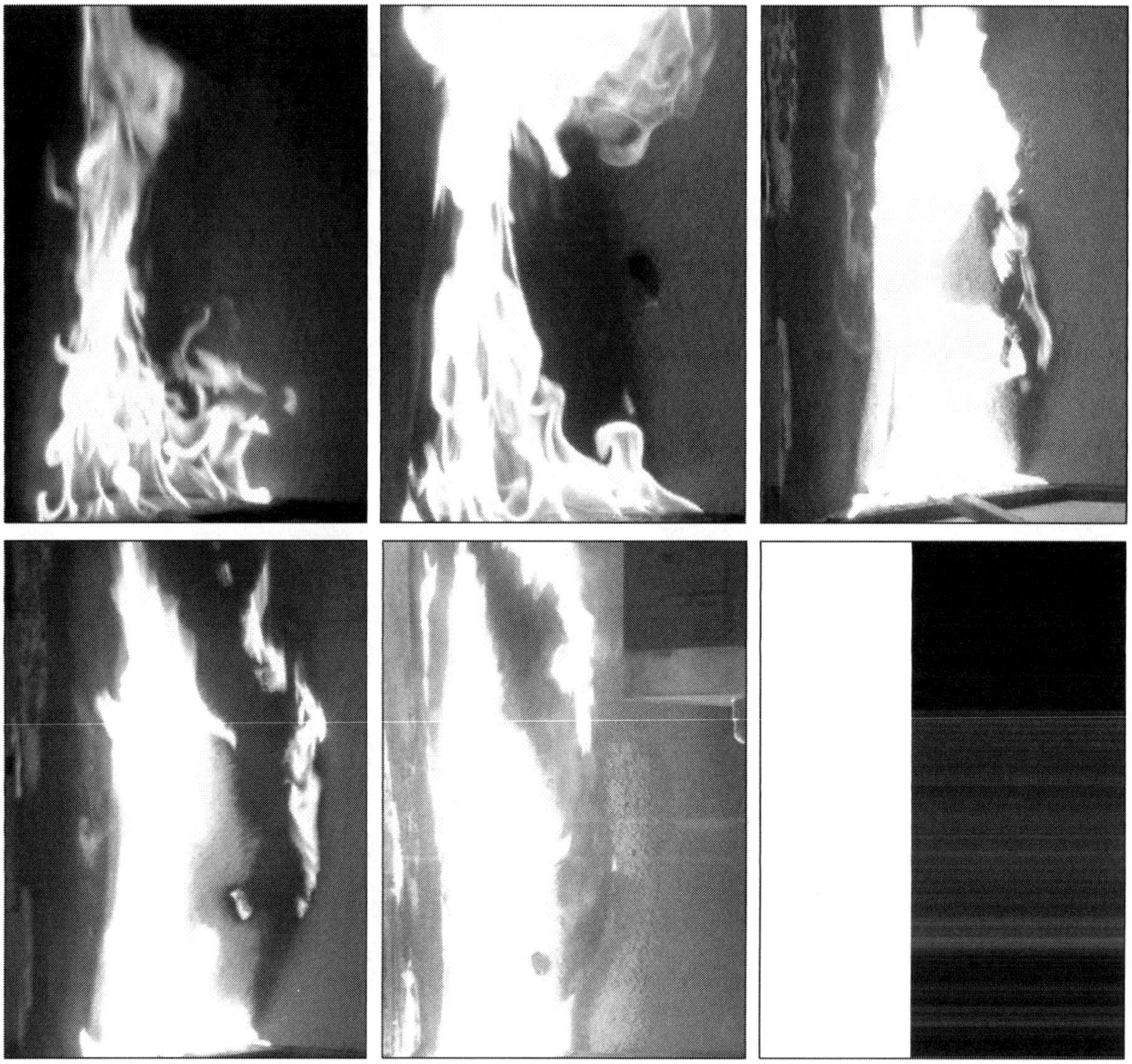

**Figure 17.** Sample E: pictures of flame propagation, due to the presence of damage, taken at different time steps of the test ($t-t_0$ 15, 120, 261, 378, 690, and 1263 s.

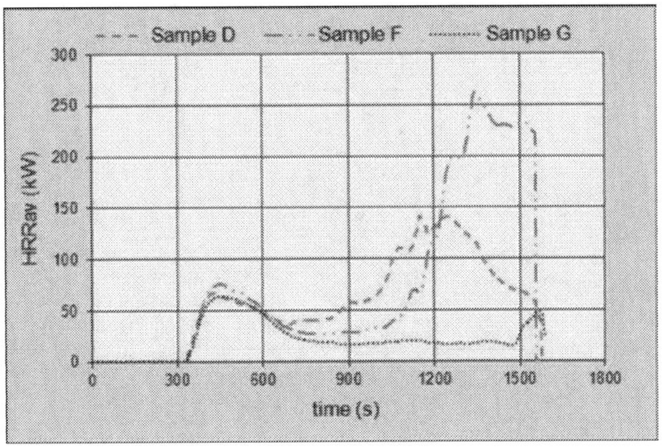


**Figure 18.** $HRR_{av}$ of samples D, F and G.

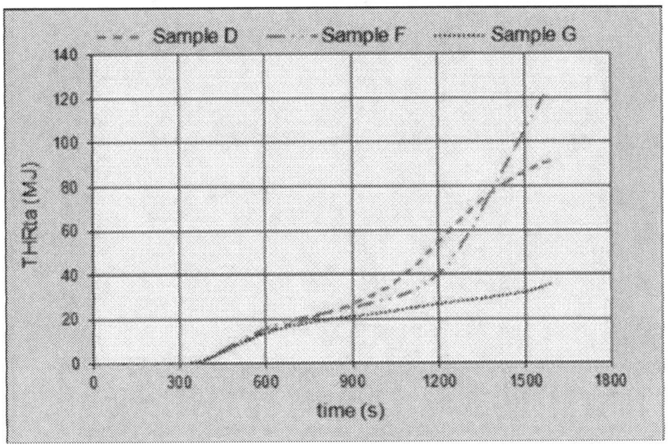


**Figure 19.** THR of samples D, F and G.

(only contribution to the organic part of the coating surface struck directly by the flame of the burner) due to heating of the surface and to the flame propagation, it liquefies the insulating material up to the "damage". The flame enters and exits from the surface through the artificial cavity and directly attacks the insulating material, in areas which would otherwise not affected, liquefying. Most ventilation in the gap that has developed between rendering and substrate feeds the flame. It is also interested the substrate of wood particles. It has a sudden increase of HRR (and of smoke) which it is greater even than that obtained in the case of sample D (wood particle boards only) in view of the presence of most combustible material (EPS). The comparison between sample E and D in the initial stage, shows once again how the rendering layer although damaged, performs protective function of the insulation and, more so, of the substrate, leading to recorded values for the heat and smoke very different from each other.

## 6. RESULTS: INSIDE FIRE ATTACK (i → o)

The following results have been processed from tests of samples "F" and "G" in comparison with "D" and aimed to point out the different fire behaviour of ETICS, with different thickness of the insulation layer (40 and 200 mm) but reversed assembled on wood particle boards (fire reaction class D-s1, d0) to be subjected at the fire attack from inside.

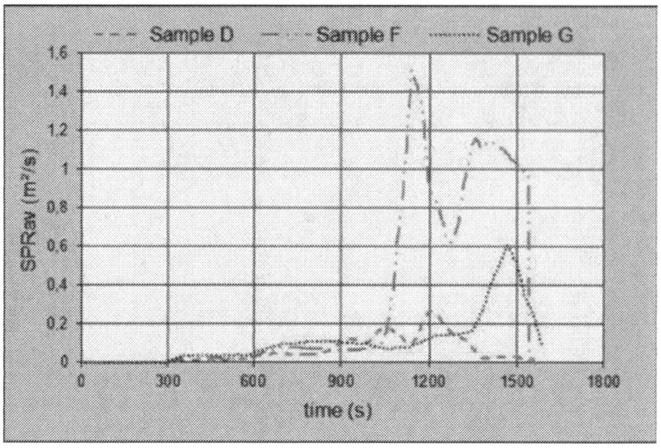


**Figure 20.** $SPR_{av}$ of samples D, F and G.

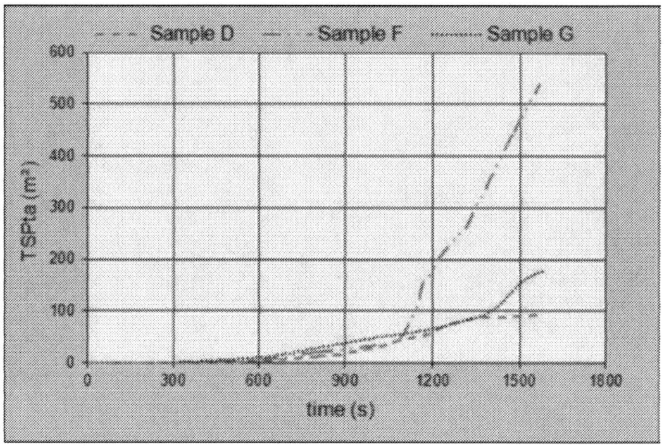


**Figure 21.** TSP of samples D, F and G.

**Table 5.** Main data from SBI tests of samples D, F and G.

| Sample | Peak HRRav (kW) | t-$t_0$ (s) | $THR_{ta}$ (MJ) | Figra (W/s) |
|---|---|---|---|---|
| D | 141.7 | 849 | 26.8 | 598.93 |
| F | 262.0 | 1044 | 25.2 | 635.13 |
| G | 63.8 | 156 | 21.0 | 544.92 |

## 6.1 Considerations

These tests have been done to check how it can be modified the reaction to fire of a wall of wood-based that is externally coated with an insulation system (ETICS). The analysis of the data shows that in a first initial phase (t–$t_0$ 400 s) the behaviour is substantially equal, being interested only the same material (wood particle board) in the combustion. Following the results deviate between them, in particular the one with insulation layer of 40 mm behaves in a completely analogous way to that of the same thickness with damage and attacked by the part of the rendering layer (sample E) that we have seen previously. When the combustion of the wood based board allows to directly attack the insulation layer, is retracts exposing the rendering layer. The relatively short distance between the flame (wood board is burning) and rendering layer and insulation system allows the latter to invest directly: It begins the combustion of its organic part. The opening of even small passages in the rendering layer increases oxygenation and leads to the almost complete combustion of the sample.

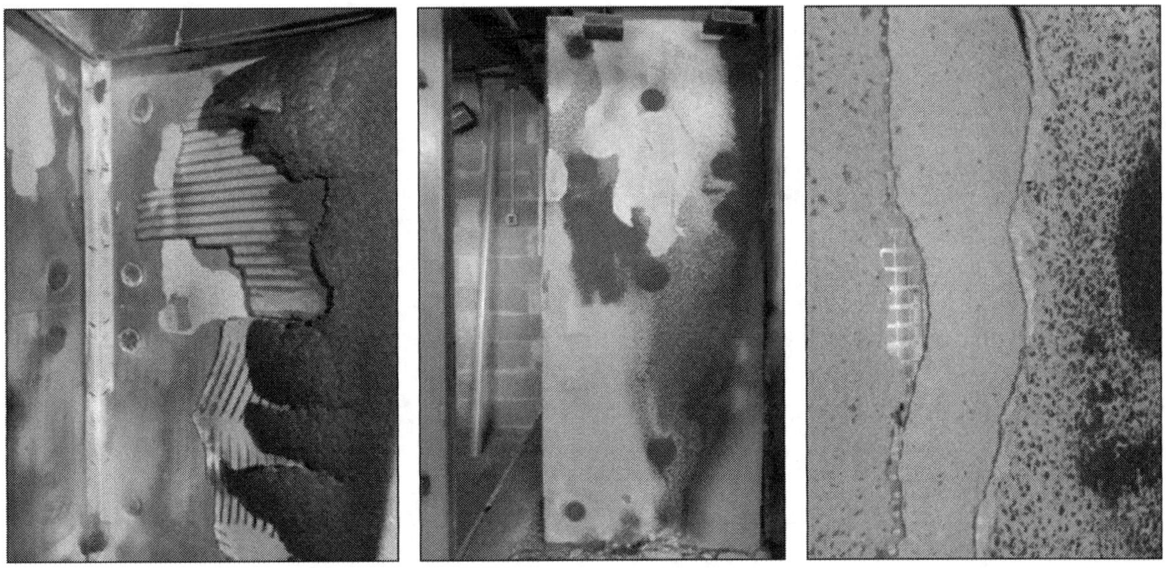

**Figure 22.** Sample F with thickness 40 mm at the end of test: inside and outside view.

**Figure 23.** Sample G with thickness 200 mm at the end of test. inside and outside view.

The situation is different for the sample with insulation layer of 200 mm: the most amount of insulation material, and the distance between the wood board and rendering layer, do not allow the complete combustion of the sample. The wood board remains somewhat "protected" by the Non-Melting of insulation material which is not allowing in this way the oxygenation and cause suffocation of the flame.

## 7. CONCLUSIONS

It is well known that it is necessary to resort to large-scale fire testing to determine the fire performance of external cladding system; despite this and the outdated determination of the European Commission to develop a harmonized test (harmonized test specification) we are still far from the final result. While the EOTA Technical Report N.073 [7] is still under discussion, the market does not stop and the use of high-performance systems, from the point of view of energy saving as the ETICS, continues to be employed and encouraged from member states to achieve objectives set by international environmental protocols. Consequently, the study presented is based on the results obtained on samples that have been tested with a shared test method (EN 13823 SBI) while knowing the limits, acting rather on the definition of mounting (samples reversed) and sample preparation in order to highlight the performance obtainable in specific end uses and this, could also resort to the use of the European fire classification system without any alteration. A first result highlighted by the work here presented is, how the different type of substrate, to which is applied the insulation system considered does not necessarily imply a change in performance in terms of reaction to fire as well as the increase of thickness of insulating material fuel (EPS). It emerges clearly from the results of tests performed on samples with artificial damage as the true discriminating role is played by the rendering layer, or by its nature, integrity, ability to perform a protective function. Hence the need to study the possible consequences, not only in the case of evident accidental damage, such as that reproduced, but also when caused by incorrect assembly (wrong overlaps of the mesh, thin thickness of the coating layers, etc.) or due to aging phenomena (i.e. cracks formation as a result of hygrothermal cycles). Without such data are not surprising the specific decisions, in some Member States, to regulate the reaction to fire class of insulation products, or alternatively (such as in Italian [8] or Danish state [9]) to require the addition of protective coating systems, separately defined about the specific reaction to fire, and so distorting the concept of system, as a whole, of the product ETICS.

## References


[1] EN 13501-1:2007+A1:2009 Fire classification of construction products and building elements. Classification using test data from reaction to fire tests.

[2] EN 13823:2010+A1:2014 Reaction to fire tests for building products. Building products excluding floorings exposed to the thermal attack by a single burning item.

[3] EN 13238:2010 Reaction to fire tests for building products. Conditioning procedures and general rules for selection of substrates.

[4] EOTA GUIDELINE FOR EUROPEAN TECHNICAL APPROVAL of External Thermal Insulation Composite Systems (ETICS) with Rendering – Edition 2000 Amended February 2013.

[5] Drysdale, D., *An Introduction to Fire Dynamics*, John Wiley & Sons, 1992.

[6] Antonatus, E *"Fire safety of Etics with EPS material properties and relevance for fire safety during transport, construction and under end use conditions in external thermal insulation component system"*, Proceedings of 1st International Seminar for Fire Safety of Façade, Matec **9**, 01001, 2013

[7] Technical Report N073 *"Large Scale Fire Performance Testing of External Wall Cladding Systems"*, EOTA PT4 Task group, 2013.

[8] ITALIAN FIRE CODE, *"Guida Tecnica Requisiti di Sicurezza Antincendio delle Facciate negli Edifici Civili"*, Ministero dell'interno, circolare n. 5043, 2013.

[9] Dragsted, A. and Vestergaard, A. B., "A new approach to the Danish guidelines for fire protection of combustible insulation". Proceedings of 1st International Seminar for Fire Safety of Façade, Matec **9**, 01001, 2013.

MATEC Web of Conferences **46**, 05009 (2016)
DOI: 10.1051/matecconf/20164605009


# Experimental study of window-ejected flame and plume on glass curtain walls

Lei Peng and Zhaopeng Ni

Tianjin Fire Research Institute of the MPS Tianjin, 300381, China

**Abstract.** This research presents an experimental study to investigate the performance of glass curtain walls exposed to exterior flame and plume ejected from windows. A test facility with 3-storey in height was constructed using steel frame to perform full-scale tests. Ventilation-controlled fire scenarios were designed to generate exterior flame and plume ejected from the burning room through a window opening. To characterize potential threats from window-ejected flame and plume to the glass curtain walls of upper floors, temperature, heat flux and air velocity at different heights above the opening upon curtain walls were measured during each test. The effects of window aspect ratio, horizontal projection were studied. It was found that the window with large aspect ratio provided more severe threat to the curtain walls on upper floors. A horizontal projection of 0.5m in depth was able to protect the glass on the upper floor from breaking.

## 1. INTRODUCTION

As more and more tall buildings are constructed with glass curtain walls, fire safety concern has started to be raised [1]. Most of the building fires are ventilation-controlled. After flashover, flames and hot gases are vented to the exterior of the building, and threaten glass curtain walls above the fire room. Once the temperature of the exterior fire has increased enough or the exterior face of the glass facade has experienced sufficient heat fluxes, the glass curtain wall would break, resulting in the spread of the fire to the upper floor [2, 3].

A lot of work has been carried out to investigate the window-ejected flame and plume upon the exterior walls above the opening [4–8]. Correlations were also developed to characterize flame heights, plume temperatures and heat fluxes upon the facade based on a large amount of small-scale tests [4–6]. Although some full scale tests were also carried out [7, 8], the attention of effects of plume temperatures and heat fluxes to the glass curtain walls on the upper floors were not well received.

In glass curtain walls, toughened glass (or tempered glass) are more and more used because of its high strength. When exposed to fire, toughened glass appears to survive higher gas temperatures and heat fluxes than float glass. It was reported that a single-layer of 10 mm thick toughened glass broke at above 600°C [9], and a double-layer of 6 mm thick toughened glass broke at about 600°C– 800°C [10]. The breaking temperatures of toughened glass are much higher than those of float glass [11].

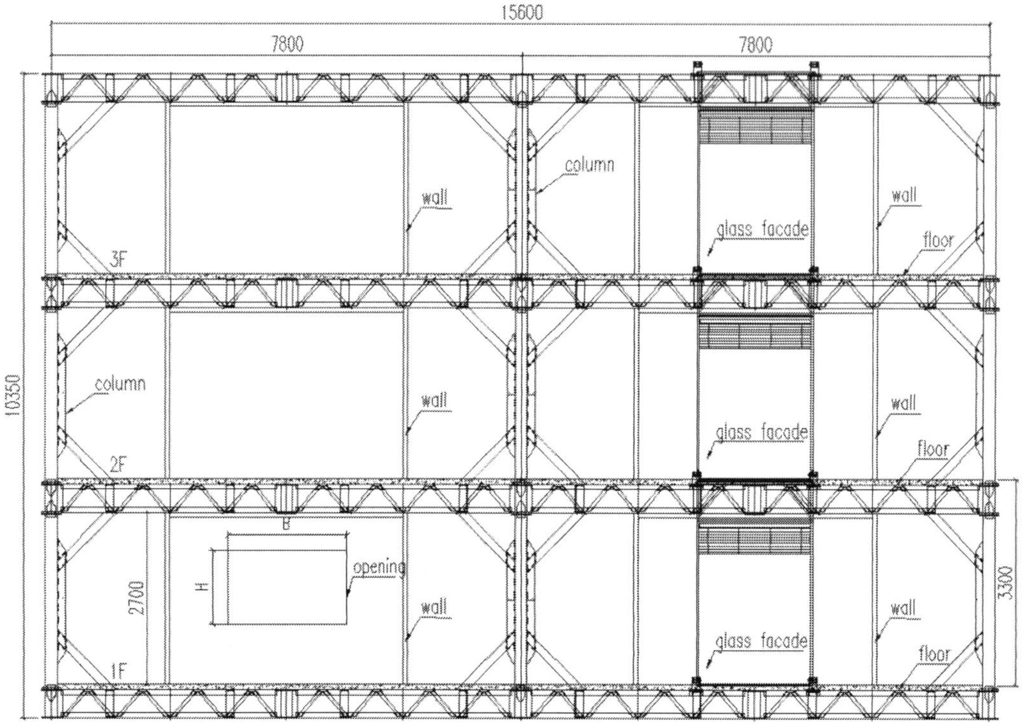


(a) front view

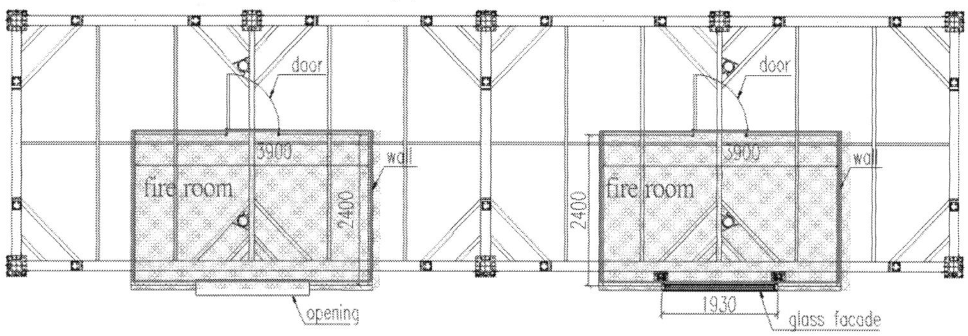


(b) plan view

**Figure 1.** Schematic of the test frame.

In this research, a series of full-scale tests were carried out to investigate the performance of toughened glass curtain walls exposed to exterior flame and fire plume ejected from windows. Both window aspect ratio and horizontal projection were considered.

## 2. EXPERIMENTAL SETUP

A test facility with 3-storey in height was constructed using steel frames to perform full-scale tests, shown in Fig. 1. The test frame has dimensions of 15.6 m wide ×3.9 m deep ×10.35 m high. Two testing setups were made on both left and right part of the test frame, using light-frame steel partition walls. Curtain walls could be installed as facade on both sides for testing.

Two identical fire rooms were located on the first floor the test frame. The walls and ceiling were covered with ceramic fibre insulation blanket. The interior dimensions of the fire rooms were 3.9 m

**Table 1.** Test Scenarios.

| Test # | Window size (m) | Wood crib weight; wood m.c. | Setup of the exterior wall | Notice |
|---|---|---|---|---|
| 1 | 1.9 × 1.1 | 337 kg; 14% | exterior walls all covered with fire boards | to study effects of window size |
| 2 | 1.4 × 1.4 | 349 kg; 15% | | |
| 3 | 0.9 × 1.9 | 354.2 kg; 14% | | |
| 4 | 1.9 × 1.1 | 340 kg; 13% | | to study effects of horizontal projection |
| 5 | 1.9 × 1.1 | 344 kg; 14% | | |
| 6 | 1.9 × 1.1 | 344.4 kg; 14.5% | glass panel installed on the 2nd and 3rd floor | – |
| 7 | 1.9 × 1.1 | 342 kg; 12% | | |

(a) Exterior walls all covered with fire boards

(b) Glass curtain walls installed on the facade

**Figure 2.** The test frame with facade installed.

**Table 2.** Window aspects in the tests.

| Test # | Window aspect (m) | | Ventilation factor | Maximum heat release |
|---|---|---|---|---|
| | Width | Height | $A\sqrt{H}$ ($m^{5/2}$) | rate expected (KJ) |
| 1 | 1.9 | 1.1 | 2.2 | 2750 |
| 2 | 1.4 | 1.4 | 2.32 | 2900 |
| 3 | 0.9 | 1.9 | 2.36 | 2950 |

wide ×2.4 m deep ×2.7 m high. A door was setup at the back of fire rooms to enable installing fire source and instrumentations.

Wood cribs conforming to BS 8414-1 [12] were used as fire source in the research. The cribs were made of pine wood with an average denity of 450 kg/m$^3$ and an moisture content in the range of 12% to 15%. The cribs were 1.5 m ×1.0 m in plane and 1.0 m high, and had an average weight of 345 kg. The fire load density in the fire rooms was about 36.8 kg/m$^2$, or 626 MJ/m$^2$ (assuming the heat of combusiton of wood is 17 MJ/kg).

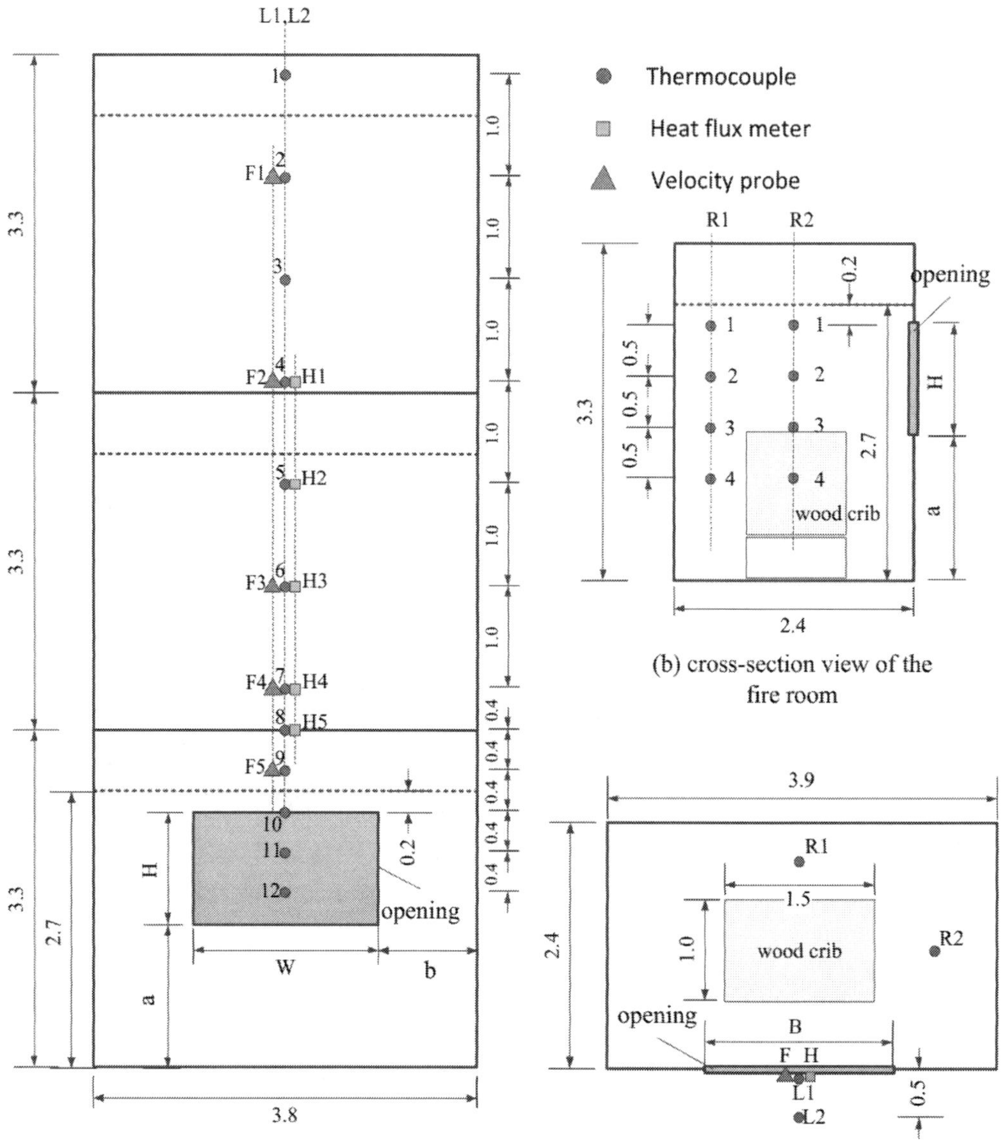


**Figure 3.** Instrumentations for Test #1 – #3.

## 3. TEST PROGRAM

In total, 7 tests were performed, shown in Table 1. The test curtain walls in Test #1–#5 were made of light-frame steel studs covered with calcium-silicate fire boards, shown in Fig. 2(a). In Test #6 and #7, glass panels were installed on the 2nd and 3rd floor, shown in Fig. 2(b).

The effects of window aspect ratio, horizontal projection were studied. In Test #1 to Test #3, three different window sizes were selected to represent typical window types: wide, square and narrow. The windows were supposed to produce similar ventilation factors and maximum heat release rates expected, shown in Table 2. In Test #4 and #5, the horizontal projection with 0.5 m and 1.0 m in depth at the top of opening were also tested.

The glass panels used in Test #6 and Test #7 were steel frame supported and assembled in prefabricated units. Each glass curtain wall unit had a size of 1.9 m wide ×3.3 m high ×0.12 m thick.

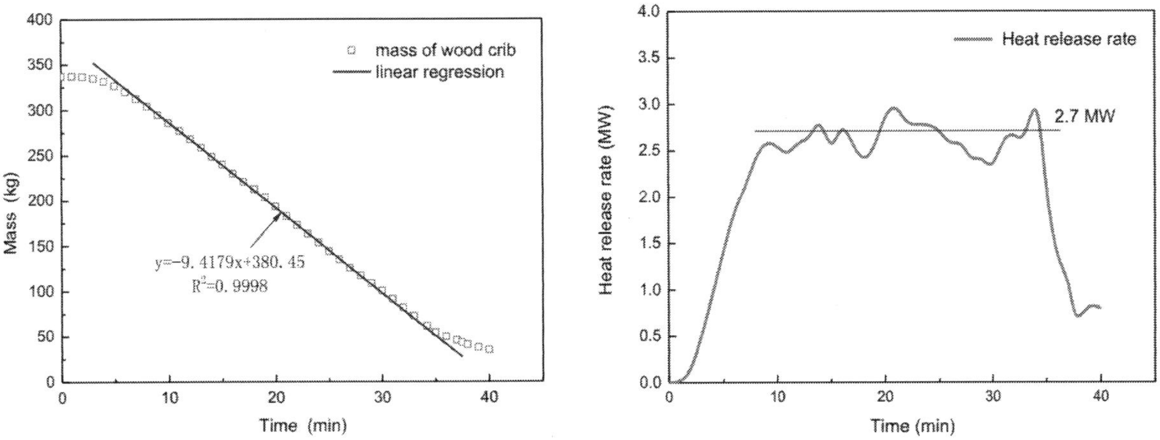


**Figure 4.** Mass loss of wood cribs and calculated heat release rate in Test #1.

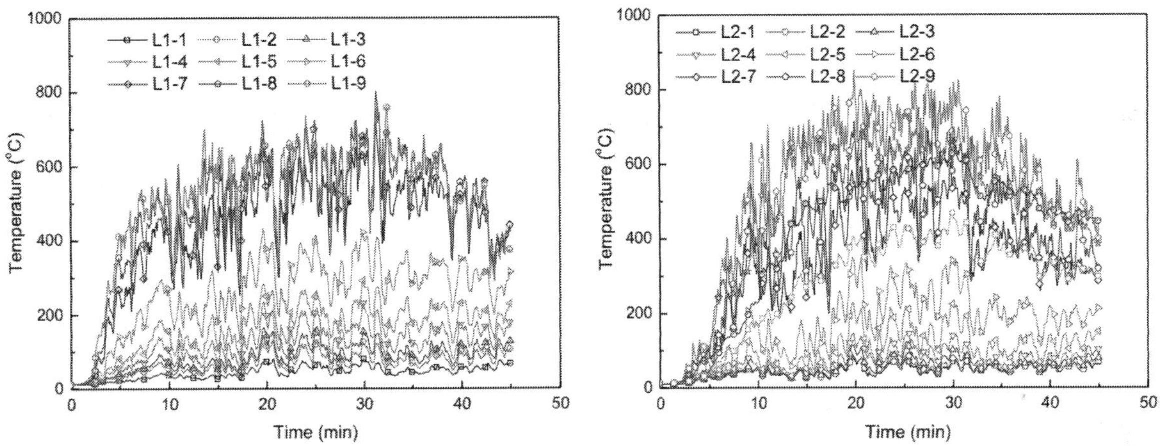


**Figure 5.** Temperatures of thermocouple tree L1 and L2 in Test #1.

The unit had 4-layer glass panels. The panels were all made of toughened glass. The outer layer glass was 10 mm thick, and the inner three layers were 6 mm thick each. The spandrel height between the window and glass panels was 0.8 m.

## 3.1 Instrumentations

To characterize threats from window-ejected flame and plume to the glass curtain walls of upper floors, temperature, heat flux and vertical gas velocity at different locations upon the curtain walls above the window were measured during tests.

The instrumentation setups in Test #1 to #3 are shown in Fig. 3. Despite horizontal projections installed, Test #4 and #5 have the same setups of instrumentations as those in Test #1 to #3. In Test #6 and #7, however, only thermocouples were used to measure the temperatures.

Metal-sheathed and mineral-insulated (MSMI) type K thermocouples (2.0 mm in diameter) were used to measure temperatures. Two thermocouple trees R1 and R2 were installed in the fire room. R1 was located in the middle of the wood cribs and back wall, while R2 was located in the middle of the wood cribs and left wall. Thermocouple R1-1 and R1-2 were set 0.2 m beneath the ceiling, and other thermocouples were located 0.5 m beneath one by one. Two thermocouple trees L1 and L2 were

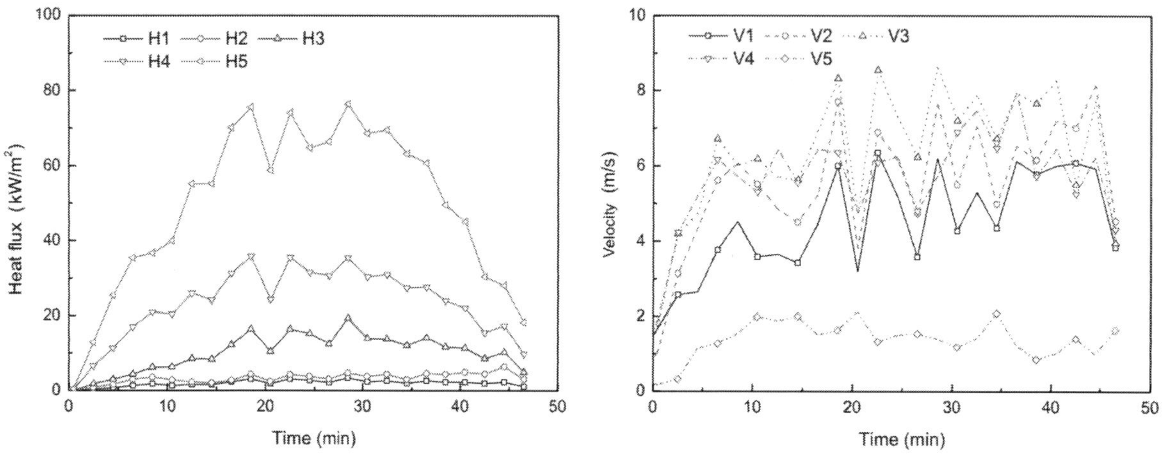


**Figure 6.** Heat flux and smoke vertical velocity upon the façade in Test #1.

installed along the central line of the facade. L1 was located close to the facade surface with a distance of 0.05 m, while L2 was located 0.5 m away from the facade surface. In L1 and L2, the thermocouple #10 was located at the top of the window opening. The spacing between two thermocouples of #1– #7 was 1.0 m, while the spacing between two thermocouples of #7–#12 was 0.4 m.

Five Gardon heat flux meters (H1 to H5) were installed along the central line of the facade for Test #1 – Test #5. The heights of H1 to H5 above the top of the opening were 4.2 m, 3.2 m, 2.2 m, 1.2 m, 0.8 m, respectively. Five bi-directional velocity probes [13] (F1 to F5) were also installed along the central line of the facade. The heights of F1 to F5 above the top of the opening were 6.2 m, 4.2 m, 2.2 m, 1.2 m, 0.4 m, respectively.

A scale with careful insulation was placed under the cribs to measure the mass loss of the wood during each fire test. Therefore, the heat release rate could be calculated by multiplying the mass loss rate by the heat of combustion of wood.

## 4. TEST RESULTS

As the mass loss was measured during the test, the heat release rate could be calculated. As shown in Fig. 4, the wood cribs generated a steady burning stage over about 30 minutes with an average heat release rate at about 2.7 MW for Test #1 (assuming the heat of combustion of wood is 17 MJ/kg).

The measured temperature profiles of thermocouple tree L1 and L2 are shown in Fig. 5 for Test #1 as an example. Thermocouple tree L1 was located close to the facade surface with a distance of 0.05 m, while L2 was located 0.5 m away from the facade surface, as shown in Fig. 3.

The measured heat flux and vertical gas velocity at different heights upon the facade along the central line are also shown in Fig. 6 for Test #1.

### 4.1 Effect of window aspects

From Test #1 to #3, it was observed that the window-ejected flame and plume from wider opening were more likely to move towards the facade above the window opening. As shown in Fig. 7, the attachment of external flame to the facade was observed in Test #1 (W:B = 1.9); the flame moved very close to the facade but no attachment occurred in Test #2 (W:B = 1); the flame ejected away from the facade and no attachment occurred either in Test #3 (W:B 0.526).

The measured temperature profiles at the central line as a function of height above the top of opening are shown in Fig. 8 for Test #1 to #3. Note that L1 was located close to the facade surface with

(a) Test #1 (b) Test #2 (c) Test #3

**Figure 7.** Window-ejected flame upon the facade for Test #1 to Test #3.

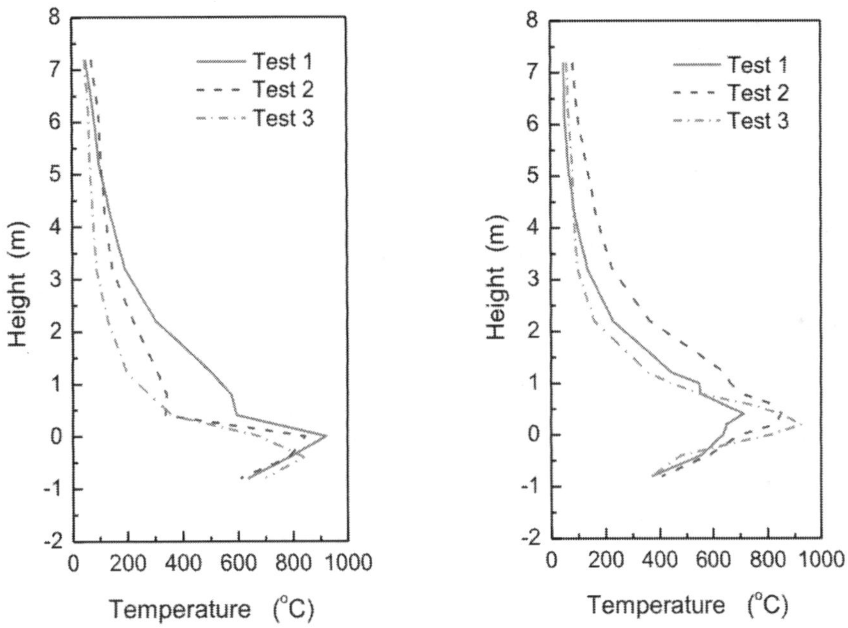


(a) Temperature profiles of L1 (b) Temperature profiles of L2

**Figure 8.** Temperature profiles along the central line of facades.

a distance of 0.05 m, while L2 was located 0.5 m away from the facade surface. It is shown that the temperatures of L1 in Test #1 are higher than those at the same height in Test #2 and #3. This suggests that the flame and plum ejected from wider window opening provide severe threats to the curtain walls on upper floors.

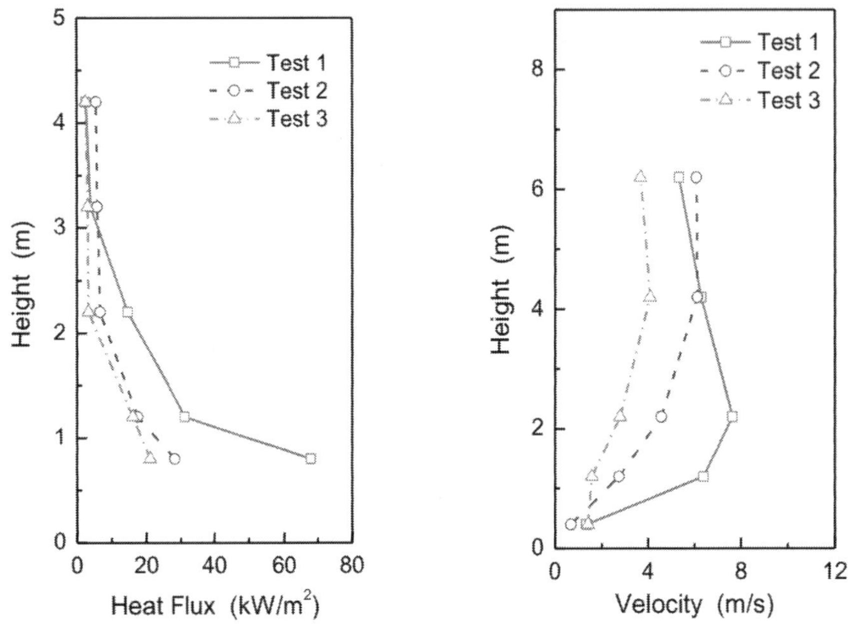


**Figure 9.** Heat flux and vertical air velocity profiles along the central line of facades.

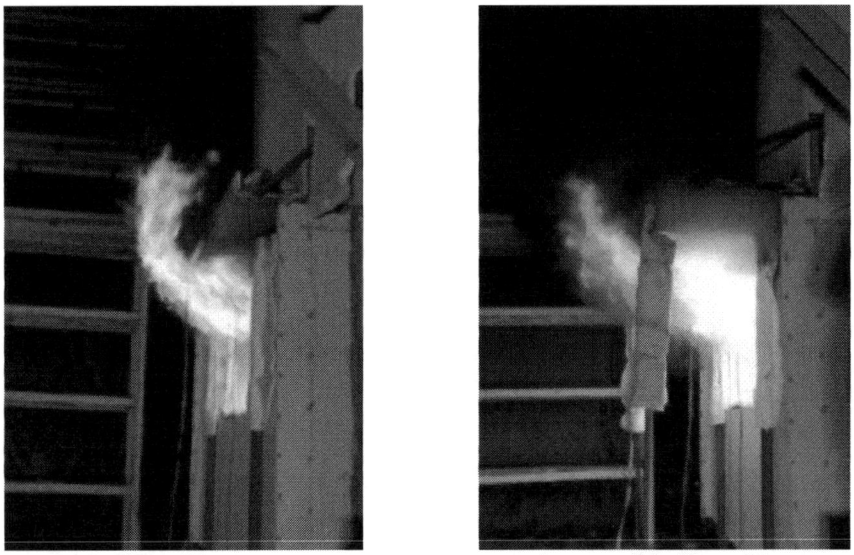

(a) Test #4 (b) Test #5

**Figure 10.** Window-ejected flame upon the facade for Test #4 and Test #5.

Figure 9 shows the measured heat flux and vertical air velocity profiles at the central line as a function of height above the top of opening for Test #1 to #3. Both heat flux and vertical air velocity were higher in Test #1 as the opening was wider.

## 4.2 Effect of horizontal projection

Figure 10 shows the ejected flame in Test #4 and #5. Due to the effect of horizontal projection, the flame could attach to the facade above the projection.

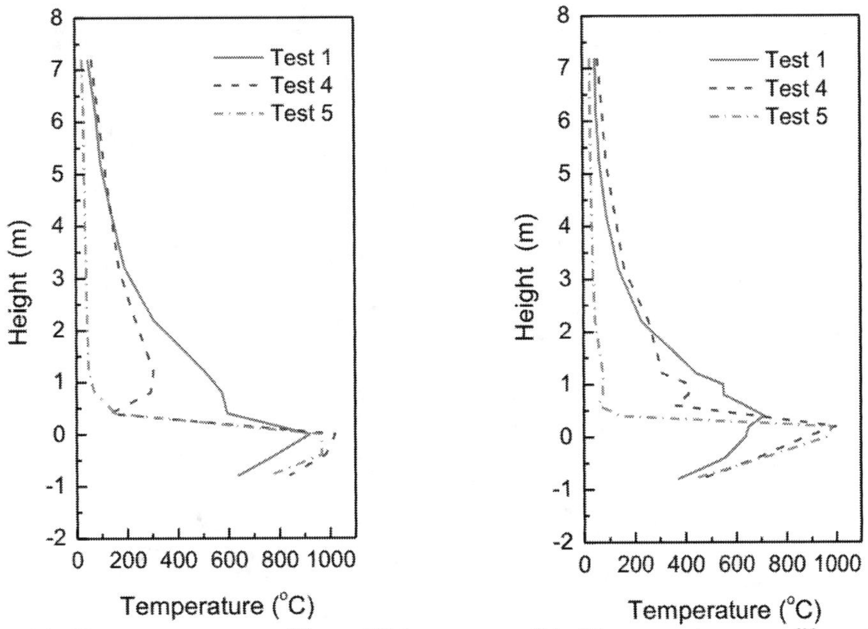


(a) Temperature profiles of L1 (b) Temperature profiles of L2

**Figure 11.** Temperature profiles along the central line of facades.

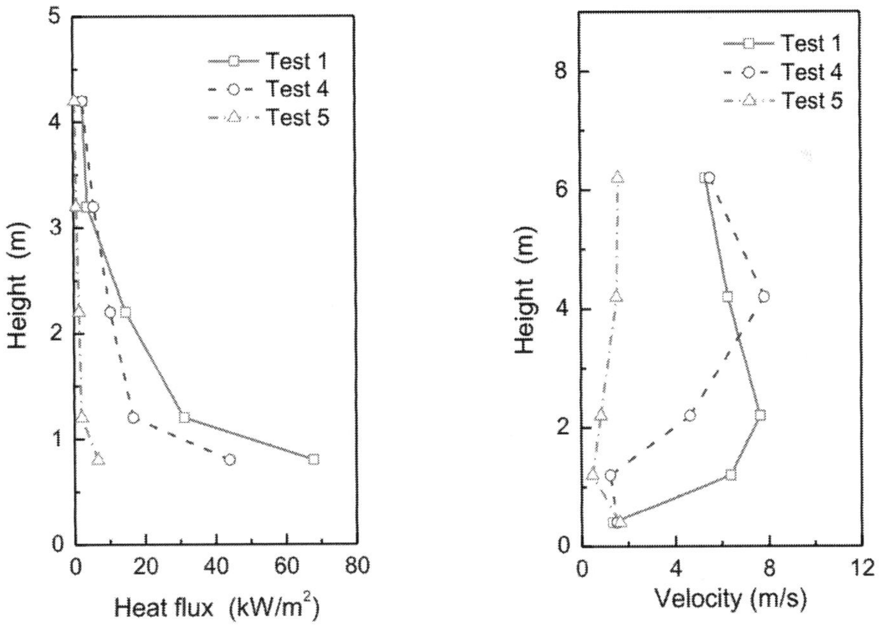


**Figure 12.** Heat flux and vertical air velocity profiles along the central line of facades.

Figure 11 shows the comparison of temperature profiles at the central line as a function of height above the top of opening for Test #1, #4 and #5. It is shown that as the depth of projection increased, the temperatures of L1 decreased significantly. The maximum temperature at the central line above the projection was about 330° in Test #4, while it was about 150° in Test #5. A horizontal projection of 0.5 m in depth was able to protect the glass on the upper floor from breaking.

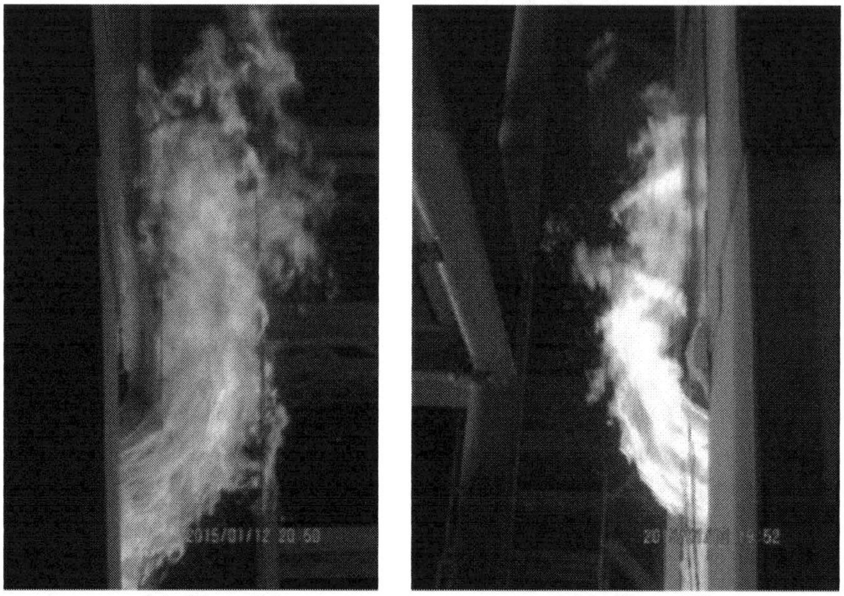

(a) Test #6 (b) Test #7

**Figure 13.** Window-ejected flame upon the facade for Test #6 and Test #7.

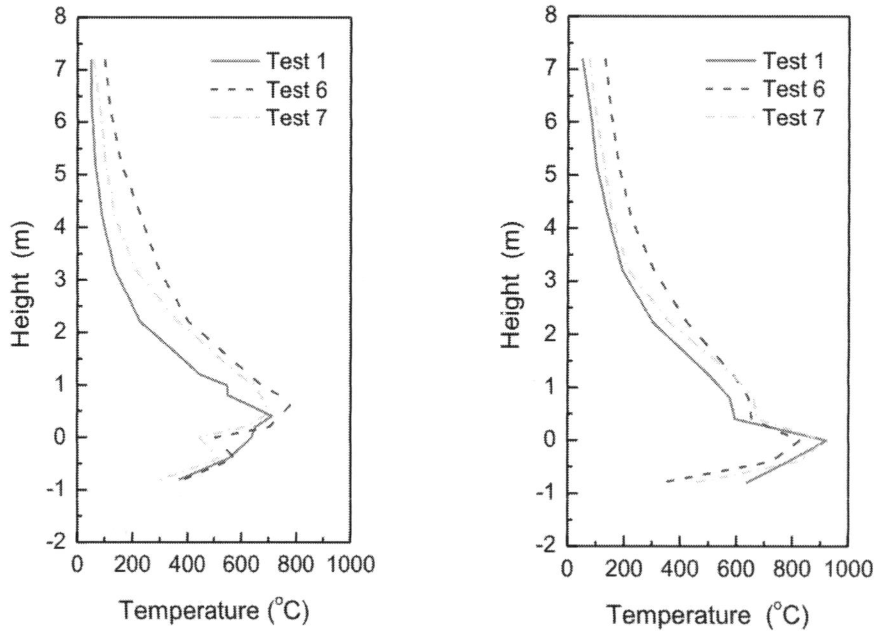


(a) Temperature profiles of L1 (b) Temperature profiles of L2

**Figure 14.** Temperature profiles along the central line of facades.

The comparison of heat flux and vertical air velocity profiles at the central line as a function of height above the top of opening is also shown for Test #1, #4 and #5. The heat flux and air velocity profiles confirm that the horizontal projection plays an important role in reducing the threat to the curtain walls on upper floors.

### 4.3 Performance of the glass curtain walls

It was observed that the breaking of glass panels did not occurred in Test #6 and #7. Figure 13 shows that the attachment of flame to the facade above the opening occurred. The spandrel with a height of 0.8 m successfully prevented the fire spread to upper floors for glass curtain walls made of 4-layer tempered glass.

Figure 14 shows the comparison of temperature profiles at the central line as a function of height above the top of opening for Test #1, #6 and #7. Test #6 and #7 had slightly higher temperatures than Test #1 at the same height. This is probably because the measured maximum heat release rates of Test #6 and #7 were slightly higher than Test #1. Therefore, more heats were discharged in the ejected flame and resulted in higher temperatures. Test #6 and Test #7 had similar temperature profiles.

## 5. CONCLUSIONS

Based on the experimental work conducted in this research, the following conclusions can be drawn:

1. The window aspect has an important impact on fire performance of the curtain walls on upper floors. The larger width-height ratio of the window (B:H), the more likely flame and plume tend to attach to the facade above the opening. This would lead to higher surface temperatures, heat fluxes, and vertical gas velocity upon the facade. As the width-height ratio is close to 1:1, the external flame and plume tend to move apart from the facade, and no attachment to the facade was observed.
2. The horizontal protection above the window also showed its important impact on the fire performance of curtain walls on the upper floors. As the projection depth increased, the temperatures, heat fluxes, and vertical gas velocities upon the facade decreased. When the depth of projection was larger than 0.5 m, no attachment to the facade was observed during the test, and the threats to the curtain walls on upper floors were reduced significantly.
3. Test #6 and #7 show that a spandrel with a height of 0.8 m might be necessary to prevent the toughened glass curtain panels on upper floors from breaking.

This research received supported from the project 2012SJ-A-31 and 2012SJ-A-32.

### References

[1] Chow, C.L., Chow, W.K., Han, S.S., and So, A.K.W. "Vertical air temperature profiles in a single skin glass façade with a 'jumping fire' scenario", *Journal of Applied Fire Science*, **17**(2), 107–130, 2007–2008,

[2] LPC. "Fire Spread in Multi-Storey Buildings with Glazed Curtain Wall Facades", Loss Prevention Council, Borehamwood, England, 1999.

[3] Xie, Q., Zhang, H., Wan, Y., Zhang, Q., and Cheng, X., "Full-Scale Experimental Study on Crack and Fallout of Toughened Glass with Different Thicknesses", *Fire and Materials*, **32**, 293–306, 2008.

[4] Yokoi, S., "Study on the Prevention of Fire Spread Caused by Hot Upward Current", Report 34, Report of the Building Research Institute, Japan, 1960.

[5] Lee, Y.P., Delichatsios, M.A., and Silcock, G.W.H., "Heat fuxes and flame heights in facades from fires in enclosures of varying geometry", *Proc. Combust. Inst.*, **31**, 2521–2528, 2007.

[6] Tang, F., Hu, L.H., Delichatsios, M.A, Lu, K.H., and Zhu W., "Experimental study on flame height and temperature profile of buoyant window spill plume from an under-ventilated compartment fire", *Int. J. Heat Mass Transfer*, **55**, 93–101, 2012.

[7] Oleszkiewicz, I., "Vertical separation of windows using spandrel walls and horizontal projections", *Fire Technology*, **27**, 334–340, 1991.
[8] Klopovic, S. and Turan, Ö.F., "Flames venting externally during full-scale flashover fires: two sample ventilation cases", *Fire Safety Journal*, **31**, 117–142, 1998.
[9] Xie, Q., Zhang, H., Wan, Y., Zhang, Q., and Cheng, X. "Full-Scale Experimental Study on Crack and Fallout of Toughened Glass with Different Thicknesses", *Fire and Materials*, **32**, 293–306, 2008.
[10] Ni, Z., Lu S., and Peng, L., "Experimental study on fire performance of double-skin glass facades", *Journal of Fire Sciences*, **30**, 457–472, 2012.
[11] Babrauskas V., "Glass breakage in fires", Issaquah, WA: FireScience and Technology Inc., 2010.
[12] BS 8414-1, "Fire performance of external cladding systems - Part 1: Test method for non-loadbearing external cladding systems applied to the face of the building", 2002.
[13] McCaffrey, B.J., and Heskestad, G., "A robust bidirectional low - velocity probe for flame and fire application", *Combustion and Flame*, **26**, 125, 1976.

MATEC Web of Conferences **46**, 05010 (2016)
DOI: 10.1051/matecconf/20164605010


# Fire behaviour of EPS ETICS on concrete or masonry facades

Afipeb, Sipev and Snmi

Paris, France

**Abstract.** In France, fire safety of facades for establishments open to the public has been governed by the Technical Instruction n°249 since 1982; its current version was made regulatory on May 24$^{th}$, 2010. Considering this revision, the Ministry of Interior requested industrials firms to revalidate the fire behaviour of their external insulation systems through testing. AFIPEB, SIPEV and SNMI decided therefore to launch a test campaign.

After a preliminary analysis of Single Burning Item (SBI) data related to EPS ETICS and some full-scale orientation tests, the polyvalent, representative and reliable LEPIR 2 test protocol was implemented. Three EPS ETICS configurations under various renders were tested: one under thick hydraulic renders, another under thin mixed renders and a third under thin organic renders.

These tests were successful. For the first 30 minutes of each test: there was no ignition nor vertical spread of fire to the facade at its second level, there was also no lateral spread of fire along the facade width. Moreover, after 60 minutes, the outer skin above the horizontal line at 5.2 m was not damaged.

The campaign confirms that EPS ETICS under thin, thick or mixed renders perfectly comply with fire safety regulations. AFIPEB, SIPEV and SNMI wrote a recommendation guide based on approved laboratories assessment. It describes construction solutions limiting fire propagation on a concrete or masonry facade insulated by EPS ETICS. This guide currently supplements IT 249 (2010).

## 1. CONTEXT

In France, since 1982, fire safety of facades for establishments open to the public has been governed by the Technical Instruction n°249, IT 249 (1982), published by the Ministry of Interior. The current version was made regulatory on May 24$^{th}$, 2010. The fire safety regulation for some residential buildings (3$^{rd}$ and 4$^{th}$ families) also refers to IT 249 (1982). Awaiting the revision of this regulation, the Ministries of Housing and of Interior have recommended the implementation of IT 249 (2010) since September 30$^{th}$, 2015.

IT 249 (2010) deals with different types of facades, systems and materials. External insulation systems on concrete or masonry with openings are addressed in paragraph 5 of which the most important parts for EPS ETICS are the following:

- Paragraph 5.1 describes the requirements on EPS ETICS component characteristics and ancillary protections. Every system satisfying all these requirements can be used without further testing.

- Paragraph 5.3 applies for EPS ETICS not described in paragraph 5.1. These systems shall be subject to a favourable recommendation delivered by CECMI *(Comité d'Etudes et de Classification des Matériaux contre le risque Incendie)* about their fire behaviour. This recommendation shall be based on an assessment delivered by an approved laboratory for fire reaction and resistance.

Driven by European and French regulations, in particular Directive 2010/31/EU on Energy Performance of Buildings and the French 2012 Thermal Regulation n°2010–1269, external thermal insulation has grown a lot in France. In this respect, products and systems have evolved to improve thermal performances and to provide a wide choice of finishing. These evolutions led the Ministry of Interior to start the revision of IT 249 (2010). In this context, the Ministry of Interior also requested industrial firms to revalidate the fire behaviour of their external insulation systems through testing.

Thus, since 2010, research and development programmes were implemented by render and insulation manufacturers. These R&D programmes resulted in fire testing campaigns carried out according to LEPIR 2 test protocol, in accordance with IT 249 (2010).

## 2. OBJECTIVE OF THE STUDY

The objective of the study was to confirm the fire behaviour of EPS ETICS under thin or thick renders on concrete or masonry facades.

## 3. STAKEHOLDERS

The test campaign has been jointly carried out by three professional organisations:

- AFIPEB *Association Française pour l'Isolation en Polystyrène Expansé dans le Bâtiment.*
- SIPEV, *Syndicat National des Industries de Peinture, Enduits et Vernis.*
- SNMI, *Syndicat National des Mortiers Industriels.*

The laboratories involved in the study were:

- CREPIM, approved fire reaction laboratory.
- CSTB, *Centre Scientifique et Technique du Bâtiment.*
- EFECTIS France, approved fire resistance laboratory.

Programme specifications were established by the three professional organisations in collaboration with the laboratories. Before implementation, they were agreed by the Ministry of Interior.

EFECTIS conducted orientation tests leading to select EPS ETICS configurations and ancillary protections. EFECTIS also established the specific characteristics of the renders, based on a statistical analysis of Single Burning Item (SBI) results related to EPS ETICS conducted by CSTB.

The LEPIR 2 tests were carried out by EFECTIS in collaboration with CREPIM.

## 4. LEPIR 2 TEST PROTOCOL

LEPIR 2, *Local Expérimental Pour Incendie Réel à 2 niveaux,* stands for "Experimental premises for real fire on 2 levels".

This polyvalent, representative and reliable protocol is a proven test in France for over 40 years. It was first published in a regulation dated September 10$^{th}$, 1970 then updated on June 11$^{th}$, 2013 after validation by notified laboratories and the Ministry of Interior. It precisely assesses the fire behaviour of facades enabling to differentiate the various systems and construction provisions.

The method applies to every type of external thermal insulation whatever the nature of the insulating material. It takes into account the whole facade, on two levels, with superimposed openings. It is

**Table 1.** Test facility dimensions and characteristics.

| Levels | Dimensions and characteristics |
|---|---|
| Each room (L & L+1) | 4.30 m in length<br>3 m in depth<br>2.50 m high |
| Spandrel (L+2) | About 1.50 m high, from the top of the openings on the 2nd level |
| Total height of the insulated façade | 6.55 m, acroterion included |
| Openings on 1st level (L) | - Openings: 1.00 × 1.50 m<br>- Width between the openings: 1.20 m<br>- One window fixed to the bare interior of the structure and one window fixed to the bare exterior<br>- PVC frame without glazing |
| Openings on 2nd level (L+1) | - Openings: 0.85 × 1.35 m<br>- Width between the openings: 1.40 m<br>- One window fixed to the bare interior of the structure and one window fixed to the bare exterior (same position as at the first level)<br>- Complete windows: PVC joinery fitted with double glazing of type 6/12/laminated44.2 PVB |

**Figure 1.** Frontal view of a LEPIR 2 test facility.

therefore representative of real constructions. The fire is fed with two important sources, representing a total load of 600 kg of calibrated wood.

## 5. Test facility

Tests are carried out on a construction composed of two rooms on two levels, each with two superimposed windows. A spandrel represents level L+2. Side walls are closed. On the rear wall (opposite to the facade), on the ground floor, an adjustable opening is installed to control air input.

### 5.1 Characteristics of the fire sources

The lower level contains two fire sources, each of 1 × 1 × 1.8 m, representing a total combustible mass of 600 kg. They are composed of slats with 3 complementary dimensions, made of softwood with a density of $480 \pm 50\,kg/m^3$. Each fire source is put on a non-combustible base, at 100 mm from the facade and 200 mm above the ground.

**Figure 2.** Three quarters view of a LEPIR 2 test facility.

**Figure 3.** Fire source implantation at the lower level.

## 5.2 Climate conditions

The wind speed is measured during 15 min at mid-height of openings at the lower level, at 1 m from the facade. A maximum mean wind speed of 3 m/s is allowed.

There must not be any rain nor snow.

## 5.3 Equipment of the test facility and observations during the test

Thermocouples are placed at all points where it is considered necessary to know temperatures reached, in particular in front of the building, at the second level on the facade and inside.

During the test, all incidents such as destructions, ignition of the external render, flame passage between the facade and the floor are registered.

From outside, a video records the entire test. Snapshots of a test are given in the Figs. 4 to 7.

## 5.4 Evaluation criteria

The test is successful under the following conditions.

- For the first 30 minutes of the test:

- o No ignition nor vertical spread of fire, for more than 20 seconds, to the facade at L+2.
- o No lateral spread of fire along the facade width.

**Figure 4.** Observation at 1 min.

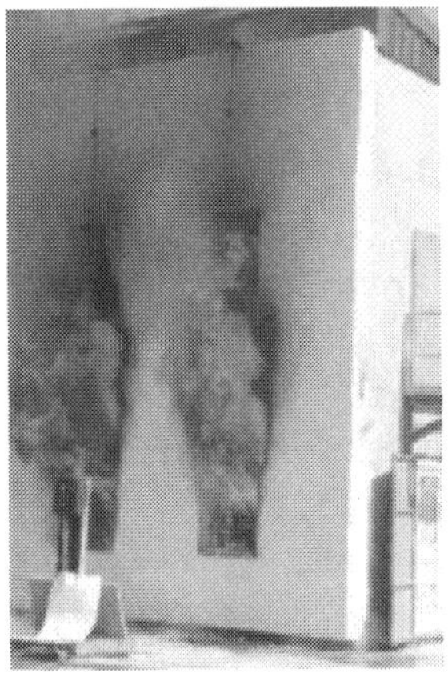

**Figure 5.** Observation at 10 min.

- For integrated construction systems, at level L+1:
  - No spread of flames on the floor through the facade/floor junction.
  - No transmission of hot gas (> 350°C) through the facade/floor junction.
  - No temperature rise greater than 180°C on non-exposed side of floor.

– For ETICS, after 60 minutes, the outer skin above the horizontal line at 5.2 m must not be damaged. The video of the facade at L+2 can prove that criteria are met.

## 5.5 Test report

At the end of the test, two documents are issued:

- A test report;
- An assessment from the laboratories in form of a conformity report for 5 years.

This test report includes a detailed description with drawings of the element, its components and the fire load. It provides temperature and radiation records, qualitative observations and results. It presents photos taken before, during and after the test.

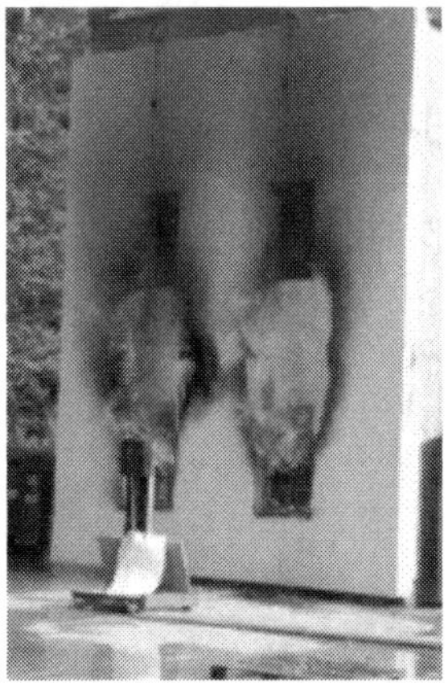

**Figure 6.** Observation at 20 min.

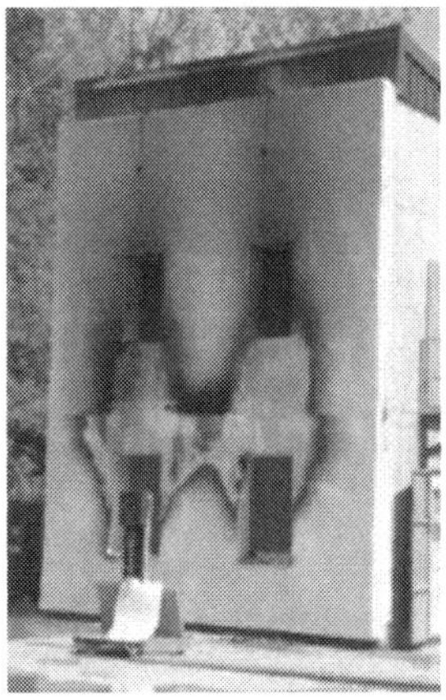

**Figure 7.** Observation at 30 min.

## 6. TEST CAMPAIGN

**Preliminary analysis of SBI data related to EPS ETICS**

The objective of this study was to identify the parameters that may affect reaction to fire of EPS ETICS systems. CSTB analysed about 300 reaction to fire test (SBI) reports, performed between 2005 and 2015. The analysis focused on:

- The Euroclass distribution by nature and chemical composition of the various renders.
- The impact of thickness and flame retardant rate of each render layer on Euroclass.
- The impact of the ratio "flame retardant rate / organic matter rate" on Euroclass.

The study highlighted that the majority of EPS ETICS systems reach Euroclass B or C. It confirmed that the fire behaviour of EPS ETICS under organic render is clearly improved when both basecoat and finishing are flame-proof.

### 6.1 LEPIR 2 orientation tests

Real scale orientation tests were carried out to assess the impact of the system components, as well as the ancillary protections.

**Figure 8.** SBI test.

These tests confirmed that horizontal fire barriers at each level fully play their role and no additional protections on openings are needed.

### 6.2 Tested configurations

Considering the conclusion of the analysis of the SBI test results and that of the preliminary test phases, three EPS ETICS configurations under renders were selected:

- EPS ETICS under thick hydraulic renders; total thickness greater than 10 mm.
- EPS ETICS under thin mixed renders (hydraulic basecoat, organic finishing); total thickness lower or equal to 10 mm.
- EPS ETICS under thin organic renders (flame-proof organic basecoat and finishing); total thickness lower or equal to 10 mm.

These configurations fulfil the thermal regulatory requirements and are at least Euroclass *B-s3,d0.* They are covered by a valid *Document Technique d'Application* (DTA) or *Avis Technique* (AT).

EPS boards, with a maximum density of 20 kg/m$^3$, are Euroclass E and ACERMI certified. They are bonded on masonry with a hydraulic mortar described in the European Technical Agreement or European Technical Evaluation of the system.

Ancillary protections consist of horizontal fire barriers made of stone wool strips. They are 200 mm high and have the same thickness as that of EPS boards. These strips run over the whole width of the facade. They are fully bonded on masonry with the same hydraulic mortar and then fixed with dowels every 500 mm.

## 7. TEST RESULTS

Tests were successful and met LEPIR 2 evaluation criteria. EFECTIS and CREPIM consequently issued the following assessment.

- The fire behaviour of EPS ETICS under renders is validated for every EPS thickness lower than those being subject to these tests: 300 mm for thick renders and 200 mm for thin renders.
- No minimum nor maximum distance is required between the lintels of the lower level and the sill of the upper level. The facades shall however meet the regulatory requirements for each type of construction.

**Table 2.** Components and features of tested configurations.

| Type of render | | Thick hydraulic | Thin mixed | Thin organic flame-proof |
|---|---|---|---|---|
| **Bonding mortar** | | Hydraulic | Hydraulic | Hydraulic |
| **Insulation** | **Nature** | EPS | EPS | EPS |
| | **Thickness** | 300 mm | 200 mm | 200 mm |
| | **Fixing system** | Bonding | Bonding | Bonding |
| **Render system** | **Reinforced basecoat** | Hydraulic render | Hydraulic render | Flame-proof organic render |
| | | Thickness 3 mm | Thickness 3 mm | Thickness 2 mm |
| | | TMO 7% | TMO 7% | TMO 7.5% |
| | | | | TFR 10% |
| | **Finishing coat** | Hydraulic render | Organic render | Flame-proof organic render |
| | | Thickness 8 mm | Thickness 2 mm | Thickness 2 mm |
| | | TMO 3% | TMO 10.7% | TMO 10% |
| | | | | TFR 8.8% |
| | **Total system** | Thickness 11 mm | Thickness 5 mm | Thickness 4 mm |
| | | FMO 4,5% | FMO 9% | FMO 8.5% |
| **Protection strips** | | Between each level one at the base | Between each level & one at the base | Between each level & one at the base |

*TMO: Taux de Matière Organique / Organic matter rate.*
*TFR: Taux d'ignifugeant / Flame retardant rate.*
*FMO: Fraction Massique Organique / Total organic mass fraction.*

**Table 3.** Render systems and protections: minimum allowed characteristics.

| Type of render | | Thick hydraulic | Thin mixed | Thin organic flame-proof |
|---|---|---|---|---|
| **Minimum fire reaction** | | B-s3,d0 | B-s3,d0 | B-s3,d0 |
| **Minimum thickness** | **Reinforced basecoat** | - | 3 mm | 2 mm flame-proof |
| | **Finishing coat** | - | - | Flame-proof |
| | **Total system** | >10 mm | 4 mm | 4 mm |
| **Maximum organic matter rate** | **Reinforced basecoat** | - | 7% | $R_{fo}$* >1 |
| | **Finishing coat** | - | 12% | $R_{fo}$* > 0,9 |
| | **Total system** | 5% | - | 9% |
| **Expanded polystyrene** | **Maximum thickness** | 300 mm | 200 mm | 200 mm |
| **Protection strips with same thickness as EPS** | **Minimum height** | 200 mm | 200 mm | 200 mm |
| | **Location** | Between each level & one at the base | Between each level & one at the base | Between each level & one at the base |

* $R_{fo}$ = (% flame retardant) / (% organic matter).

- The render systems whose finishing include an organic matter mass fraction lower than 12% do not contain enough organic load for parietal or lateral spread of fire along the render.
- Expanded polystyrene insulation materials have a CE marking according to EN 13163 standard. They have a density lower or equal to 20 kg/m$^3$. They shall be flame-proof and at least Euroclass E. The EPS boards supplier shall provide the proof that the raw material producer performs continuous monitoring of the fire proof characteristics (Euroclass D for a conventional 60 mm thick EPS board).
- Stone wool for protection strips shall have a nominal density greater or equal to 90 kg/m$^3$, have a CE marking according to EN 13162 standard, be Euroclass A1 and be ACERMI certified or equivalent.

# 8. RECOMMENDATION GUIDE

AFIPEB, SIPEV and SNMI wrote a recommendation guide based on EFECTIS and CREPIM assessment. This guide practically describes construction solutions intended to limit fire propagation on a concrete or masonry facade insulated by EPS ETICS.
This guide recalls that:

- Facades may have openings or may be blind walls.
- Openings may be equipped with all type, nature and dimension of joinery and can be installed in every position, from the bare interior to the bare exterior of the structure, and without constraints of minimum distance between superimposed openings.
- EPS ETICS can be installed as first insulation or as extra insulation.

The guide also describes:

- EPS ETICS characteristics and related technical justifications.
- EPS insulation boards characteristics, related justifications and installation instructions.
- Render systems composition and specific features.
- Protection strips and related construction provisions.
- Installation instructions around openings.

In addition to EPS ETICS solutions resulting from this campaign, the guide includes technical solutions, previously tested by other actors, for EPS ETICS under thick hydraulic renders, without protection strips, but with reinforced frame protection at openings. This guide supplements IT 249 (2010); it will be available on the Ministry of Interior website (`www.interieur.gouv.fr`).

# 9. CONCLUSION

The LEPIR 2 test protocol provides a relevant analysis about the fire behaviour of facades thanks to its main assets: test facility dimensions and representativeness, testing procedures and laboratories dual competence in fire reaction and resistance.

Our LEPIR 2 test campaign confirms that EPS ETICS under thin or thick renders perfectly comply with fire safety regulations for buildings. Thus, actors in the construction sector can still design and construct buildings fulfilling the highest thermal requirements, at reasonable cost, while keeping the good fire behaviour of the facades.

Industrial firms positively answered the request from public authorities. EPS ETICS solutions are published in a recommendation guide that will be available on the Ministry of Interior website. This should allow to update IT 249 in the coming future.

## Reference

*Internal reports and assessments issued by CREPIM, CSTB and EFECTIS for AFIPEB, SIPEV and SNMI, campaign 2014/2015.*

MATEC Web of Conferences **46**, 05011 (2016)
DOI: 10.1051/matecconf/20164605011


# Reaction-to-fire performance of fire-retardant treated wooden facades in Japan with respect to accelerated weathering

Miki Nakamura[1], Hideki Yoshioka[2,a], Manabu Kanematsu[1], Takafumi Noguchi[3], Shinji Hagihara[4], Akio Yamaguchi[5], Ken Shimizu[5], Toshiyuki Sugita[6], Yumi Matsumoto[3], Yuhei Nishio[3] and Tetsuya Hayakawa[7]

[1] Tokyo University of Science (TUS), Japan
[2] National Institute for Land and Infrastructure Management (NILIM), Japan
[3] The University of Tokyo, Japan
[4] Japan Testing Center for Construction Materials (JTCCM), Japan
[5] Koshii & Co, Ltd, Japan
[6] Misawa Homes Co., Ltd., Japan
[7] Tokyo System Vac. (TSV), Inc., Japan

## 1. INTRODUCTION

Wood has been used for building facades to enhance the aesthetic design of buildings since the revision to the evaluation method associated with the amendment of the Building Standard Law of Japan in 2000. In response, wood that has been pressure-impregnated with fire retardants (fire-retardant treated wood) is often used to ensure it is safe in the event of a fire. Currently, when fire-retardant-treated wood is tested for certification of reaction to fire performance, a cone calorimeter test is conducted in Japan. This test applies radiant heat to the surface of a square specimen, 100 mm each side, immediately after it has undergone fire-retardant treatment. However, when applying fire-retardant treatment to wood, aqueous chemical injection is the standard procedure. When wood is actually used to construct a building, there is a concern about environmental forces such as wind and rain that could cause the wood to deteriorate, and concerns about performance degradation associated with aging. One of the past studies in Japan [1] conducted a cone calorimeter test after an outdoor exposure test and accelerated weathering test, compared the post-test performance with the initial performance and confirmed the amount of remaining fire retardant in the treated wood had been reduced. However, no comparison of the fireproof performance of fire-retardant wood in actual use in a building facade had been conducted in Japan. There have been already valuable researches [e.g. 2, 3] on this issue internationally, but this paper is the first step in Japan and authors hope to focus on the wooden façade construction technique and the standard façade test in Japan.

---

[a]e-mail: yoshioka-h92te@nilim.go.jp

**Table 1.** Combination of accelerated-weathering and reaction-to-fire tests.

| Method | Accelerated-weathering test | | Reaction-to-fire test | |
|---|---|---|---|---|
| | Standard | Specimen size | Standard | Specimen size |
| 1 | JSTM J 7001 | 1.980 × 1.980 mm | JIS A 1310<br>Façade test | 1.820 mm (width)<br>4.095 mm (height) |
| 2 | | | ISO 5660-1<br>Cone test | 100 mm × 100 mm |
| 3 | NT FIRE 053 | 130 × 500 mm | | |

**Table 2.** Overview of specimens.

| Specimen | | Wood | Thick | Fire retardant | | Coating |
|---|---|---|---|---|---|---|
| | | | | Type | Amount | |
| A | Fire-retardant treated | cedar | 18 mm | Phosphoric acid | 172 kg/m$^3$ | Paraffin |
| B | Fire-retardant treated, and JSTM J 7001 | | | | 131 kg/m$^3$ | |
| C | Fire-retardant treated, and NT FIRE 053 | | | | 139 kg/m$^3$ | |
| D | No fire-retardant treated or accelerated-weathering | | | - | - | |

Under these circumstances, this study aims to discuss a method to evaluate the reaction-to-fire performance of fire-retardant-treated wood used as building facades with aging deterioration taken into consideration. To be specific, a building facade fire-spread test in Japan (JIS A 1310) [4, 5] and cone calorimeter test (ISO 5660-1) [6] were conducted on specimens that had undergone an accelerated weathering test so as to compare their performance and make fundamental reviews.

## 2. OUTLINE OF THE EXPERIMENTS

The combination of accelerated-weathering and reaction-to-fire tests in this study is shown in Table 1. Two kinds of accelerated-weathering test were conducted – NT FIRE 053 [7], one of the generally accepted standard tests especially in North Europe, and the Japanese standard test method (JSTM J 7001) [8]. Two kinds of reaction-to-fire test were conducted – a facade test that assumes an actual fire (JIS A 1310) and a small-scale test (cone calorimeter test, ISO 5660-1). Assuming aging deterioration of fire-retardant-treated wood, an accelerated weathering test was conducted, which was then followed by a reaction-to-fire test. The fire properties were compared among the two tests, and the burning condition of the fire-retardant wood after the accelerated weathering test was compared with that of non-treated wood.

### 2.1 Outline of specimens

The specimens were as described as in Table 2. The wood used was 18-mm-thick cedar lumber pressure-impregnated with a fire retardant. Phosphoric acid was used as a fire retardant to increase its fire retardant performance with the thermal insulation effect in the solid phase. Paraffin was used as the coating and it is insoluble in water. The concentration of the fire retardant impregnated in specimens (A) was 22.9%. Specimens (A) were only treated with a fire retardant and not subject to an accelerated weathering test. The fire retardant agent with a concentration of 23.5% was impregnated in JSTM J7001 specimens (B) and NT FIRE 053 specimens (C). The weight of the specimens was measured after impregnation of the chemical to calculate the amount of the impregnated chemical agent. The average amount of the agent was 172 kg/m$^3$ for A, 131 kg/m$^3$ for B, and 139 kg/m$^3$ for C. The impregnated amount of A was the greatest among them.

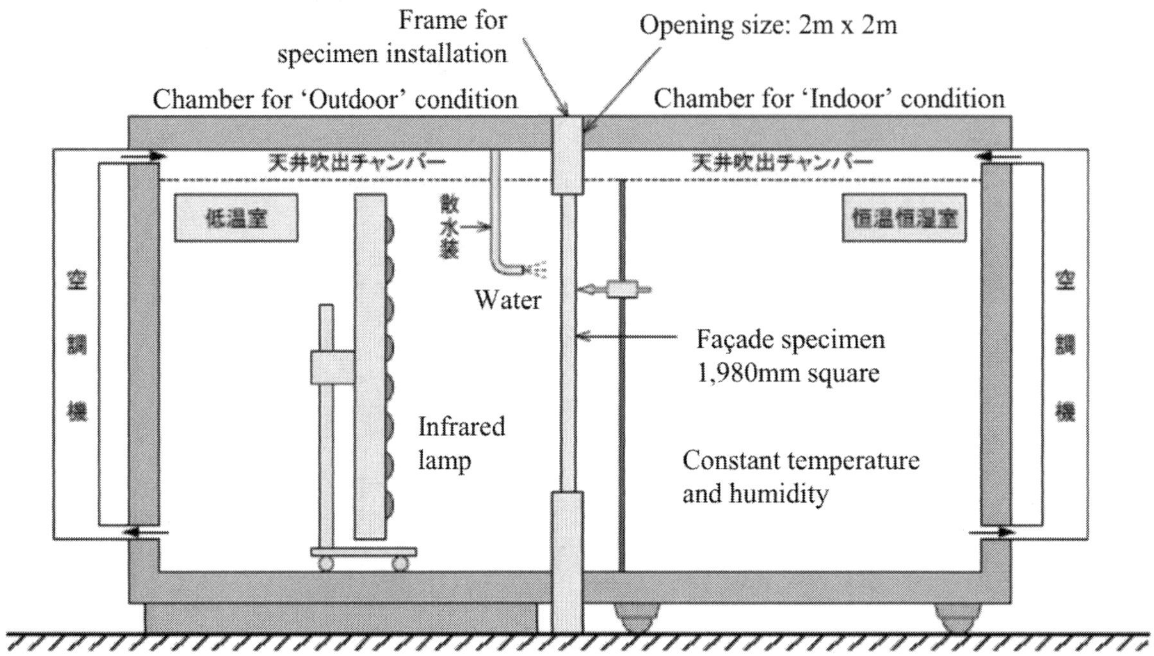


**Figure 1.** JSTM J 7001 test apparatus.

## 2.2 Accelerated weathering test – 1 (JSTM J 7001)

An accelerated weathering test was conducted on two specimens, each with surface dimensions of 1980 × 1980 mm, according to the Japan Testing Center for Construction Materials standard (JSTM J 7001) durability test method. It is usually used as durability test for exterior facades in general, but in this research it is applied to be used as accelerated-weathering for fire-retardant treated wooden facades. An outline of the test equipment is shown in Fig. 1. Specimens were installed on a specimen installation frame set up between the two chambers, one for outdoor condition and another for indoor condition, according to the actual installation procedure. Specimens were heated for 4 hours and then cooled for 4 hours. This eight-hour process was set as one cycle. A total of 90 cycles were conducted for the specimens. For heating, an insolation device with an infrared lamp was used to irradiate the surface of a specimen until the surface temperature rose to 80 °C. For cooling, water was sprinkled at a rate of 1 litter/($m^2$· min) at the beginning, and the temperature was lowered to −20 °C.

## 2.3 Accelerated weathering test – 2 (NT FIRE 053)

An accelerated weathering test was conducted according to Northern European standard NT FIRE 053 Method A. To be specific, four specimens, each measuring 130 × 500 mm, were placed with the longer side slanted to 18 °. Water was sprinkled for 96 hours and the specimens were dried for 72 hours as one cycle. Four cycles were conducted in the test. The temperature of the water sprinkled ranged from 2 °C to 16 °C. Water was sprinkled over the surface through a spray nozzle from above the specimen. To make conditions more equal in the water sprinkling for all specimens, the layout of each specimen was changed every cycle so that the average sprinkling volume on the specimen's surface remained at 0.32 L/($m^2$· min). To prevent water from being sprinkled on the specimen edges and the fire retardant from leaching from the edges, the specimen surfaces were coated with sealant. The inside of the test equipment was kept between 57 °C and 60 °C during drying, and wind with an average velocity of 19.3 m/min was blown over the specimen surfaces.

**Table 3.** Details of accelerated-weathering tests.

| | | JSTM J 7001 | NT FIRE 053 |
|---|---|---|---|
| Time for one cycle | | 8 h | 168 h |
| Number of cycle for conducting | | 90 cycles | 4 cycles |
| Heating | Total time | 360 h | 288 h |
| | Temperature | 80 °C | 60 °C |
| Cooling | Total time | 360 h | 384 h |
| | Temperature | −20 °C | 2 ∼ 16 °C |
| Watering | Total time | 15 h | 384 h |
| | Amount of water per minute | 1 LTR/(m$^2$ · min) | 0.3 LTR/(m$^2$· min) |
| | Total amount of water | 900 LTR/m | 6912 LTR/m$^2$ |

The method of conducting the accelerated weathering test used is described in Table 3. The temperature difference that occurred between heating and cooling was 100 °C for JSTM J 7001 and 50 °C for NT FIRE 053. During the former method, the surface temperature dropped to below 0 °C, and dew condensation eventually occurred. The amount of water sprinkled per minute was greater for JSTM J 7001, whilst the total amount of sprinkling water that the specimen surface received throughout the test was greater for NT FIRE 053.

### 2.4 Reaction-to-fire test – 1 (cone calorimeter test)

The cone calorimeter test was conducted according to ISO5660-1. The specimen surface was cut to a size of 100 × 100 mm after the accelerated weathering test, and the specimens were kept intact and dried for about one week until the water content before the cone test reached about 10%. A cone test was conducted by heating a specimen at a position 25 mm away from the electric heater at a heating intensity of 50 kW/m$^2$ for 20 min. The mass of each specimen was measured during heating to calculate the mass reduction ratio. The oxygen concentration was also measured to calculate the heat release rate using the oxygen consumption method.

### 2.5 Reaction-to-fire test – 2 (facade test)

An outline of the equipment used for the facade test JIS A 1310 is shown in Fig. 2. After accelerated weathering test (JSTM J 7001), each specimen was modified to have surface dimensions of 1820 × 4095 mm and opening dimensions of 910 × 910 mm. Facade specimens were set in place so that the burning chamber matched the opening of the specimen, and the burner in the chamber was ignited to let the flames gush out from the opening. The heating intensity in the chamber was set to 1200 kW, the heating time was 20 min, and thermocouples were installed in a total of 19 positions in the vent layer between the specimen and the substrate and a total of 5 positions on the specimen surface (Fig. 3). Heat flux meters were installed on the top surface of the specimen, the center of the opening on the surface facing the specimen 2000 mm away, and heights of 0, 500, 900, 1500, 2000, and 2500 mm from the top of the opening. The oxygen concentration in the exhaust duct during the experiment was measured, and the heat release rate was calculated using the oxygen consumption method.

## 3. EXPERIMENT RESULTS AND DISCUSSION

### 3.1 Results of cone calorimeter tests (ISO5660-1)

The damage condition of the burned surface before and after the cone calorimeter test is shown in Fig. 4. The results of JSTM J 7001 specimens B showed almost the same tendency between the two specimens. Thus, the result of B-1 is shown as a representative specimen. No major change occurred to the size of specimen A, which was only treated with a fire retardant, and B-1 from before and after the cone test.

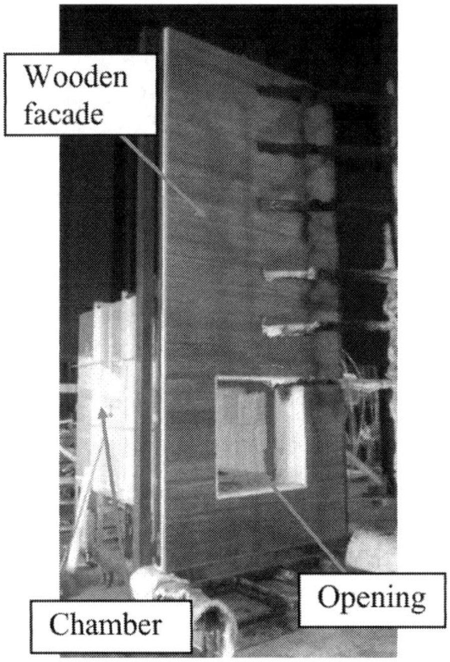


**Figure 2.** JIS A1310 facade test equipment.

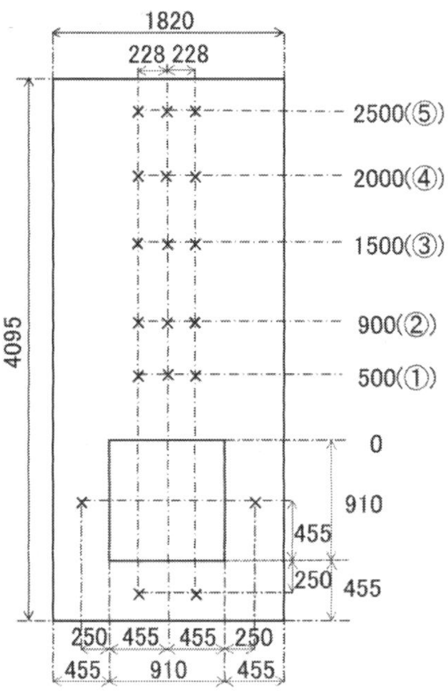


**Figure 3.** Layout of thermocouples (unit: mm).

The NT FIRE 053 specimen C and non-treated specimen D became smaller than before the cone test, carbonized after the test, and showed gloss on their rear surface.

The measurement results and the results of heat release rate and total heat released are shown in Table 4, Fig. 5, and Fig. 6, respectively. Although no difference occurred between A and B in terms of heat release rate and total heat released, C ignited 9.7 min after the start of heating, and its heat release rate became a maximum 13.6 min after the start of heating. D ignited, its heat release rate peaked

**Figure 4.** Damage on the burned surface before and after the cone calorimeter test. (left: before test, right: after test).

immediately after the start of heating, and the peak occurred again after 12 min. B still maintained a sufficient amount of fire retardant to prevent ignition even after the accelerated weathering test. C had much of its fire retardant leached in the accelerated weathering test and ignited. It is speculated according to the condition of the rear of specimen C after the cone test that the specimen was burned to its rear in a burning process that reached the peak. For the burning condition of D, it is speculated that specimen burning became moderate after the first peak because the carbonized layer that formed on the wood surface shut off the burning process, and then another peak occurred because the rear of the specimen began burning.

The resultant mass reduction rate and comparison of specimens A, B, and C that all contained fire retardants indicated that the larger the mass before the heating test, the larger the mass reduction rate would likely become. This is presumably because the more fire retardant remains in a specimen before the heating test, the more is leached during burning.

In order for a building material/product to be certified in Japan regarding reaction-to-fire performance, the total heat released must be 8 MJ/m2 or less, in 5 minutes for flame retardant material, 10 min for quasi-noncombustible material, and 20 min for noncombustible material. Specimens A, B, and C exceeded 8 MJ/m2 in 15 min, 13.4 min and 9.6 min after the start of heating, respectively. These results show A and B have the performance of quasi-noncombustible material and that C has the performance of flame-retardant material. As for the result of the accelerated weathering test, the condition of B showed just leaching of some fire retardant, while the condition of C showed much more leaching of fire retardant. These results confirmed the deteriorated performance of the specimens subjected to the accelerated-weathering test.

The dry densities of specimens before the cone test were 0.51 g/cm$^3$ for A, 0.43 g/cm$^3$ for B-1, 0.40 g/cm$^3$ for B-2, and 0.37 g/cm$^3$ for C. B and C had a smaller dry density than A, which suggests that the fire retardant of these two specimens was leached as a result of the accelerated-weathering tests. Comparing B and C, C had a smaller dry density than B. This is presumably because C had a lot of its fire retardant leached as a result of a large amount of water being sprinkled and a long duration of water sprinkling during the accelerated-weathering test and its dry density before the cone test eventually was reduced.

## 3.2 Results of facade tests (JIS A 1310)

The damage condition and burned area of each specimen after completion of the facade fire test are shown in Fig. 7. A 100 × 100 mm lattice metal net was put over each specimen after completion of the test, and the burned condition such as burn-through or carbonization was checked for each space. The burn-through height was 1,500 mm for A, 2,000 mm for B, and 2,700 mm for C, while the burn-through area was 0.84 m$^2$ for A, 1.35 m$^2$ for B and 4.23 m$^2$ for C (see Table 5). Comparing the specimens after burning, specimen D suffered a wider area of burning than A or B. A comparison between A and B

**Table 4.** Cone calorimeter test results.

| Specimen | | Condition before test | | | THR 20 min | Peak HRR | Mass decrease rate | Time to reach each condition | | |
|---|---|---|---|---|---|---|---|---|---|---|
| | | Mass | Dry density | Moisture content | | | | THR 8 $MJ/m^2$ | Peak HRR | Ignition |
| Fire retardant treated | A | 97.1 g | 0.51 $g/cm^3$ | 7.0% | 12.3 $MJ/m^2$ | 15.6 $kW/m^2$ | 6.2 $g/s/m^2$ | 15.0 | 18.9 | No |
| Fire retardant treated and JSTM J 7001 | B-1 | 81.8 g | 0.43 $g/cm^3$ | 7.7% | 14.7 $MJ/m^2$ | 19.3 $kW/m^2$ | 5.6 $g/s/m^2$ | 13.4 | 19.6 | No |
| | B-2 | 75.8 g | 0.40 $g/cm^3$ | 7.0% | 13.5 $MJ/m^2$ | 19.4 $kW/m^2$ | 5.4 $g/s/m^2$ | 14.4 | 19.1 | No |
| Fire retardant treated and NT FIRE 053 | C | 71.5 g | 0.37 $g/cm^3$ | 8.3% | 41.7 $MJ/m^2$ | 92.8 $kW/m^2$ | 4.9 $g/s/m^2$ | 9.6 | 13.6 | 9.7 |
| No treatment | D | 53.1 g | 0.28 $g/cm^3$ | 7.9% | 76.6 $MJ/m^2$ | 116.0 $kW/m^2$ | 6.4 $g/s/m^2$ | 2.2 | 0.9 | 0.3 |

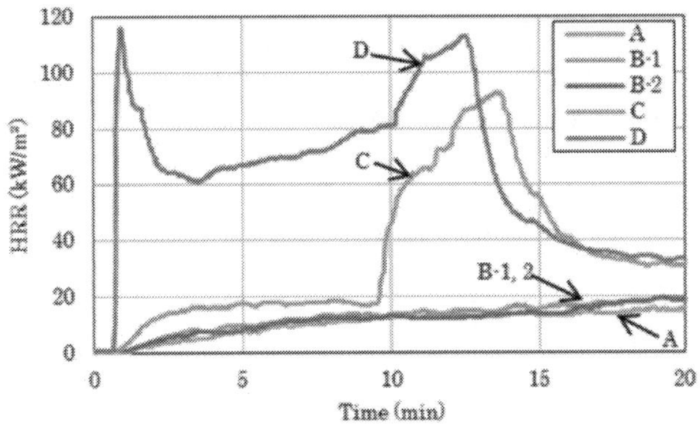


**Figure 5.** Cone calorimeter test: heat release rate.

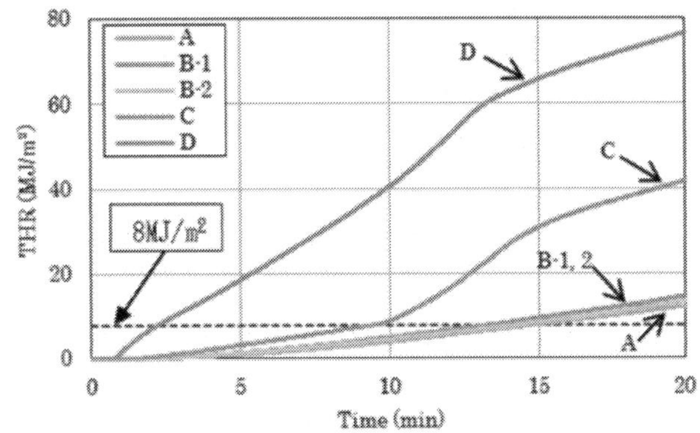


**Figure 6.** Cone calorimeter test: total heat released.

indicated the burning had spread slightly more for B than A. The fire retardant contained in A and B was leached on the surface and ran down after the facade test, and it was seen even from the lower part of the opening that was not exposed to the flame during the test. It is shown that the temperature rise causes leaching of the fire retardant. In addition, part of the surface at the upper part of specimen A fell off after the test, just as if it was shaved off. This is a sign that the amount of the leached fire retardant was greater for A than B and that the area beside the opening was more intensely carbonized for A than B (see Fig. 8). Specimens A and B share a similar burning tendency. Immediately after the start of the test, they both gave off a burning sound, the surface wood sprang up, and the fire was automatically extinguished after completion of heating. On the other hand, specimen D had all its surface burned immediately after the start of test, the burning sound started in 3 minutes, ash started to fly off from the surface thereafter, and self-burning continued even after the end of the test (end of heating from chamber).

Temperature changes on the specimen surface and at the vent layer (right-side from centreline) are shown in Fig. 9. Temperature changes at the vent layer located right side from centreline are shown here, because the thermocouples situated at the centreline of the vent layer, which was installed down the vertical furring strip, did not show any major temperature change during the façade test. The trend of temperature rise on the surface is almost the same for A and B. The temperature rose soon after the start of heating at 500 mm and 900 mm above the opening, while the temperature softy rose at 1,500 mm, 2,000 mm and 2,500 mm above the opening. On the other hand, all the surface temperatures started to rise for D immediately after the start of heating, and only the temperature rise at 500 mm shows almost the same trend for D, as A and B. Regarding temperature changes at the right-side

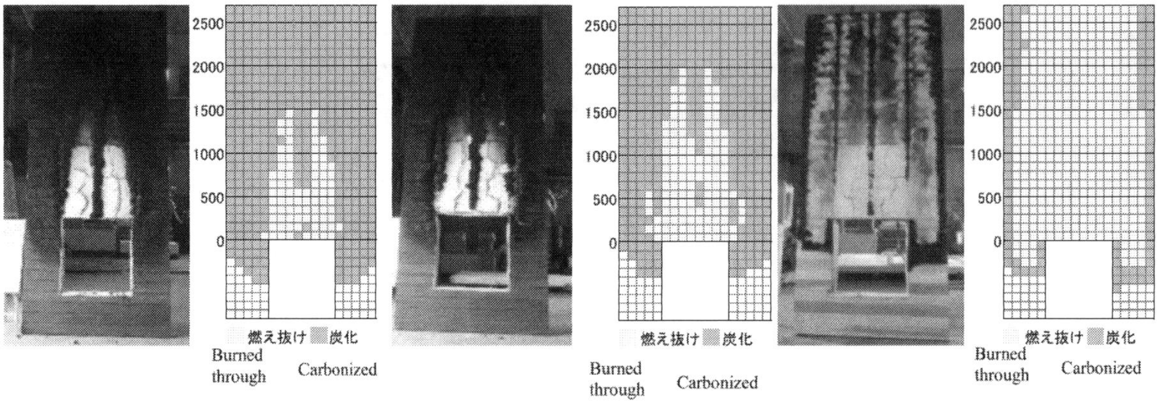


(a) Façade specimen A:
Fire-retardant treatment only.

(b) Façade specimen B:
Firstly, fire-retardant treated,
and then, JSTM J 7001 conducted.

(c) Façade specimen D:
No treatment at all (cedar).

**Figure 7.** Damage condition and the burned area after the facade test.

**Table 5.** Overview of JIS A 1310 façade test results.

| Specimen | | Portion of burned-through area at façade after the test | | Time when the façade specimen was burned through from surface to substrate | | | | | Peak Heat Flux at top of facade | Peak HRR* | THR* 20 min |
|---|---|---|---|---|---|---|---|---|---|---|---|
| | | Height | Area | 500 mm | 900 mm | 1500 mm | 2000 mm | 2500 mm | | | |
| Fire retardant treated | A | 1.5 m | 0.84 m² | 13 min | 15.5 min | - | - | - | 11.7 kW/m² | 1,361 kW | 1,350 MJ |
| Fire retardant treated and JSTM J 7001 | B | 2.0 m | 1.35 m² | 10 min | 12.5 min | 14.5 min | 19.5 min | - | 11.0 kW/m² | 1,306 kW | 1,348 MJ |
| No treatment | D | 2.7 m | 4.23 m² | 8 min | 8.5 min | 10.5 min | 11.5 min | 12.5 min | 32.9 kW/m² | 1,796 kW | 1,788 MJ |

* Heat release values (peak and total) include the contribution of gas burner.

vent layer, the temperature rise time at each measurement point decreases in descending order from A to B to D. Assuming the time when the temperature at the right-side vent layer almost matches the surface temperature is equal to the time when a façade specimen surface is burned through, then the burn-through time at 500 mm is 13 min, 10 min, and 8 min for A, B and D, respectively, while that at 900 mm is 15.5 min, 12 min and 8.5 min for A, B and D, respectively. Table 5 tabulates those differences among the specimens. Burn-through was visually confirmed up to 1,500 mm with specimen A, while the temperature at 1,500 mm at the right-side vent layer did not reach the surface temperature. B was burnt through up to 2,000 mm, and C was burnt through up to 2,500 mm. These results confirmed that fire-spread further expanded after the accelerated weathering test was conducted to the façade specimen, but it is still more moderate than the cedar specimen without fire-retardant treatment.

### 3.3 Comparison of reaction-to-fire test results

The cone calorimeter results were compared among the specimens A, B and D. While a difference was confirmed for D from the other two, no major difference was observed between A and B. And regarding the facade tests, their results are almost the same as those of the cone calorimeter test for A, B and D in terms of the visual records and surface temperature measurements, which means that D showed far more intense fire propagation than A or B. However, differences were observed in temperature changes at the vent layer even between A and B, in that B had higher temperatures than A. These results confirmed that

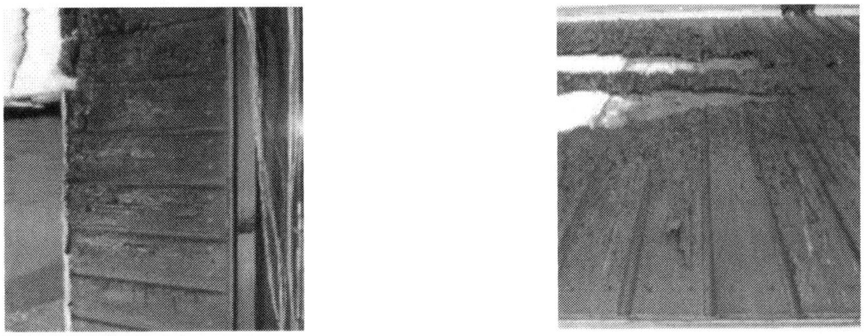

(a) Condition of leaching (b) Condition of peeling

**Figure 8.** Damage condition of specimen A (only with fire-retardant treatment).

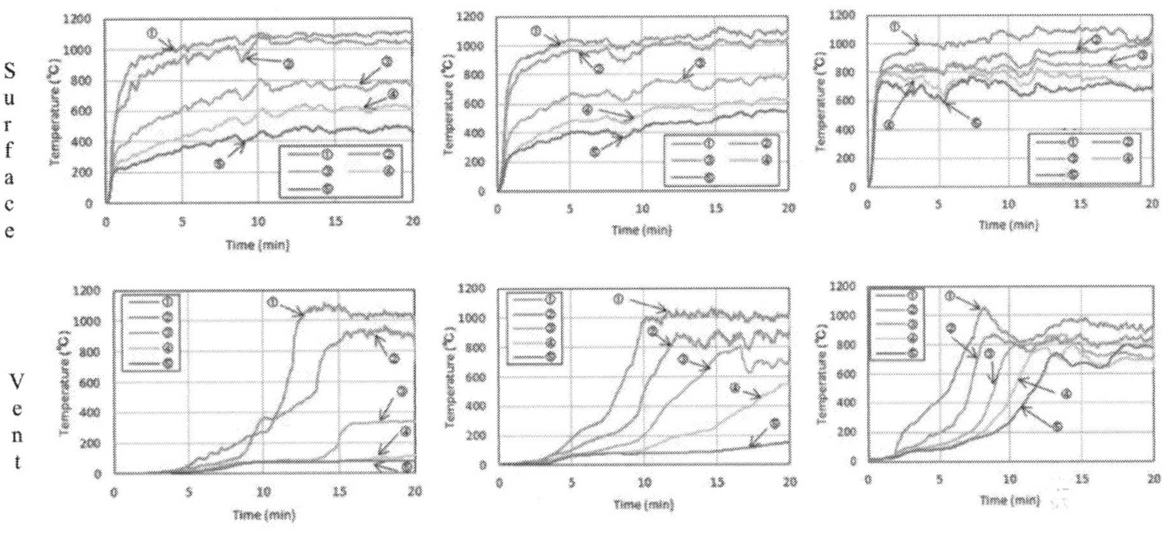


(a) Façade specimen A: Fire-retardant treatment only.

(b) Façade specimen B: Firstly, fire-retardant treated, and then, JSTM J 7001 conducted.

(c) Façade specimen D: No treatment at all (cedar).

①500 mm, ② 900 mm, ③1500 mm, ④ 2000 mm, ⑤ 2500 mm (each height from the opening top)

**Figure 9.** Temperature changes during the facade tests.

it is feasible to determine differences in fire-spread performance among the specimens by façade tests, while they showed no major difference by the cone calorimeter tests.

The result of the cone calorimeter tests showed that the mass reduction of A and B depends on the leached amount of fire retardant during the cone test. It turned out that A had a larger mass reduction than B. A visual record from the facade test also confirms that the amount of leached fire-retardant after the façade test is larger for A than B. This indicates that the mass reduction resulting from a cone calorimeter test can be used to judge the leaching condition of the fire-retardant in a facade test.

Regarding the facade tests, the area of fire-spread over the specimen was larger with the specimen which underwent JSTM J 7001 test (one of the accelerated weathering tests), compared with the specimen without weathering test. Considering the cone calorimeter test results, it would be expected that fire will spread even more intensely for the specimens that underwent NT FIRE 053 test (another accelerated weathering test).

# 4. SUMMARY

The findings obtained from this study on reaction-to-fire performance of fire-retardant treated wood for building facades with respect to accelerated-weathering are summarized as follows:

1. Regarding the cone calorimeter tests, the specimen that underwent the JSTM J 7001 test (accelerated-weathering) resulted in the total heat released for twenty minutes (14 MJ/m$^2$), which was just slight increase compared with the fire-retardant treated specimen without accelerated weathering (12 MJ/m$^2$). On the other hand, the specimen that underwent the NT FIRE 053 test (different type of accelerated-weathering) was burned to a much greater degree (42 MJ/m$^2$), which was still better than the normal cedar specimen without fire-retardant (77 MJ/m$^2$).
2. It is tentatively concluded that the NT FIRE 053 test method produces more deterioration effect on fire-retardant treated wood than the JSTM J 7001 test, as an accelerated weathering test. This result indicates that the water sprinkling time and amount of water sprinkled during the accelerated weathering test affect the status of leaching of the fire retardant, in that total amount of watering is more with NT FIRE 053 test (6912 LTR/m$^2$ in total) than JSTM J 7001 test (900 LTR/m$^2$ in total).
3. Regarding the JIS A 1310 façade tests, specimen with JSTM J 7001 test (accelerated-weathering) showed not so much different temperatures on the façade surface from the specimen without accelerated-weathering, while the former specimen had higher temperature increase at vent layer than the latter specimen. Fire spread slightly earlier with the former façade specimen, because fire-retardant leached out after the JSTM J 7001. But the difference is not very significant, and both the specimens showed far more moderate fire behaviour than cedar façade specimen without fire-retardant treatment.

With respect to future works, considering the specimen size of façade, NT FIRE 053 test is not very realistic option as accelerated weathering test because there is not a physically large enough laboratory in Japan for conducting with façade test size specimen, even though it is already accepted to some extent internationally, and no problem from technical viewpoint. Therefore, authors will intend to modify the test based on JSTM J 7001 with more water to be sprinkled, until cone calorimeter result will be similar to that with NT FIRE 053. And after that, the façade specimen which underwent the modified JSTM J 7001 with more watering, will be tested with JIS A 1310 façade test, to be compared with the façade specimen without accelerated weathering.

Authors greatly appreciate Dr. Tamás Bánky (ÉMI), Dr. Dhionis Dhima (CSTB) and Dr. Simo Hostikka (Aalto university), etc. on their kind help in gathering relevant information especially from European countries. Technicians and students at the University of Tokyo and Tokyo University of Science are also appreciated for their dedicating support in conducting experiments.

## References

[1] Harada T., et al.: Weatherability and combustibility of fire-retardant-impregnated wood after accelerated weathering tests, Journal of Wood Science, **55**(5), pp. 359–366, 2009.
[2] Östman B. and Tsantaridis L.: "Fire performance of multi-storey wooden facades," 1st International Seminar for Fire Safety of Facades, November 2013, Paris, France.
[3] Belloncle C., et al.: "Artificial and natural weathering of fire proofed wood cladding," 1st International Seminar for Fire Safety of Facades, November 2013, Paris, France.
[4] JIS A 1310: 2015, Test method for fire propagation over building façades.
[5] Yoshioka H., et al.: "Façade Tests on Fire Propagation along Combustible Exterior Wall Systems", Fire Science and Technology, **33** No. 1, pp. 1–15, 2014.

[6] ISO 5660-1, Reaction-to-fire tests – Heat release, smoke production and mass loss rate – Part 1: Heat release rate (cone calorimeter method) and smoke production rate (dynamic measurement).
[7] NT FIRE 053: Accelerated Weathering of Fire-Retardant Treated Wood for Fire Testing, 2003.
[8] JSTM J 7001: 1996, Test method for thermal deformation and durability of real-scale exterior wall (in Japanese).

MATEC Web of Conferences **46**, 06001 (2016)
DOI: 10.1051/matecconf/20164606001


# A case study of fires in structural elements

Nils Johansson and Patrick van Hees

Department of Fire Safety Engineering Lund University, Sweden

**Abstract.** Several severe fires in structural elements, or so-called "structural fires", have occurred in Sweden in recent years. In order to study this phenomenon, a dozen fires in structural elements have been studied in a recent project. This paper provides a description of the characteristics of "structural fires" and examples of how such events can be managed and limited. Constructional fire protection, rescue service response and fire development are the three main factors that control the outcome of structural fires. However, there are several underlying and more specific factors affecting the outcome of each fire.

## 1. INTRODUCTION

Fires in buildings can be divided into three different categories depending on the characteristics of the building and the area in which the fire starts and develops (see Fig. 1) [1]. The first category: "Fires in structural elements" involves fires that start in, or spread to, a structural element e.g. inside walls, attic space, facade or roof. The second category: "Compartment fire" involves fires that start in, or spread to, enclosures where a hot-gas-layer can form as the fire develops. The temperature throughout the hot-gas-layer will be rather homogenous due to the stirring of the hot gas by the fire. The enclosure can, however, be so large that it is not reasonable to assume that a homogenous hot-gas-layer will form, and this is the last category is referred to as "Fires in large enclosures". The terminology might create a bit of confusion because "Fires in large enclosures" are also fires in compartments but the "Compartment fire" is considered to be an established concept that refers to a model of fires in small- or medium-sized enclosures (e.g. residential rooms, cellular offices, or small industrial units) and therefore, that terminology is used in this paper.

The three categories constitute an idealization and the categories will of course overlap and it is possible that several categories may be included in the same fire, e.g. a compartment fire can spread to a structural element, like the façade, or vice versa.

Several devastating fires in recent years have created an attention to fires in structural elements, so-called "structural fires", in Sweden. Fires in structural elements can lead to an extensive damage and in many cases a totally destroyed building. In a Canadian study [2] the severity of 120 fires in apartment buildings were studied and categorized based on where they started as: compartment fires, concealed fires and exterior fires (which included façade fires). Ten per cent of the compartment fires in the study spread to another fire compartment or to the exterior of the building through window openings. The concealed fires did spread in 56% of the cases to other concealed spaces or to a compartment.

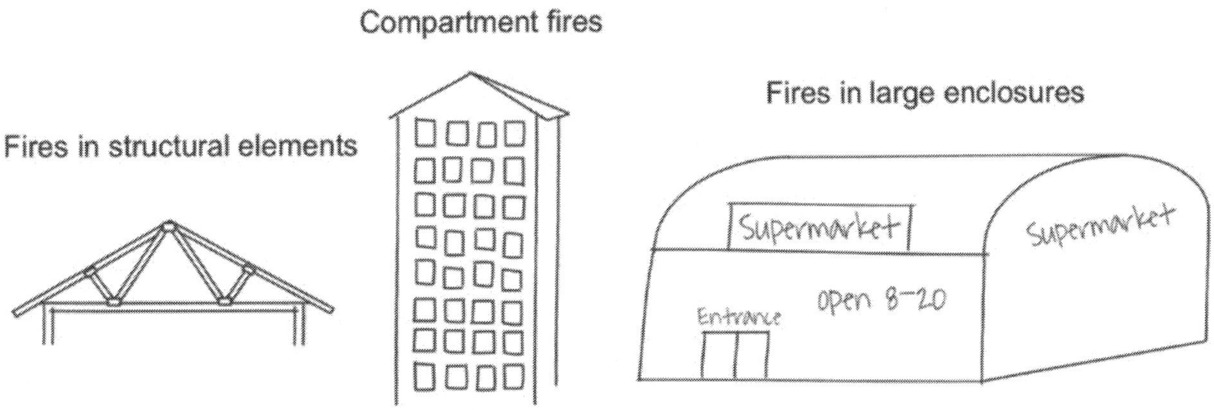


**Figure 1.** Different categories of fires in buildings. Based on an illustration in Johansson [1].

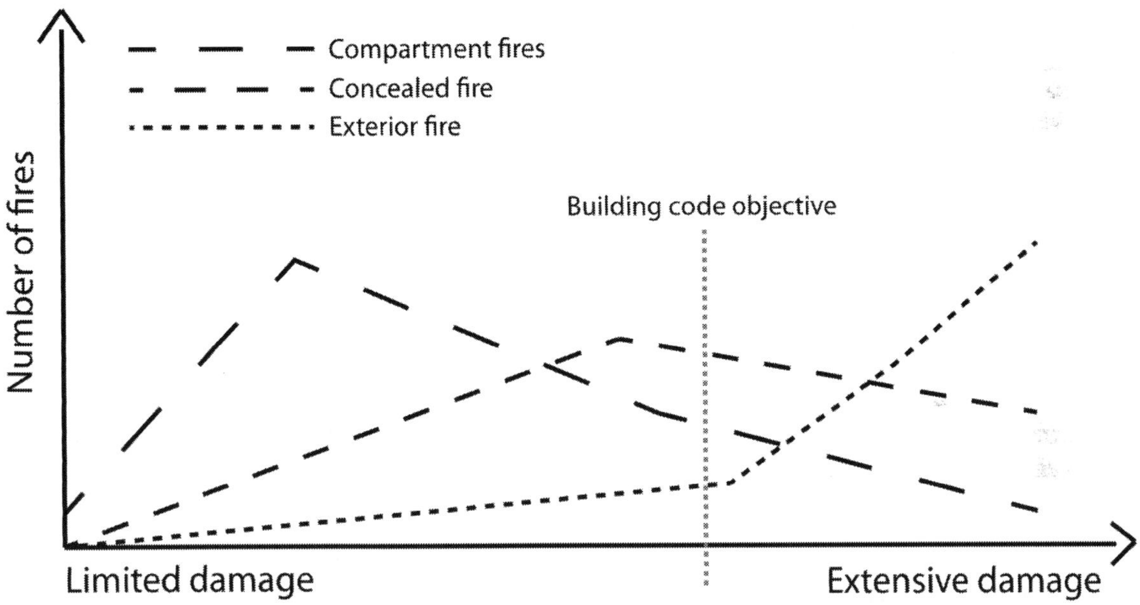


**Figure 2.** Schematic representation of the results from the study by Senez, Calder & Li [2].

The exterior fires did spread in 92% of cases to more than one floor, and in several cases all the way up to the roof of the building. The Canadian study is limited to 120 fires, but it indicates that the concealed and exterior fires (both are included in the fires in structural elements category in Fig. 1) more often will result in larger fire damage and property losses compared to fires in compartments.

The concept of "fires in structural elements" is not clearly defined but it has been used to describe fires inside walls, floors, crawl attics, roof structures etc., i.e. in concealed spaces areas where it is difficult easily detect and observe the fire. This type of fire is radically different from the well-ventilated compartment fires. A lot of effort has been put into research on how various objects burn, e.g. the size of the heat release rate and the combustion products [3, 4] much of this research has also been focused on fires with good supply of oxygen [5]. Fires in concealed spaces are different, since the fire is restricted, the volume is smaller and oxygen supply is limited. A fire in a normal room usually burns with flames while a concealed fire generally smoulders. If fresh air for some reason, flows to the concealed fire (e.g. if parts of a roof collapse) it can flare up. Well-ventilated compartment fires are different from ventilation limited concealed fires; consequently, the models that are used to describe compartment fires cannot be used for concealed fires.

Exterior fires can also be included in the term "structural fires". Although it is not as common in Sweden as concealed fires but there are several international examples of fires in tall buildings where vertical flame spread occurred rapidly along the façade [6]. In a comprehensive case study [6] funded by the Fire Protection Research Foundation has some twenty facade fires in high-rise buildings been included. Some of the conclusions from the case study were:

- Extensive facade fires often result in extensive property damage, but rarely in any fatalities.
- The events occur more frequently in countries with insufficient or no rules on combustible materials in facades at times when the buildings were erected.
- An internal fire (compartment fire) that spread to the outside of the building is the most common initiating scenario for exterior fires.
- Falling burning materials can pose a great danger and contribute to the spread of fire downwards.
- U-shaped facades (such as glass-enclosed balconies that creates a corridor vertically along the facade) presents a risk of fire spread faster than on flat facades.
- Extensive facade fires occur more frequently during construction and before the building is taken fully into operation.

The damage after a fire is related to direct fire and smoke damage, and also to water damage. The water damage may be due to the rescue effort but can also have other causes, e.g. rainwater after a roof fire or water from water pipes that for some reason burst during the fire [7].

There is no good overview of the phenomenon of "structural fires" in Sweden. One reason for this can be that the collection of fire statistics are based on the compartment fire concept that the fire starts in a object inside a room and spreads to other objects in the room of origin or other rooms in the building, "structural fires" fall outside of this description. The circumstances differ between these two categories of fires and the models that are used to describe and predict compartment fires cannot be used on fires in structural elements.

In summary, there is a lack of knowledge about "structural fires" and it is important to study such fires in order to clarify why fires in structural elements in general causes larger property losses than compartment fires. There is a lot of data, in terms of fire investigation reports, that can be studied, and a structured study of previously occurred fires in structural elements could be used to identify measures to prevent or limit the impact of future fires in structures. Therefore, a number of actual fires in structural elements are studied in this paper with the objective to clarify what factors that contribute to the extensive fire damage in some of these fires.

The research presented in this paper has been conducted within a project on fires in structural elements. The paper is a summary of a report [8] written in Swedish.

## 2. METHOD

A number of occurred fires in structural elements in various types of buildings are studied in the paper with a case study methodology developed previously by the authors [9]. The basis of the methodology is to conduct a systematic review of several actual events in order to identify underlying problems or faults that have led to a certain unwanted event (see Fig. 3). The method consists of the following six steps:

1. Define the problem
2. Interviews and literature review
3. Specify problem
4. Select events

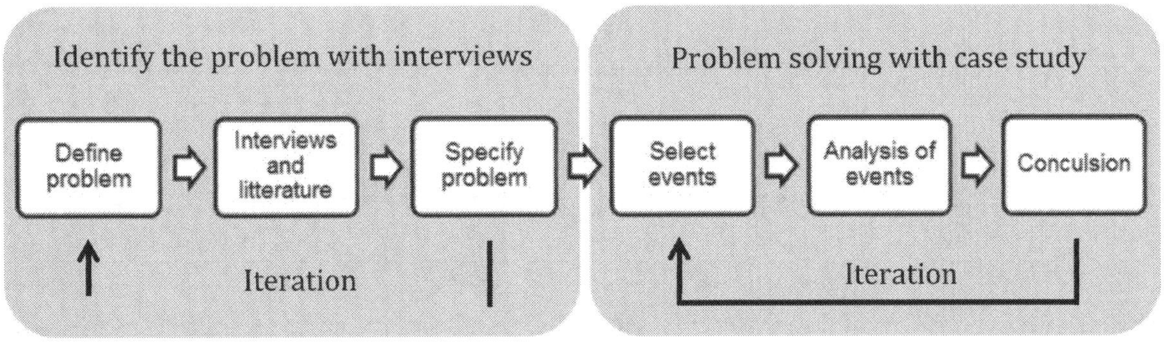


**Figure 3.** The six different steps in the method used in the paper. The method is based on a method presented by Johansson et al. [9].

5. Analysis of events
6. Conclusions.

The first three steps are essentially used to refine and specify the characteristics of the events that are relevant to study. In the first step the problem is defined. How the problem is formulated depends on the information that is available, and it can be rather vague when little information is available. In this paper the problem is related to "structural fires": how they are defined, begins and develops. In step 2 more information is gathered in order to refine the problem. In this study interviews with different stakeholders was considered as the best option in order to gather more information. Semi-structured interviews were performed with people at the Rescue Service in Gothenburg, a representative for the Swedish Contingency Agency and a representative from The Swedish Fire Protection Association. The results from the interviews made it possible to specify the problem (step 3). In this study the first three steps are used to clarify and define what a fire in a structural element is and how it behaves.

In step 4 a number of events are selected in accordance with the problem description. In step 5, *Analysis of events*, the material for each event is studied. The material consists of fire investigations from the rescue service and information (newspaper articles and photos) about the event found on the Internet. Each event is analysed in regard to the fire development and the behaviour of the building in a fault tree. In a fault tree analysis a deductive systems analysis is conducted [10], based on a certain outcome of a system (successful or unsuccessful). Fault tree analysis starts with specifying the outcome and then identifying factors or events that have or may have contributed the outcome. Two types of so-called gates can be used in the fault tree, namely "and"-gates (more than one component needs to occur) and "or"-gates (only one of several components are necessary). An example of a fault tree is presented in Fig. 4.

The results of the analysis of the individual events are combined in Step 6 to find common underlying factors. The common parts for the events are included in a new fault tree where elements common to all studied the fault tree are presented (see Fig. 5). If the conclusions are not considered to be sufficiently substantiated step 4 and 5 can be repeated. Additional events may reinforce or complement the conclusions and this iteration is repeated until the findings are considered to be sufficiently robust.

## 3. RESULTS AND DISCUSSION

People from three different stakeholder organisations in Sweden was interviewed and a dozen events, which occurred between 2012 and 2014, was studied and analysed thoroughly with the case study approached presented in the method section.

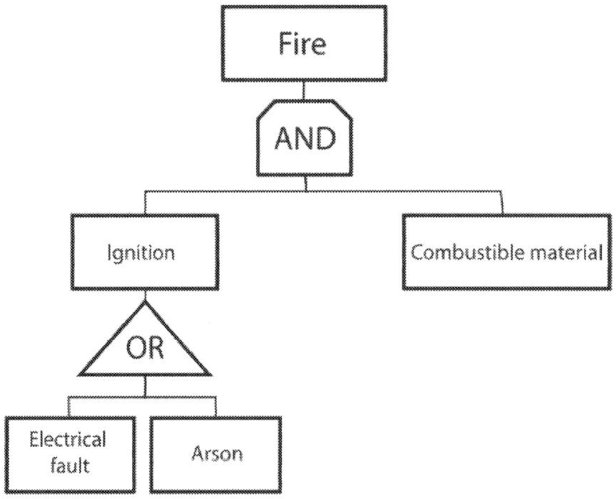


**Figure 4.** Example of a fault tree.

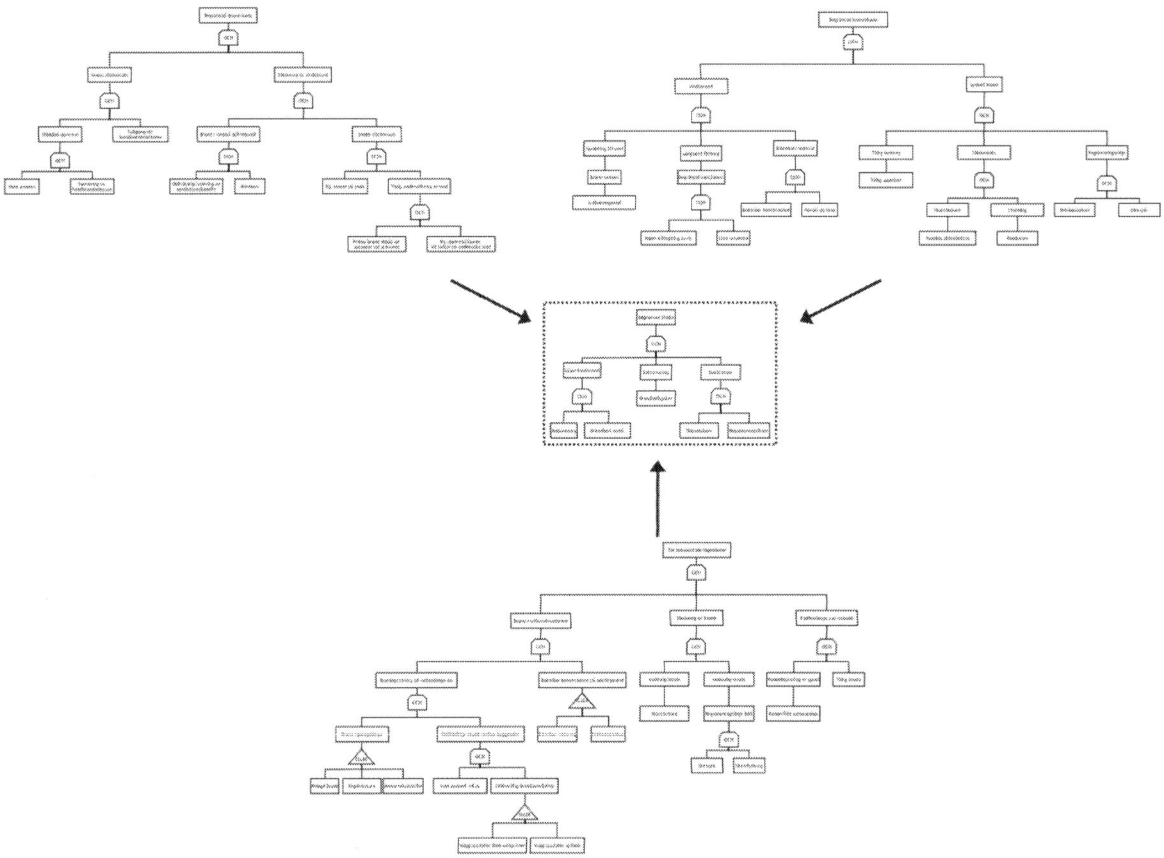

**Figure 5.** The end result of step 6 is a fault tree (centre) which provides a schematic representation of the features common in the studied events.

## 3.1 Characteristics of structural fires

Based on the conducted interviews conducted and the case study the following two types of fires can be included in the term structural fires:

- Hidden fires
- Fires in building materials.

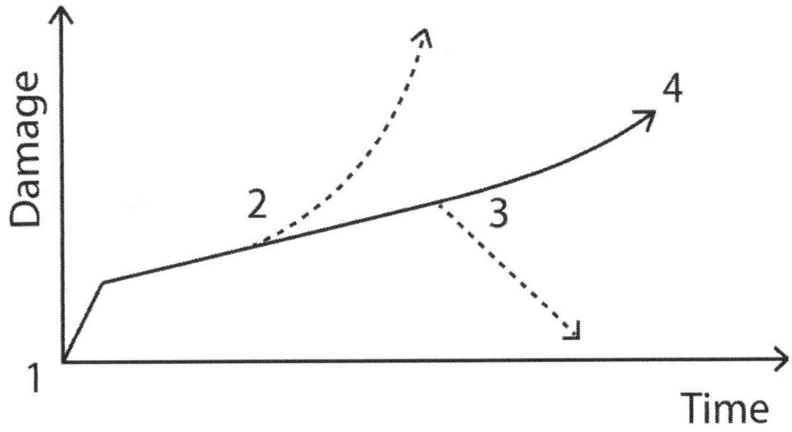


**Figure 6.** Illustration of fire damage over time for a fire in a concealed space.

Concealed fires include fires in composite structures such as walls, ceilings and roofs. In a composite wall with an air gap, a so-called chimney effect can occur if the ventilation is good [11]. The concealed fires are difficult for rescue services to access and extinguish. Fires in building materials are fires in linings inside the building and materials on the exterior, i.e. fires on the walls, ceilings, floors, facade and roof.

Fires in building materials in a room (e.g. a wall) differ from fires in loose furnishing but the fire development will go through similar stages as the typical compartment fire (growth, flashover and flashover), this type of fire are therefore not covered by what is commonly referred to as structural fire. When it comes to fires outside the building such as roof or facade fires there will be a good supply of air, the fire will also be influenced by external conditions (e.g. wind speed and direction). External fires are therefore very different from compartment fires and fires in concealed spaces. These different types of fires should not be lumped together because they are different due to the different conditions and fire dynamics.

Based on the results from the interviews it is considered that fires in concealed spaces is what usually is referred to when the term structural fires is used. A fire in a concealed space can be schematically described as fire damage as a function of time as presented in Fig. 6.

In Fig. 6 is a general description given of a fire in a concealed space. The fire starts in the concealed space (point 1); it can be initiated by e.g. heat transfer from a hot pipe or duct, from electrical installations or due to fire spread from a compartment to the concealed space. The concealed space is small and the amount of air is limited, which means that the fire will become ventilation controlled quickly. The extent of the damage is increasing, but not as fast as if there would be a good supply of oxygen. If the enclosure breaks (e.g. if there is no fire rating on a supporting structure) or openings to the fire are created by the rescue services, air can flow in and the fire can flare up (point 2). This can lead to a chimney effect in the construction and fire spread to more areas in the building, causing the amount of fire damage to increase rapidly. If the space is intact, the rescue services can intervene by supplying a extinguishing agent, e.g. with a cutting extinguisher or piercing nozzle, without the fire flares up (point 3). If the fire is allowed to grow, the enclosure will eventually collapse and the more air will flow to the fire, which will mean that the extent of the damage will increase faster (point 4).

The time scale of fires in concealed spaces is difficult to assess because it will be dependent on several factors such as the supply of air (oxygen) and the material burning. But, the fire development and damage progression can in general be expected to be slower than what in a compartment fire.

Based on the above reasoning, structural fires are considered characterized by the following:

- Fire inside a building structure (e.g. attic, ventilation gap)
- Difficult for the rescue services to access the fire
- Under-ventilated fire
- In many cases, slower fire development compared to compartment fires.

### 3.2 Case study of structural fires

A feature in the studied events is that structural fires with extensive fire damage occur if the fire development is severe and there have been shortcomings in fire barriers (fire separations) or in the rescue operations. Based on the events with extensive damage there are clear indications that problems in the structural fire protection or rescue response often are an underlying reason for the extensive damage. In several of the events the rescue service had problems getting the situation under control.

In cases where there were problems with fire separations in the building it has been due to the building did not met the building regulations in some aspects or that the building regulations were not sufficient to avoid extensive fire damage. The latter is mainly because the main objective of the Swedish building regulations primarily is and has been on life safety and secondary on property protection.

In all studied cases of structural fires with extensive fire damage, the fire started as a compartment fire (normally on the stove in the kitchen or in a living room). The fire then spread to a concealed space, e.g. through a window to the attic or to the roof structure through inadequately insulated ventilation duct. In cases with extensive fire damage there is often an uncertainty about the functionality of the fire separations, i.e. if the defect was caused by incorrect installation or if a tested and properly installed fire separation according to the regulations was insufficient.

A prerequisite for the occurrence of a structural fire is that there are flammable construction materials in the building structure. A structural fire would not occur in incombustible structure. But it is often not an alternative to construct a building of flats in totally non-combustible materials, e.g. wooden roof structures are often used in concrete buildings.

If the structural fire has spread to a second fire compartment when the fire service arrives on site, it will often take more time before the rescue service gets the situation under control. A rapid fire growth or a late call to the rescue service will result in that the there is less time take measures. The size and growth of the fire is controlled by the amount and type of combustible material and the ventilation conditions. The situation when the rescue service arrives will determine the possibilities to tackle the fire. If search and rescue of missing people is required the structural fire can spread further before full attention can be given to fire fighting measures. In about half of the studied structural fires with extensive fire damage lifesaving measures were taken initially.

In a majority of the studied events the rescue service was undersized. More personnel or other equipment (such as cutting extinguishers) in the early stage would most likely have reduced the fire damage. It is explicitly stated in some of the fire investigations reports that if cutting extinguisher had been in used, the outcome would have been different. Another aspect is that the rescue service sometimes failed to interpret the building construction, which may have affected how the resources (personnel and equipment) were allocated and used during the operation.

Knowledge and experience of the rescue service is something that emerges as an important part of a successful operation. This is not identified on the basis of the individual events but it is something that is perceived as an important success factor in the structural fires with limited damage. Good knowledge of the building construction and fire protection (e.g. fire separations) is important. Knowledge of fire dynamics is also a essential for a successful rescue service operation; such knowledge can be used to e.g. predict the fire behaviour and thereby to apply the most effective actions to extinguish the fire.

The factors and actions behind a successful management of a structural fire are in many cases not present in the studied structural fires were the outcome was an extensive damage. That the identified success factors are missing in the events with extensive damage are considered to strengthen the results.

## 4. CONCLUSIONS

A schematic representation of fires in concealed spaces is presented in this paper. The model is simple, but it constitutes a first step to represent structural fires. However, further work to create models, which can lead to a better understanding of structural fires, are considered necessary.

The events studied in this work can be divided into two categories: structural fires that resulted in limited damage and structural fires that resulted in extensive damage. Conclusions have been drawn in regard to each category. Constructional fire protection, rescue service response and fire development are the three main factors that determine the outcome of structural fires and these appears to be important in all the studied fires. However, there are several underlying and more specific factors affecting the outcome and these have been identified for each studied fire. The result is based a dozen events and it is not possible to generalize the result completely, but the results are considered to have a good validity because the identified failures as well as success factors reappear in the studied fires.

The Swedish Fire Research Board (Brandforsk) has funded the research presented in this paper.

### References

[1] Johansson, N, Fire Dynamics of Multi-Room Compartment Fires, Report 1055, Department of Fire Safety Engineering, Lund University, 2015.

[2] Senez, P L, Calder, K D & Li, H H, Fire loss statistical consideration in relating failure and building damage to the building code objectives, Proceedings of the 13th International Fire Science and Engineering Conference, 2013.

[3] Särdqvist, S, Initial Fires, Department of Fire Safety Engineering, Lund University, 1993.

[4] Hertzberg, T, Sundström, B & van Hees, P, Design Fires for Enclosures, Brandforsk projekt 314-001, SP Report 2003:02, SP, Borås, 2003.

[5] Karlsson, B & Quintere, J G, Enclosure fire dynamics, CRC Press, Boca Raton, 2000.

[6] White, N & Delichatsios, M, Fire Hazards of Exterior Wall Assemblies Containing Combustible Components, Fire Protection Research Foundation, National Fire Protection Association, 2014.

[7] Lundberg, O, Brand i flerbostadshus – En rapport om skadekostnader vid brand relaterade till de branddrabbade husens byggnadsmaterial, Betongforum, 2005 (in Swedish).

[8] Johansson, N, A case study of fires in structural elements, Report 3191, Department of Fire Safety Engineering, Lund University, 2015 (in Swedish).

[9] Johansson, N, van Hees, P & Särdkvist, S, "Combining Statistics and Case Studies to Identify and Understand Deficiencies in Fire Protection", Fire Technology, **48**(4) DOI: 10.1007/s10694-012-0255-z, 2012

[10] Vesely WE, Goldberg FF, Roberts NH, Haasl DF, The fault tree handbook. NUREG-0492, US Reglatory Commission, 1981.

[11] Svensson, S, Byggnadstekniskt brandskydd i MSB:s utbildningar, Report 3182, Department of Fire Safety Engineering, Lund University, 2014 (in Swedish).

MATEC Web of Conferences **46**, 06002 (2016)
DOI: 10.1051/matecconf/20164606002


# Post incident analysis report: Lacrosse Docklands, 25 November 2014

Greg Badrock

Metropolitan Fire and Emergency Services Board, Melbourne, Australia

## 1. EXECUTIVE SUMMARY

This Post Incident Analysis (PIA) provides a detailed account of the fire incident that occurred at the Lacrosse building on 25 November 2014.

It includes information compiled by MFB Fire Safety Officers investigating the sequence of fire events, the suitability of building materials used in construction, performance of installed fire safety equipment, evacuation of the building and fire causation.

The main observations are:

- External wall cladding (Alucobest) rapid fire spread.
- Use of combustible external wall cladding on Type A construction.
- Building material design, selection and installation.
- High occupancy rate.
- Mass evacuation necessary due to fire development and spread.
- Emergency Warning and Intercommunication System (EWIS) was compromised.
- Sprinkler system operated well beyond its designed capability.
- Sprinklers were not required on the balconies under the Building Codes of Australia (BCA).

## 2. FIRE CALL 15657

At 02:24 hours on Tuesday morning 25 November 2014, MFB fire crews responded to an exchange call for a reported apartment fire at 673 – 675 La Trobe Street, Docklands. When the first fire crews arrived on scene at 02:29 hours, they observed that the fire had already extended up the external walls and balconies over approximately 6 levels. At 02.35 hours, only 6 minutes later, crews reported back that fire had reached the roof of the building above the $21^{th}$ floor.

The fire scenario and fire behaviour encountered by the attending MFB fire-fighters on that morning is not a scenario commonly encountered by MFB crew attending high-rise buildings. Rapid vertical fire spread up the building appeared to be directly associated with the external façade of the building, rather than associated with the internal parts or extensive fuel loads stored on many of the balconies.

Due to rapid fire spread and penetration into internal parts of the building over many levels, the entire building was evacuated resulting in more than four hundred evacuees assembling in La Trobe Street. It appears the rapid fire spread caused the EWIS to be compromised on most fire affected levels, preventing it from operating as designed on those levels. Fire crews were therefore forced to enter every level and alert occupants of each apartment to ensure total evacuation.

After the fire, it was observed that many apartments contained bedding arrangements indicating a higher occupancy level than what would normally be expected. This resulted in increased combustible fuel loads due to the greater amount of personal belongings. It was fortunate that the installed fire sprinkler system operated well above its designed capability preventing further internal spread.

MFB fire investigators concluded that the fire was started by a cigarette butt disposed of in a plastic container located on the top of a timber topped outdoor table, positioned towards the southern end of apartment 805 balcony.

## 3. FIRE INCIDENT EVENTS

The following information was compiled after MFB fire safety officers interviewed a number of fire-fighters and occupants who were present during the incident. The Bureau of Meteorology (BoM) records indicate that the temperature during the night was around 12 degrees, with a westerly wind of 20 to 30 kilometres per hour. At approximately 01:30 hours on the morning of Tuesday 25 November 2014, an occupant from Apartment 805 of the building, claims he investigated the smell of smoke. After checking the kitchen and making sure the gas stove was turned off, he returned to bed. Sometime later, the same occupant was woken by two other house mates who had discovered the fire burning on the balcony.

From inside the apartment, he could see a fire on the right hand side (south) of the balcony. The occupants of the apartment unsuccessfully attempted to extinguish the fire using a container of water. All of the occupants from Apartment 805 then evacuated the apartment via the buildings northern end isolated fire stairs.

At 02:24 hours, MFB fire crew responded to an exchange call for a reported apartment fire at 673 La Trobe Street, Docklands. The fire indicator panel history log shows activation of the Level 8 fire sprinkler flow switch at 02:25 hours, which also generated an alarm to the MFB. Several exchange calls followed confirming that the building was well alight and the fire was spreading rapidly up the building. Refer to call history Appendix 2 of the full report.

When the first fire crew arrived on scene at 02:29 hours, they observed fire travelling upwards rapidly and involving about six floors. They also observed that the fire was burning up the external wall cladding and spreading onto the balcony on each level. By this time a number of people had already evacuated and had congregated outside the building entry in La Trobe Street.

The occupants from Level 6, Apartment 605 reported later to fire-fighters, seeing fire embers and flaming debris falling from levels above their apartment and igniting materials on their balcony. They then evacuated the building (see Fig. 1).

At 02:30 hours, the Senior MFB Officer in attendance provided the following word back; "Structure Fire, Respond 3rd Alarm". This was followed by a message that crew wearing breathing apparatus and equipped with hose lines were entering the building to evacuate all occupants and investigate the extent of fire spread. By 02:35 hours, it was reported that the fire had spread to Level 21 via the external face of the building. At 02:38 hours, the status of the alarm was upgraded to a 4th alarm.

Fire-fighters confirmed that the sprinkler system operating within the apartments had held the fire in check, and was preventing further internal spread and fire development. Fire-crew used hose lines connected to internal hydrants and portable fire extinguishers to totally extinguish fires on Levels 10 and 19.

An MFB aerial appliance referred to as a "Ladder Platform" was set up on the La Trobe Street overpass and at approximately 02:46 hours, was operational and had water onto the fire. The water stream from the water monitor on this appliance was able to reach all levels on the building, making extinguishment of the burning façade more efficient. With several hundred civilians from the building assembling on the north side of La Trobe Street, MFB Officers arranged their evacuation to a sheltered area at Southern Cross Station.

At the height of the fire, MFB committed 122 personnel, 22 appliances, 3 aerial appliances and 4 specialist vehicles.

**Figure 1.** Picture of the fire on level 6 (red arrow) and fire on level 8 (point of origin – blue arrow) extending up to level 14 (02:29).

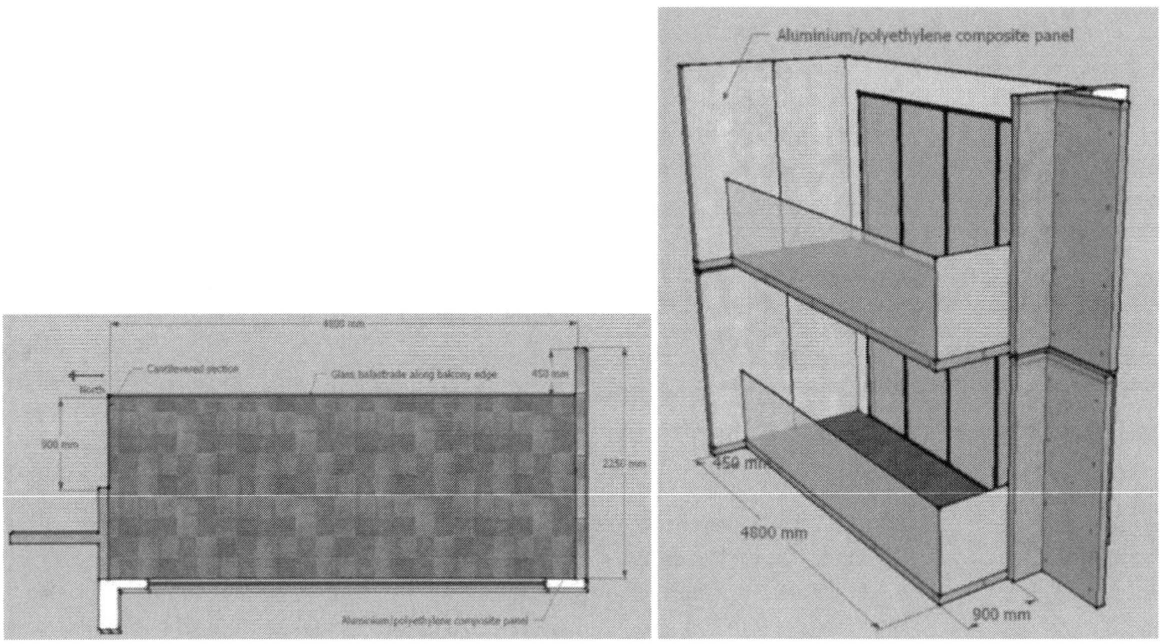


**Figure 2.** Scheme of balcony and external wall.

# 4. BUILDING CONSTRUCTION ASSESSMENT

## 4.1 Balcony and external wall construction and characteristics

The apartments known as Apartments 605 to 2105 are located in vertical alignment from the 6th floor to the 21th floor on the east façade of the building. They include suspended concrete balconies of approximately 1.8 metres deep × 4.7 metres wide. A double-glazed sliding door assembly is fitted in the external wall of the building and provides access to the balcony.

The northern end of these balconies are bounded by 900 mm concrete panels, which leaves the remaining 900 mm balcony depth cantilevering out beyond the building face. A 1 metre high glazed balustrade is fitted to the remaining northern edge of the balconies and returns along the length of the eastern edge and butts into the southern wall.

In contrast to the northern end of these balconies, the walls at the southern end of the balconies extend out approximately 2.25 metres and some 450 mm from the external face of the balconies.

These walls are built of lightweight steel stud construction. The internal face of the walls are lined with two layers of 13 mm standard grade gypsum plasterboard, contain insulation batts, along with a combustible PVC stormwater downpipe and several combustible electrical/television cabling and input face plates. The external face is lined with a 4 mm aluminium/polyethylene composite panel façade containing a polyethylene core.

MFB Fire Investigators removed a large sample of the aluminium/polyethylene composite panel façade, fitted to the southern end of the balconies, for further investigation. The removed section of panel contained manufacturer labelling and serial identification on the internal face indicating the following: "ALUCOBEST 11060167 HY 103 4 mm 2011/06/17 20:51:45".

The product is believed to incorporate a Polyethylene core material. The Alucobest Technical Manual provided on the web link at `www.alucobest.com`, indicates that it is a product manufactured by a China based company titled Shanghai Huayuan New Composite Materials Co. Ltd. The MFB reasonably believes this to be the product installed to the southern end of the balconies of Rooms 605–2105 and is likely to be the same product installed throughout the remainder of the façade of the building. Alucobest Aluminium/Polyethylene Composite Panel – Fire Behaviour Properties. As mentioned above, it is assumed the Alucobest panel taken from the Lacrosse building is the standard grade, and not the "Fire Resistant" model, detailed on the Alucobest Technical Manual. All references made to the "Fire Resistant" range is referred to as Alucobest FR. The sample of panel does not include manufacturer labelling to indicate the sample is from the Alucobest FR range. There is no discussion or reference made to fire behaviour tests of the standard grade Alucobest aluminium/polyethylene composite panel in the available Alucobest technical manual. It appears that this product has not been tested in accordance with AS1530.1:1994 – Combustibility Test for Materials.

### 4.2 Testing of Alucobest to 1530.1

MFB forwarded a sample of the "Alucobest" (ACP) taken directly from the façade of the Lacrosse building to the CSIRO test facility in North Ryde, NSW. On April 1, 2015 the CSIRO determined the following indicative test outcomes:

" ***Observations***: *Sustained flaming was observed on the specimen at 55 seconds into the test. The test was terminated at 93 seconds due to excessive flaming and smoking.*

***Designation:*** *The material is deemed COMBUSTIBLE according to the test criteria specified in Clause 3.4 of AS 1530.1:1994.*"

For the full CSIRO test report see Appendix 3, of the full PIA report

## 5. ISSUES

### 5.1 External wall cladding (Alucobest) rapid fire spread

### 5.2 MFB comment:

First-hand accounts from attending MFB fire-fighters and residents of the building, describe the fire as appearing to be associated directly with the façade of the building rather than the combustible contents and storage on the external balconies. Burning and flaming facades on high-rise buildings is not a common phenomenon witnessed by the MFB and is of genuine concern. Of even greater concern is the speed and intensity of the fire spread.

The MFB Fire Investigation team have determined that the fire originated on the balcony of the 8th floor. The FIP history log shows activation of the Level 8 sprinkler flow switch at 02.25 hours, one minute later than an exchange call received by the MFB. Fire crews arrived on scene at 02.29 hours, some five minutes later and reported vertical fire spread to approximately the 14th floor. At 02.35 hours, MFB crew reported that the vertical fire spread had reached the roof. From the timeline described above, it is reasonable to derive external vertical fire spread occurred from the 8th floor to the roof above the 21th floor within 10 to 15 minutes, penetrating the adjacent internal rooms on all floors. In the case examined in this report, the upward vertical spread of fire was restricted only by the height of the building. If the building and the construction of the external walls continued to a greater height of upward of 21 storeys, it is highly probable fire spread would have continued beyond 21 storeys.

## 5.3 Use of combustible external wall cladding on Type A construction

## 5.4 MFB comment:

Due to the use and number of storeys, Stage 1 of The Lacrosse Apartment Building is considered a building requiring Type A construction when determined under C1.1 of the BCA. In accordance with the deemed-to-satisfy requirements of Specification C1.1 of the BCA, external walls of Type A buildings must be non-combustible, notwithstanding any requirement for fire rating. Non-combustible is a defined term in the BCA and is defined as the following:

*Applied to a material – not deemed combustible as determined by AS1530.1 – Combustibility Test for Materials; Applied to construction or part of a building – constructed wholly of materials that are not deemed combustible.*

Additionally, a material may be considered non-combustible under C1.12 of the BCA, if it meets the defined criteria within that clause. Standard grade Alucobest aluminium/polyethylene composite panel does not meet the criteria and nor is it likely that it has been successfully tested in accordance with AS1530.1. Therefore, a building permit application specifying the use of standard grade Alucobest aluminium/polyethylene composite as an external wall cladding system, proposes an alternative solution to the deemed-to-satisfy requirements of the BCA. Evidence of suitability for the material and form of construction must be obtained in accordance with A2.2 of the BCA, to demonstrate it meets the relevant performance requirements. This may be in the form of a Certificate of Conformity/Accreditation. The MFB have not been able to gain such documentation for the Alucobest range and these products are not included in the ABCB – Register of CodeMark Certified Products.

## 5.5 Building material design, selection and installation

## 5.6 MFB comment:

The MFB has been unable to obtain the complete building approval documentation as the total of the approved drawings, specifications was not available at the Council Offices. The MFB was therefore unable to substantiate if Alucobest panel was specified in the approved drawings or a competitor product.

Anecdotally, many of the same/similar products in Australian competitors range have obtained Certificates of Conformity under the ABCB – CodeMark Scheme. The conditions on the certificates generally require mechanical fixing systems for the panels on Type A and B buildings, in lieu of adhesive flat tape as used on the Alucobest product on the Lacrosse Building. The MFB have been informed by industry representatives that they believe this requirement is to prevent delamination of the aluminium face sheet, which can expose the combustible core material to the effects of fire.

Brigade member accounts, along with video footage, details large flaming flat sheets falling from the building façade.

Additional to the combustible cladding, the lightweight external walls at the southern end of balconies include combustible 100 mm PVC stormwater downpipes and associated lagging within the core of the wall. The downpipes are connected to the drains housed in the balcony floors to drain stormwater collected on the surface of the balconies. A fire collar is installed around the PVC downpipe penetration, through the concrete balcony, to retain the required fire resistance level.

Inspections conducted after the day of the subject fire incident revealed that many fire collars failed to operate as designed i.e. did not close sufficiently to prevent fire spread to connecting levels.

### 5.7 High occupancy rate

### 5.8 MFB comment:

The occupancy of many of the apartments appeared to be in excess of what would normally be expected in a two bedroom apartment and what a two bedroom apartment is designed for. During investigations it was identified that some apartments had sleeping arrangements for up to eight people. This led to a greater level of storage of personal belongings within the apartment and on the apartment balcony.

As a result, the higher fire fuel load has allowed for a more intense fire to develop on the balcony. In this instance the sprinkler system did not cover the balcony so the fire was able to develop to the point of causing the glass panels located between the apartment's living room and the balcony to break and expose the internal rooms to fire. Fortunately in this instance, the sprinkler system installed within each apartment prevented major internal fire spread.

Additionally, high occupancy of apartments can impact on safe evacuation of occupants in the event of an emergency. Some of the apartments in this building had temporary structures assembled around the bed to provide privacy. These light weight structures, along with other furnishings and contents, may impede clear egress from the apartment making it difficult for occupants to exit safely.

### 5.9 Mass evacuation necessary due to fire development and spread

### 5.10 MFB comment:

External wall construction and materials used in this building allowed for rapid vertical fire spread, involving a relatively large portion of the high-rise building as opposed to a single level. In light of this fire, Officers had no choice but to evacuate the entire building. Over 400 evacuees were assembled in La Trobe Street before the MFB identified that a large sheltered evacuation centre would be needed.

Initially the evacuees were escorted from La Trobe Street to the Southern Cross Station bus centre, where they were monitored by Ambulance Victoria and provided with water and blankets. The MFB Incident Management Team (IMT) called for the response of the Municipal Emergency Response Officer (MERO) and the establishment of an Emergency Relief Centre, (ERC). During the morning it was decided to use Etihad Stadium as a recovery and information centre and all evacuees were transferred to this location.

The care and management of the displaced occupants also presented a challenge for the MFB and other agencies. During the MFB debrief fire-fighters indicated that it was a difficult process to register and account for all occupants, as some evacuees did not speak English and some did not understand the reason for providing their name and address.

Due to the fire damage and the fire systems being inoperable, the occupants were unable to return to their apartment for an extended period of time. Some occupants were displaced for a longer period while the building underwent structural repair.

### 5.11 Emergency Warning and Intercommunication System (EWIS) was compromised

### 5.12 MFB comment:

In the hours after the fire, MFB personnel conducted numerous interviews with occupants and fire-fighters that were involved in the initial fire response. Many of the occupants spoken to said they did not hear alarms. These occupants said they were awoken to the fire by "screaming, banging and other loud noises".

The remaining witnesses fell into two different groups. One group commented that the alarms came on for a few seconds only or they could hear alarms but they were very quiet and distant. The other group said they heard the alarms and evacuated. Approximately 10 minutes after the first appliance arrived on scene a fire-fighter used the EWIS PA facility to make an evacuation announcement. No witnesses questioned reported hearing this message. Subsequent investigations found that the following scenario resulted in the EWIS"s failure.

Directly above the balcony air-conditioner compressor units (south wall), is a metal exhaust grill. The grill is connected to a formed sheet – metal collection box located in the ceiling space above Bedroom 2. The collection box is approximately 600 mm wide, 250 mm high and 600 mm deep. Two 150 mm non-insulated aluminium flexi ducts were connected to the collection box from the bathroom exhaust fans. A EWIS sounder was located in the ceiling of Bedroom 2 and directly adjacent to the above exhaust collection box. This design arrangement was typical for all apartments in vertical alignment from 605 to 2105.

The EWIS system is designed and installed as per Australian Standard AS1670.4. This Standard requires all wiring between the EWIS"s main panel and the evacuation zone to be fire rated. The zone wiring itself is not required to be fire rated, instead non-fire rated thermal plastic sheath (TPS) wire is used.

As identified earlier, the fire started on the balcony of Apartment 805 on Level 8. Hot gases as a result of the fire, entered the ceiling space over Bedroom 2 via the external wall grill and compromised the wiring and designated sounder of the EWIS (Appendix 5) full report. This resulted in a fault in the speaker loop and subsequent failure of the entire sounder system on the entire 8th level. This has been confirmed by Representatives from the installation/maintenance company.

Based on witness statements and subsequent investigations, it is believed that the EWIS on Level 8 and 9 operated for approximately 30 seconds after the FIP received its first activation transmission from the smoke detector outside Apartment 805. It then failed due to the fire compromising the system. It is also believed that the EWIS operated on Level 7 for approximately 5 minutes before it too failed.

As described earlier the fire burnt up the side of the building extremely fast. It is believed that the fire caused the EWIS"s system to fail on most of the levels ahead of the cascading EWIS evacuation sequence that was ascending the building. This would suggest that those who said they heard the alarm and evacuated would have been located below Level 9.

### 5.13 Sprinkler system operated well beyond its designed capability

### 5.14 MFB comment:

In total, 26 fire sprinkler heads activated over 16 floors during the fire incident. As the fire spread to each level, fire sprinkler heads generally activated within the lounge and bedroom 2 and prevented internal fire spread and development into apartments. This put a significant demand on the installed sprinkler system and associated water supply. Additionally, two internal fire hydrants were used by fire-fighters to extinguish fires not extinguished by the sprinklers.

The installed combined fire hydrant/fire sprinkler system, compliant with AS2118.6, was designed to facilitate simultaneous operation of four sprinkler heads and two fire hydrants. See Appendix 8.(full report) It is possible that not all sprinkler heads along with the two fire hydrants were operating at the

**Figure 3.** Picture of typical combustibles present on balconies.

same time; however due to the time-line of events, it is reasonable to conclude that the system operated significantly beyond its designed capability.

Had the combined fire hydrant/fire sprinkler system not exceeded its designed capability, it is likely that significant fire development and spread would have occurred in some of the subject apartments on Levels 6 to 21. Spread beyond the subject apartments to adjacent apartments and common areas may also have occurred.

This would have presented an extremely difficult scenario for fire-fighters and occupants of the building, and may have resulted in serious injury and/or death. Significantly increased property damage and loss would have occurred along with the negative impact on occupant displacement and emergency service/recovery agency resource.

### 5.15 Sprinklers were not required on the balconies under the BCA

### 5.16 MFB comment:

The balconies connected to Apartments 605 to 2105 were approximately 1.8 metre in depth × 4.7 metre long. The sprinkler system did not extend beyond the apartments" internal areas to the balconies and it is acknowledged by the MFB that this was not a deemed-to-satisfy requirement under the Building Code of Australia (AS2118.1 Sect. 5.7.10).

*AS 2118.1 Sect. 5.7.10 – Covered balconies*

*Portions of covered balconies that exceed 6 $m^2$ floor area and have a depth in excess of 2 m shall be sprinkler protected.*

The Chief Officer Report and Consent pursuant to Regulation 309 of the Building Regulations 2006, dated 29 March 2011, contains notification under 309(3) for the deletion of sprinklers to balconies (not Apartments 605 to 2105) and indicated the balconies in question would have low fuel loads. MFB recommendations included a comment that balconies were not to be used for storage and requested that this measure was to be included as part of the Essential Safety Measures for the building.

In this instance, typical combustibles identified to be present on balconies throughout the building consisted of clothing, bedding, bicycles, electrical appliances and other miscellaneous combustible materials, notwithstanding the air conditioner compressor units and other combustible furniture items (see Fig. 3).

High occupancy rates in apartments can lead to reliance on balconies for additional storage space, increasing the fire load.

Had the sprinkler system extended to the balcony area of each apartment, fire would have most likely been contained to the level of fire origin.

## 6. CONCLUSION

In this instance, fire quickly extended up the external wall involving the cladding and fuel loads on each balcony. Had the external wall cladding been of a non-combustible type, the likelihood of fire spread beyond the level of ignition would have been greatly reduced.

Furthermore, if a sprinkler head had been installed, it too would have reduced the chance of fire spread.

In the words adopted within the Building Code of Australia, a building must have elements to the degree necessary to avoid the spread of fire in a building. Amongst other considerations, the degree necessary is relative to:

- The function and use of the building – residential building with sleeping occupants not in a ready and aware state.
- The fire hazard – the installed cladding material appeared to be a readily combustible material, which upon ignition sustained and supported rapid spread of fire. Additionally, there was excessive storage of combustibles on many of the balconies.
- The number of storeys – the building contains 23 storeys (high-rise).
- Fire brigade intervention – High-rise, 400+ occupants and multiple internal ignitions over 15 floors.

The fire behaviour on the morning of the 25 November 2014, clearly demonstrated to all concerned, that the elements installed to the external walls of this building did not avoid the spread of fire to the degree necessary.

An improved system and understanding for ensuring appropriate material selection, approval and installation is of major importance to the MFB.

There were many contributory factors leading to the events on 25 November 2014. In this instance, increased combustible fuel loads on the unsprinklered balcony aided in the ignition and fire propagation to such a degree, as to allow ignition of the external wall cladding. This incident in isolation may have resulted in a far more manageable outcome with damage on one level only. However, as previously described in this report, rapid fire spread directly related to the involvement of the external cladding material, occurred to such an extent that the only likely fact preventing further vertical fire spread was the limitation of the building height to only 21 storeys.

Clearly, the external cladding material on this building did not to the degree necessary avoid the spread of fire as required by the Building Code of Australia. Simultaneous internal fire ignition events over multiple floors are simply an unacceptable fire safety solution for a residential high-rise building, or any other occupiable building for that matter.

If not for the excellent performance of the internal fire sprinkler system and the quick and professional response of the MFB fire-fighters to prevent further internal fire spread and development, the consequences of this fire would likely have resulted in greater impact on occupants, the surrounding community amenity, resources and infrastructure. There could have been a greater likelihood of serious injury or even loss of life.

Whilst damage and injury were minimised, the social impact of this fire was considerable and cannot be ignored. All occupants of the building were displaced for some days during building refurbishment and reinstatement of the fire safety systems, whilst the occupants of the fire affected apartments were displaced for considerably more time. Prevention of similar incidents in new and existing developments should be a priority for the entire construction industry. This must start with ensuring an improved process and/or understanding for appropriate material selection, approval and installation. In particular, the MFB would urge all stakeholders in the construction industry to exercise greater diligence and caution with the selection and installation of aluminium/polyethylene composite cladding panels, and encourage selection of those products with appropriate and clear product accreditation and certificates of conformity. Given the risk of fire on a residential balcony is real, and as in this instance, the sprinkler system within the apartments does not extend to protect the balcony on each level, there is the possibility for this scenario to repeat itself.

The full Post Incident Analysis (PIA) report can be located at the following link:

http://www.mfb.vic.gov.au/Media/docs/Post_Incident_Analysis_ Lacrosse_ Docklands-f699bfbc-344a-4168-bdd8-ef8aa5e61c1e-0.pdf

# AUTHOR INDEX